# SYMMI

## An Introduction to Group Theory and Its Applications

Roy McWeeny
Professor Emeritus
University of Pisa, Italy

DOVER PUBLICATIONS, INC.
Mineola, New York

### *Copyright*



### *Bibliographical Note*

This Dover edition, first published in 2002, is an unabridged republication of the work originally published in 1963 by Pergamon Press Ltd., Oxford, England as Volume 3 in Topic 1 (Mathematical Techniques) of The International Encyclopedia of Physical Chemistry and Chemical Physics. (Preliminary material related solely to the 1963 Encyclopedia as a whole has not been included in the Dover edition, and references to this volume as a component of that set have been deleted.)

*Library of Congress Cataloging-in-Publication Data*

McWeeny, R.
Symmetry : an introduction to group theory and its applications / Roy McWeeny.
p. cm.
Originally published: Oxford ; New York : Pergamon Press, 1963.
Includes bibliographical references and index.
ISBN 0-486-42182-1 (pbk.)
1. Group theory. 2. Symmetry (Physics) I. Title.

QD453.3.G75 M38 2002
512'.2—dc21

2002023722

Manufactured in the United States by Courier Corporation
42182108
www.doverpublications.com

# CONTENTS


PAGE

PREFACE .. vii

CHAPTER 1 GROUPS

1.1 Symbols and the group property .. .. .. 1
1.2 Definition of a group .. .. .. .. .. 6
1.3 The multiplication table .. .. .. .. 7
1.4 Powers, products, generators .. .. .. .. 9
1.5 Subgroups, cosets, classes .. .. .. .. 11
1.6 Invariant subgroups. The factor group .. .. 13
1.7 Homomorphisms and isomorphisms .. .. .. 14
1.8 Elementary concept of a representation .. .. 16
1.9 The direct product .. .. .. .. .. 18
1.10 The algebra of a group .. .. .. .. .. 19

CHAPTER 2 LATTICES AND VECTOR SPACES

2.1 Lattices. One dimension .. .. .. .. 22
2.2 Lattices. Two and three dimensions .. .. .. 25
2.3 Vector spaces .. .. .. .. .. .. 27
2.4 $n$-Dimensional space. Basis vectors .. .. .. 28
2.5 Components and basis changes .. .. .. .. 31
2.6 Mappings and similarity transformations .. .. 33
2.7 Representations. Equivalence .. .. .. 38
2.8 Length and angle. The metric .. .. .. .. 41
2.9 Unitary transformations .. .. .. .. .. 47
2.10 Matrix elements as scalar products .. .. .. 49
2.11 The eigenvalue problem .. .. .. .. .. 51

CHAPTER 3 POINT AND SPACE GROUPS

3.1 Symmetry operations as orthogonal transformations .. 54
3.2 The axial point groups .. .. .. .. .. 59
3.3 The tetrahedral and octahedral point groups .. .. 69
3.4 Compatibility of symmetry operations .. .. 75
3.5 Symmetry of crystal lattices .. .. .. .. 78
3.6 Derivation of space groups .. .. .. .. 85

CHAPTER 4 REPRESENTATIONS OF POINT AND TRANSLATION GROUPS PAGE

4.1 Matrices for point group operations .. .. .. 91
4.2 Nomenclature. Representations .. .. .. 95
4.3 Translation groups. Representations and reciprocal space .. .. .. .. .. .. .. 105

CHAPTER 5 IRREDUCIBLE REPRESENTATIONS

5.1 Reducibility. Nature of the problem .. .. .. 109
5.2 Reduction and complete reduction. Basic theorems .. 110
5.3 The orthogonality relations .. .. .. .. 116
5.4 Group characters .. .. .. .. .. .. 121
5.5 The regular representation .. .. .. .. 124
5.6 The number of distinct irreducible representations .. 125
5.7 Reduction of representations .. .. .. .. 126
5.8 Idempotents and projection operators .. .. 131
5.9 The direct product .. .. .. .. .. 133

CHAPTER 6 APPLICATIONS INVOLVING ALGEBRAIC FORMS

6.1 Nature of applications .. .. .. .. .. 140
6.2 Invariant forms. Symmetry restrictions .. .. 141
6.3 Principal axes. The eigenvalue problem .. .. 147
6.4 Symmetry considerations .. .. .. .. 150
6.5 Symmetry classification of molecular vibrations .. 151
6.6 Symmetry coordinates in vibration theory .. .. 159

CHAPTER 7 APPLICATIONS INVOLVING FUNCTIONS AND OPERATORS

7.1 Transformation of functions .. .. .. .. 166
7.2 Functions of Cartesian coordinates .. .. .. 170
7.3 Operator equations. Invariance.. .. .. .. 174
7.4 Symmetry and the eigenvalue problem .. .. 181
7.5 Approximation methods. Symmetry functions .. 187
7.6 Symmetry functions by projection .. .. .. 190
7.7 Symmetry functions and equivalent functions .. 195
7.8 Determination of equivalent functions .. .. 197

| | | PAGE |
|---|---|---|
| CHAPTER 8 | APPLICATIONS INVOLVING TENSORS AND TENSOR OPERATORS | |
| 8.1 | Scalar, vector and tensor properties .. .. .. | 203 |
| 8.2 | Significance of the metric .. .. .. .. | 206 |
| 8.3 | Tensor properties. Symmetry restrictions .. .. | 208 |
| 8.4 | Symmetric and antisymmetric tensors .. .. .. | 211 |
| 8.5 | Tensor fields. Tensor operators .. .. .. | 218 |
| 8.6 | Matrix elements of tensor operators .. .. .. | 224 |
| 8.7 | Determination of coupling coefficients .. .. | 231 |
| APPENDIX 1 | Representations carried by harmonic functions .. | 235 |
| APPENDIX 2 | Alternative bases for cubic groups .. .. .. | 241 |
| INDEX .. | .. .. .. .. .. .. .. .. .. | 245 |

# PREFACE

THE OPERATIONAL principles underlying the construction of symmetrical patterns were certainly known to the Egyptians. They were formulated symbolically during the nineteenth century and played their part in the development of the theory of groups. But only during the past thirty-five years has the immense importance of symmetry in physics and chemistry been fully recognized.

The value of group theoretical methods is now generally accepted. In chemical physics alone, the symmetries of atoms, molecules and crystals are sufficient to determine the basic selection and intensity rules of atomic spectra, and of the electronic, infra-red and Raman spectra of molecules and crystals. The vector model for coupling angular momenta, the properties of spin, the Zeeman and Stark effects, the splitting of energy levels by a crystalline environment, and even the nature of the periodic table may all be traced back to symmetry of one kind or another.

Many excellent textbooks on group theory and its applications are already available. They range from rigorous but formal presentations to highly condensed accounts which deal with particular applications and give only a superficial treatment of underlying concepts. The first kind of approach is unpalatable, except to the professional mathematician; the second provides a useful vocabulary and a set of working rules—but no real understanding of group theory. The present book is intended primarily for physics and chemistry graduates who possess a fair amount of mathematical skill but lack the formal equipment demanded by the standard texts (by Wigner, Weyl and others). Accordingly, the elementary ideas of both group theory and representation theory (which, incidentally, provides the basic mathematical tools of quantum mechanics) are developed in a leisurely but reasonably thorough way, to a point at which the reader should be able to proceed easily to more elaborate applications. For this purpose, emphasis is placed upon the *finite* groups which describe the symmetry of regular polyhedra and of repeating patterns. By restricting the scope in this way, it is possible to include geometrical illustrations of all the main processes. In fact, over a hundred fully worked examples have been incorporated into the text.

A discussion of the permutation group and of the rotation groups would have defeated the aim of introducing all the basic ideas in a limited space. However, there is much to be said for dealing with special groups in the context in which they occur. A study of the rotation group, for example, is essential to any full discussion of the quantum mechanical central field problem. And such applications present no real difficulty once the basic principles have been grasped.

The book is constructed so that it may be read at various levels and with emphasis on any one of the main fields of application. Chapter 1 is concerned mainly with elementary concepts and definitions; Chapter 2 with the necessary theory of vector spaces (though sections 2.8–10 may be omitted in a first reading). Chapters 3 and 4 are complementary to 1 and 2: they provide the reader with an opportunity of actually working with groups and representations, respectively, until the ideas already introduced are fully assimilated. The more formal theory of irreducible representations is confined to Chapter 5. In a first reading it would be sufficient to grasp the main ideas behind the orthogonality relations (5.3.1), passing then to section 5.4 for the properties of group characters, and finally to the construction of irreducible basic vectors according to (5.7.8) and the examples which follow. The rest of the book deals with applications of the theory. Chapter 6 is concerned largely with quadratic forms, illustrated by applications to crystal properties and to molecular vibrations. Chapter 7 deals with the symmetry properties of functions, with special emphasis on the eigenvalue equation in quantum mechanics. Chapter 8 covers more advanced applications, including the detailed analysis of tensor properties and tensor operators.

Much of the material in this book has been presented to final year undergraduates and to graduate students in the Departments of Mathematics, Physics and Chemistry at the University of Keele, and the exposition has often been guided by their response. It is a pleasure to acknowledge this enjoyable and valuable contact, and also the many conversations with colleagues both here and at the University of Uppsala. Finally, anyone writing a book in this field must be deeply conscious of his debt to those who pioneered the subject, over thirty years ago; I should like to acknowledge, in particular, that the books by Wigner and Weyl have been a constant influence.

R. McWeeny

CHAPTER 1

# GROUPS

## 1.1. Symbols and the Group Property

A large part of mathematics involves the translation of everyday experience into symbols which are then combined and manipulated, according to determinable rules, in order to yield useful conclusions. In counting, the symbols we use stand for *numbers* and we make such statements as $2 + 3 = 5$ without giving much thought to the meaning of either the symbols themselves or the signs $+$ (which indicates some kind of *combination*) and $=$ (which indicates some kind of *equivalence*). In group theory we use symbols in a much wider sense. They may, for instance, stand for geometrical operations such as rotations of a rigid body; and the notions of combination and equivalence must then be defined operationally before we can start translating our observations into symbols. We do arithmetic without much thought only because we are so familiar with the operational definitions, which are far from trivial, which we learnt as children. But it is worth reminding ourselves how we began to use symbols.

How did we learn to count? Perhaps we took sets of beads, as in Fig. 1.1, giving each set a name 1, 2, 3, . . . (the " whole numbers "). A set of cows, for example, can then be given the name 3, or said to

FIG. 1.1. Sets of objects representing the whole numbers.

contain 3 cows, if its members can be put in " one-to-one correspondence " with the beads of the set named 3 (a bead for each cow, no cows or beads left over). The same number is associated with different sets if, and only if, their members can be put in one-to-one correspondence : in this case the numbers of objects in the different sets are *equal*. If $x$ objects in one set can be related in this way to $y$ objects in another set we write $x = y$. If the numbers of fingers on my two hands are $x$ and $y$, I can say $x = y$ because I can put them into one-to-one correspondence : and I can say $x = y = 5$ because I can put the members of either set in one-to-one correspondence with those in the set named 5 in Fig. 1.1. This provides an operational definition of the symbol $=$. We observe that the sets in Fig. 1.1 have been given distinct names because none can be put in one-to-one correspondence with any other: $1 \neq 2 \neq 3 \neq 4 \ldots$. Numbers may be *combined under addition* (or " added "), for which we usually use the symbol $+$, by *putting together* different sets to make a new set. If we put together a set of 4 objects and a set of 1 object the resultant set is said to contain $(4 + 1)$ objects: but there is another name for the number of objects in this set because it can be put in one-to-one correspondence with the set of 5 objects. Hence the different collections contain equal numbers of objects and we write $4 + 1 = 5$. The whole numbers are conveniently arranged in the ordered sequence (Fig. 1.1) such that $1 + 1 = 2$, $2 + 1 = 3$, $3 + 1 = 4$, etc., so that sets associated with successive numbers are related by the addition of 1 object. Generally, we say that if the members of sets containing $x$ and $y$ objects, when put together to form a new set, can be put into one-to-one correspondence with those of a set of $z$ objects, then $x + y = z$. The operational meaning of the law of combination (indicated by the $+$) and of the equivalence ($=$) is now absolutely clear. But, of course, the terminology is quite arbitrary: instead of $2 + 3 = 5$ we could just as well write

2 combined with 3 gives 5 or 2 ! 3 : 5

What matters is that we agree upon (i) what the symbols stand for, (ii) what we shall understand by saying two of them are equal, or equivalent, and (iii) what we shall understand by combining them.

In group theory we deal with collections of symbols, $\mathsf{A}$, $\mathsf{B}$, ..., $\mathsf{R}$, ..., which do not necessarily stand for numbers and which are accordingly set in distinctive type (gill sans—instead of the usual italic letters). We refer to the members of the collection as " elements " and often denote the whole collection by showing one or more typical elements in braces, $\{\mathsf{R}\}$ or $\{\mathsf{A}, \mathsf{B}, \ldots, \mathsf{R}, \ldots\}$. The elements may, for example, represent geometrical operations such as rotations of a rigid body, and the law of combination is then non-arithmetic. Nevertheless rotations can be compounded in the sense that one ($\mathsf{R}$) *followed by* a second ($\mathsf{S}$) *gives the same final result as* a third ($\mathsf{T}$): the italic phrases describe, respectively, the law of combination (sequential performance) and the nature of the equivalence. In order not to restrict the law of combination we shall normally leave the question completely open, simply writing side by side the two symbols to be combined and, purely formally, calling the result of combination their "*product*". We adopt the = sign to denote equivalence in whatever sense may be appropriate. In dealing with any particular collection of elements, the meaning of the term "product" and the significance of the = sign will be agreed upon at the outset.

When the elements of a collection are the natural numbers and the law of combination is addition one important property is at once evident. Three numbers may be combined, by applying the law of combination twice, in either of two ways: with the general terminology these are

$$\text{(i)} \quad (abc)_1 = a(bc) \qquad \text{or (ii)} \quad (abc)_2 = (ab)c$$

*and the results are always identical.* To give an example

$$\text{(i)} \quad (2\,3\,4)_1 = 2\,(3\,4) = 2\,7 = 9 \qquad \text{(ii)} \quad (2\,3\,4)_2 = (2\,3)4 = 5\,4 = 9$$

The order in which the terms of a "product" are dealt with is therefore irrelevant and there is no need to bracket together the pairs to show how a triple product must be interpreted. The law of combination is said to be *associative.* In this case there is no ambiguity and an ordered product of any number of elements is *uniquely* equivalent to a certain single element. This is a fundamental property of all the collections we shall study and is assumed in defining the *group property*:

> Any collection of elements $\{A, B, \ldots, R, \ldots\}$ has the *group property* if an associative law of combination is defined such that for any ordered pair $R$, $S$ there is a unique product, written $RS$, which (in some agreed sense) is equivalent to some single element $T$ which is also in the collection. (1.1.1)

The condition that the combination of two elements should yield an element of the same kind (i.e. also in the collection) is itself non-trivial: the collection is then *closed* and we refer to the *closure* property. It should be noticed that the law of combination refers to an *ordered* pair, so that the possibility $RS \neq SR$ is admitted from the start. It must also be stressed that a collection with the group property is not *a group* unless further conditions are fulfilled. Before introducing these we give examples of various collections with the group property:

EXAMPLE 1.1. We first consider the natural numbers under addition. This collection has already been mentioned in defining the group property—which it evidently possesses. In dealing with such collections, in which *two* laws of combination are recognized (addition and multiplication), it is of course usual to employ the special symbol $+$ to indicate combination by addition. The associative property $a(bc) = (ab)c$ may be written without ambiguity as

$$a + (b + c) = (a + b) + c$$

A second important property is $ab = ba$ or, with the $+$ notation,

$$a + b = b + a$$

The elements of the collection are in this case said to *commute* and the law of combination to be *commutative.*

EXAMPLE 1.2. Let us now take the natural numbers under multiplication. The product under multiplication is in this case the simple arithmetic product of two numbers. Again, to avoid ambiguity, the law of combination may be indicated explicitly by the $\times$ sign. The law is again associative [e.g. $3 \times (4 \times 2) = (3 \times 4) \times 2$] and $a(bc) = (ab)c$ may in this case be written

$$a \times (b \times c) = (a \times b) \times c$$

The collection is clearly commutative under multiplication also: $ab = ba$ or, with the $\times$ notation,

$$a \times b = b \times a$$

In both the preceding examples the *order* in which the elements are combined is irrelevant. Collections in which the law of combination is commutative are termed "*Abelian*": but many of those we shall meet are *non-Abelian* and the reference to an *ordered pair* must be carefully observed. The following example introduces a non-commutative law of combination:

EXAMPLE 1.3. Let us consider *operations* which bring the lamina of Fig. 1.2 into a position indistinguishable from that which it originally occupied. Such an operation (which need not be *mechanically* feasible) is said to bring the system into *self-coincidence* and is called a *symmetry operation.* If we single out one point (+) on the lamina, and identify different operations by observing what happens to this point, it is clear that the lamina is brought into self-coincidence by the six operations indicated in Fig. 1.2(a–f). So long as we ignore the possibility of turning the lamina over, this is the full set of symmetry operations. The symbols beneath the six results are merely conventional names for the operations which produce them. We include the so-called identity operation $E$ for reasons which will presently become clear. $C_3$ and $\bar{C}_3$ stand for *rotations* through 120° anti-clockwise (+ve) and clockwise (−ve) while $\sigma^{(1)}$, $\sigma^{(2)}$ and $\sigma^{(3)}$ stand for *reflections* which send all points on one side of a broken line (including the +) into their " images " on the other side. It is not necessary to define positive rotation through 240° as *another* symmetry operation because, in

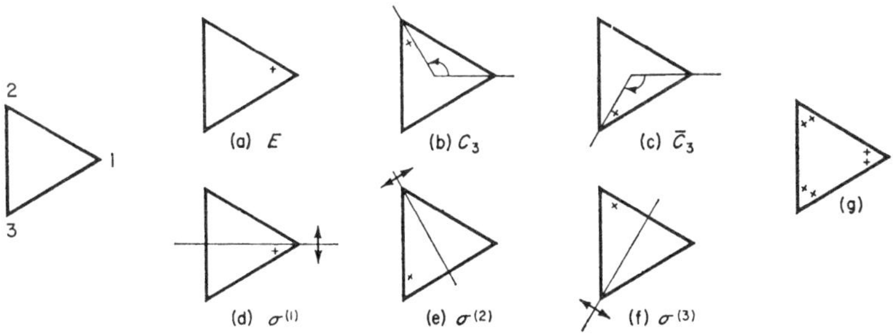


FIG. 1.2. Symmetry operations. (a)–(f) indicate operations bringing the triangular lamina into self-coincidence. (g) indicates six " equivalent points ".

terms of final effect, it is completely equivalent to $\bar{C}_3$. We note also that if the final positions of the + signs in Fig. 1.2 are superimposed in a single triangle they form a pattern (Fig. 1.2(g)) or set of "equivalent points " which has the full symmetry of the lamina itself. The number of equivalent points in such a set is the number of symmetry operations in the collection (including $E$).

We now define a law of combination by agreeing that $RS$ ($R$ and $S$ denoting any two symmetry operations) shall stand for operation $S$ followed by operation $R$: *symmetry operations may be combined by sequential performance.* The convention of putting the first operation on the *right* of a product is in accord with the accepted practice in other branches of mathematics—e.g. in $\frac{\mathrm{d}}{\mathrm{d}x}\left(x\frac{\mathrm{d}}{\mathrm{d}x}\right)$ the $x\frac{\mathrm{d}}{\mathrm{d}x}$ operates first. Once introduced, it must be carefully observed in all that follows. Since the combined effect of $R$ and $S$ is, by definition, to send the lamina into self-coincidence, it follows that $RS = T$ where $T$ is one of the six possible symmetry operations, and the = implies equivalence of effect (i.e. $RS$ and $T$ each bring the + into the same final position). To give an example, $\sigma^{(1)}C_3$ represents the operation $C_3$ followed by reflection $\sigma^{(1)}$: but $\sigma^{(1)}$ operating on the result shown in Fig. 1.1(b) gives that shown in Fig. 1.1(e)—which is also the result of the single operation $\sigma^{(2)}$. We therefore write

$$\sigma^{(1)}C_3 = \sigma^{(2)}$$

It is again stressed that writing the symbols side by side indicates *sequential performance* of the operations which they represent, and that the $=$ sign indicates equivalence, as regards final effect. It is at once clear that the symbols must now, in general, be regarded as *non-commutative*: for reversal of order shows that $C_3\sigma^{(1)} = \sigma^{(3)} \neq \sigma^{(1)}C_3$.

### 1.2. Definition of a Group

We now complete the formal definition of a group. A collection of elements $\{A, B, \ldots, R, \ldots\}$ is a group $\mathsf{G}$ if

> (a) it possesses the group property
> (b) it contains a *unit element* $E$ such that $RE = R$ (all $R$ in $\mathsf{G}$) and
> (c) it contains for every element $R$ an *inverse*, which may be called $R^{-1}$, such that $RR^{-1} = E$ (all $R$ in $\mathsf{G}$). (1.2.1)

In a non-Abelian group it is clearly important to know whether $ER$ is the same as $RE$: if $E$ is a *right unit* is it also a *left unit?* A simple argument shows that this is so: for if we let $ER = S$, then $E = E(RR^{-1}) = SR^{-1}$; but if $R^{-1}$ is the unique right inverse of $R$ this means $S = R$ and hence $ER = R$. A similar argument shows that $R^{-1}R = E$: a *right inverse* is also a *left inverse.*

We now re-examine the collections discussed in section 1.1.

Example 1.4. The collection of natural numbers, combined under addition, satisfies (a) but does not form a group because there is *no unit*—no number $x$ such that $n + x = x + n = n$. We may remedy this omission by introducing a *unit under addition* which we call 0 with this property: $n + 0 = 0 + n = n$. But the collection is still not a group because $n$ has no *inverse*, $x_n$ say, such that $n + x_n = x_n + n = 0$. To remove this defect we must include the *negative integers*, written $-1$, $-2$, etc. and defined by $n + (-n) = (-n) + n = 0$. The resultant collection $\{\ldots -n, \ldots, -2, -1, 0, 1, 2, \ldots, n, \ldots\}$ is then a group under addition.

Example 1.5. The collection of natural numbers, combined under multiplication, satisfies conditions (a) and (b), 1 being the *unit under multiplication* because $n \times 1 = 1 \times n = n$. But the collection is not a group unless we admit also the inverses written $n^{-1}$ or $1/n$ such that $n(n^{-1}) = (n^{-1})n = 1$. And if we admit $n^{-1}$ we must, according to (a), admit all numbers of the form $mn^{-1}$ (or $m/n$). The collection of all positive rational numbers (*excluding zero*) is thus a group under multiplication.

EXAMPLE 1.6. The collection of symmetry operations of an equilateral triangle (example 1.3) obviously satisfies both (a) and (b). It satisfies (c) because for every operation we can find an operation which restores the lamina to its original position; for every element there is an inverse. It should be noted that an element and its inverse need not be different members of the collection: each of the reflections, for example, is *its own* inverse. The collection $\{E, C_3, \bar{C}_3, \sigma^{(1)}, \sigma^{(2)}, \sigma^{(3)}\}$ is therefore a group. it is the symmetry group usually denoted by $C_{3v}$ and treated more fully in chapter 3.

### 1.3. The Multiplication Table

If a group G contains $g$ elements, $g$ is called its *order*. We shall be concerned mainly with groups of finite order—*finite groups*—and in this case the properties of the group are conveniently summarized in a *multiplication table* which sets out systematically the products of all $g^2$ pairs of elements.

EXAMPLE 1.7. Let us consider two finite groups, with ordinary multiplication as the law of combination:

$$G_1 = \{1, -1\}, \qquad G_2 = \{1, -1, i, -i\}$$

The multiplication tables of $G_1$ and $G_2$ are

$G_1$:

| | 1 | −1 |
|---|---|---|
| 1 | 1 | −1 |
| −1 | −1 | 1 |

$G_2$:

| | 1 | −1 | $i$ | $-i$ |
|---|---|---|---|---|
| 1 | 1 | −1 | $i$ | $-i$ |
| −1 | −1 | 1 | $-i$ | $i$ |
| $i$ | $i$ | $-i$ | −1 | 1 |
| $-i$ | $-i$ | $i$ | 1 | −1 |

The rows and columns of such tables are labelled by the group elements while the body of the table contains their $g^2$ possible products: the element equivalent to $RS$ (in that order) is placed at the intersection of the $R$-row and the $S$-column.

Generally the multiplication table is the array

| | A | B | C | ... |
|---|---|---|---|---|
| A | AA | AB | AC | ... |
| B | BA | BB | BC | ... |
| C | CA | CB | CC | ... |
| ... | ... | ... | ... | ... |

each entry being written as the single element to which it is equivalent. Such a table exhibits a certain *structure* which is sufficient to specify an *abstract group.*

EXAMPLE 1.8. Consider the group $G_2$ in example 1.7 and give the elements new names, $A$, $B$, $C$, $D$. The table then reads

| | A | B | C | D |
|---|---|---|---|---|
| A | A | B | C | D |
| B | B | A | D | C |
| C | C | D | B | A |
| D | D | C | A | B |

This table *defines* a certain group G, *irrespective of what the symbols stand for*: for it may be possible to find many different sets of quantities with exactly the same multiplication table. G is called an *abstract group.*

When the group elements are quantities of some specified kind (e.g. the collection of numbers 1, $-1$, $i$, $-i$) with a specified law of combination (e.g. multiplication), we obtain a particular *realization* of the abstract group G. Thus $G_2$ is a realization of G by a set of numbers combined under multiplication. But a study of G may be regarded as a study of *all groups* with the structure exhibited, in particular, by $G_2$. In physics and chemistry we are always concerned with realizations of abstract groups (e.g. by symmetry operations, vectors, permutations, etc.) and emphasis will naturally be placed upon such realizations. Consequently, although we must introduce certain concepts and definitions belonging to abstract group theory, we shall always provide concrete illustrations. As a final example we give the multiplication table of $C_{3v}$ (example 1.6), a group which we shall use repeatedly for illustrative purposes.

EXAMPLE 1.9. By sequential performance of the operations defined in Fig. 1.2 (exactly as in example 1.3) we obtain the following multiplication table:

| | $E$ | $C_3$ | $\bar{C}_3$ | $\sigma^{(1)}$ | $\sigma^{(2)}$ | $\sigma^{(3)}$ |
|---|---|---|---|---|---|---|
| $E$ | $E$ | $C_3$ | $\bar{C}_3$ | $\sigma^{(1)}$ | $\sigma^{(2)}$ | $\sigma^{(3)}$ |
| $C_3$ | $C_3$ | $\bar{C}_3$ | $E$ | $\sigma^{(3)}$ | $\sigma^{(1)}$ | $\sigma^{(2)}$ |
| $\bar{C}_3$ | $\bar{C}_3$ | $E$ | $C_3$ | $\sigma^{(2)}$ | $\sigma^{(3)}$ | $\sigma^{(1)}$ |
| $\sigma^{(1)}$ | $\sigma^{(1)}$ | $\sigma^{(2)}$ | $\sigma^{(3)}$ | $E$ | $C_3$ | $\bar{C}_3$ |
| $\sigma^{(2)}$ | $\sigma^{(2)}$ | $\sigma^{(3)}$ | $\sigma^{(1)}$ | $\bar{C}_3$ | $E$ | $C_3$ |
| $\sigma^{(3)}$ | $\sigma^{(3)}$ | $\sigma^{(1)}$ | $\sigma^{(2)}$ | $C_3$ | $\bar{C}_3$ | $E$ |

The multiplication table in the preceding example illustrates one general and fundamental property which we note before continuing:

> Each row (and column) of a multiplication table contains each element of the group *once* and *once only*. (1.3.1)

This is a simple consequence of the uniqueness of the law of combination. If $AB = C$ and we postulate that the $A,D$-element of the array (i.e. another product in the same row of the table) is also equal to $C$ we have $B = A^{-1}C$ and $D = A^{-1}C$: since the result of combination is unique, $D = B$, and therefore $C$ can appear only in *one place*—at the intersection of the $A$-row and $B$-column. Stated in another form, this theorem asserts that multiplication of all the elements of the group in turn by any single element simply reproduces the same collection in a different order. This is apparent in the examples already given. We write this result (sometimes called the *rearrangement theorem*) $R\mathsf{G} = \mathsf{G}$ (all $R$ in $\mathsf{G}$), the equivalent statement about the columns being $\mathsf{G}R = \mathsf{G}$ (all $R$ in $\mathsf{G}$): here, for instance, $R\mathsf{G}$ is used to denote $\{RA, RB, \ldots, RS, \ldots\}$ and the "equality" is used to indicate that two collections each contain the same set of elements, though not necessarily in the same order.

## 1.4. Powers, Products, Generators

We note that with the help of the multiplication table any product of group elements can be reduced to a single element. If, for example, we take the symmetry group $\mathsf{C}_{3v}$, the table in example 1.9 shows that

$$C_3\sigma^{(1)}\sigma^{(3)} = C_3\bar{C}_3 = E$$

Conversely, all the elements of a group of order $g$ may be expressed as products, whose factors are drawn from a limited number $(<g)$ of elements called *generators*. In the group $C_{3v}$, for instance,

$$\bar{C}_3 = C_3^2, \qquad E = C_3^3, \qquad \sigma^{(2)} = \bar{C}_3\sigma^{(1)} = C_3^2\sigma^{(1)}, \qquad \sigma^{(3)} = C_3\sigma^{(1)}$$

so that all elements may be expressed in terms of only *two*, $C_3$ and $\sigma^{(1)}$. Here we have adopted the usual algebraic terminology, writing $C_3C_3 = C_3^2$. This raises the question of how to handle " powers " of group elements. Before we can continue, it is clearly necessary to check the consistency of this terminology and to make sure of its interpretation. In general we write $RR\ldots R$ ($n$ factors) $= R^n$. Since $RR^{-1} = E$ it follows that $R^2R^{-1}R^{-1} = E$ and hence that $(R^2)^{-1} = R^{-1}R^{-1}$: similarly $(R^n)^{-1} = R^{-1}R^{-1}\ldots R^{-1}$ ($n$ factors) $= (R^{-1})^n$. The usual " laws of indices " are then obeyed if we write $(R^{-1})^n = R^{-n}$ and note that $R^0 = E$.

We now return to the discussion of generators. In general, a group may be determined by giving a *system of generators*, with specified properties and forming their powers and products until no further elements can be found. Since, if $R$ is in G, $R^2, R^3, \ldots, R^n, \ldots$ are also in G, G will be infinite unless $R^n = E$ for some value of $n$. A set of elements $\{R, R^2, \ldots, R^n\ (= E)\}$ is called a *cyclic group* of order $n$.

EXAMPLE 1.10. In the case of $C_{3v}$ the properties of the generators are

$$C_3^3 = E, \qquad \sigma^{(1)2} = E, \qquad C_3\sigma^{(1)}C_3 = \sigma^{(1)}$$

Only six *distinct* products can be found when these properties are observed and they satisfy the original multiplication table when we name the products as follows:

$$C_3^2 = \bar{C}_3, \qquad C_3^3 = E, \qquad C_3^2\sigma^{(1)} = \sigma^{(2)}, \qquad C_3\sigma^{(1)} = \sigma^{(3)}$$

No further elements arise because, for example,

$$C_3\sigma^{(1)3}C_3^2 = C_3\sigma^{(1)}C_3^2 = \sigma^{(1)}C_3 = C_3^2\sigma^{(1)} = \sigma^{(2)}$$

the last step arising from $C_3\sigma^{(1)}C_3 = \sigma^{(1)}$ on multiplying by $C_3^2$

The definition of a group by means of its generators is an exceedingly useful device. In chapter 3, for example, it provides a simple method of classifying the symmetry groups. In particular $C_{3v}$ is evidently determined by specifying one *typical* rotation and one *typical* reflection, which can be combined and repeated until all possible operations have been exhausted.

We conclude with a formal definition:

| | |
|---|---|
| A set P of elements of a group G is a *system of generators* of the group if every element of G can be written as the product of a finite number of factors, each of which is either an element of P or the inverse of such an element. | (1.4.1) |

With the convention already introduced, $(R^{-1})^n = R^{-n}$, this merely means that each element can be written as a product of (positive *or* negative) integral powers of the generators.

### 1.5. Subgroups, Cosets, Classes

Any collection of the elements of G which *by themselves* form a group H is called a *sub*group of G. We already possess examples of subgroups.

EXAMPLE 1.11. The set $\{E, C_3, \bar{C}_3\}$ is a subgroup of $C_{3v}$. It is the cyclic group $C_3$ and is determined by the *single* generator $C_3$ with the property $C_3{}^3 = E$. But $\{E, \sigma^{(1)}\}$, $\{E, \sigma^{(2)}\}$, $\{E, \sigma^{(3)}\}$ are also subgroups and are also cyclic since $\sigma^{(1)2} = \sigma^{(2)2} = \sigma^{(3)2} = E$. The group therefore contains four cyclic subgroups.

Every group contains, of course, two trivial or "improper" subgroups—the unit element and the whole group itself; other subgroups are said to be *proper* subgroups. It should be noticed that the set of subgroups does not correspond to a *partition* of the original collection, for $E$ must be a member of *every* subgroup. The possibility of usefully partitioning the group into distinct sets of elements (no element common to two or more sets) will now be examined and will enable us to establish the following basic result:

| | |
|---|---|
| The order $h$ of any subgroup H must be a divisor of the order $g$ of the group G. | (1.5.1) |

This theorem allows us to state, for example, that the group $C_{3v}$ can have proper subgroups of order 2 and 3 only: there is no point in looking for subgroups of order 4 and 5.

The proof of (1.5.1) depends on the idea of a *coset*. Let $H = \{A_1, A_2, \ldots, A_h\}$ and suppose $R_1, R_2, \ldots$ are elements of G *not* contained in H : then the collection defined by

$$HR_k = \{A_1R_k, A_2R_k, \ldots, A_hR_k\}$$

is called the *right coset* of H with respect to $R_k$. Thus with $G = C_{3v}$ and $H = \{E, C_3, \bar{C}_3\}$ the right coset of H with respect to $\sigma^{(1)}$ is (see example 1.9) $H\sigma^{(1)} = \{\sigma^{(1)}, \sigma^{(3)}, \sigma^{(2)}\}$. The coset $HR_k$ is *not* a group, although it contains the same number of elements as H. Cosets have the following properties:

(i) Every element of a group appears either in the subgroup or in one of its cosets.
(ii) No element can be common to both a subgroup and one of its cosets.
(iii) No element can be common to two different cosets of the same subgroup.
(iv) No coset can contain the same element more than once.

Each proposition may be established by assuming the contrary and showing that this leads to contradictions—a very common method of proof in group theory. As a result, it follows that the $g$ elements of G are partitioned into a number of distinct sets each containing $h$ elements: thus $g = nh$ where $n$ is an integer, and the theorem is established.

It is, of course, also possible to define left cosets. For a finite group, however, it can be shown that right and left cosets give exactly the same partitioning of the group (though the right and left cosets with respect to one particular element are not necessarily identical).

Finally, we introduce the idea of a *class*. This depends on the following definition:

An element $B$ is said to be *conjugate* to $A$ with respect to $R$ if

$$B = RAR^{-1}$$

If we form all the elements conjugate to $A$ as $R$ runs through the whole group G and collect the *distinct* results, we get a *class*—the class of all elements conjugate to $A$. We see at once that any two elements of a class are conjugate to each other (with respect to some member of the group). For if $\{A_1, A_2, \ldots\}$, is some class we can suppose $A_i = RA_1R^{-1}$ and $A_j = SA_1S^{-1}$: hence $A_i = R(S^{-1}A_jS)R^{-1} = TA_jT^{-1}$ where $T = RS^{-1}$ is some element of the group.

Arguments analogous to those employed in dealing with cosets then allow us to establish that a group may also be partitioned into *distinct classes*. The classes, however, may contain different numbers of elements. The unit element, for example, always forms a class by itself since it is sent into itself by conjugation with all the group elements: $E = RER^{-1}$ (all $R$ in G). The classes of $C_{3v}$ are readily found using the multiplication table in example 1.9. They are

$$\{E\} \qquad \{C_3, \bar{C}_3\} \qquad \{\sigma^{(1)}, \sigma^{(2)}, \sigma^{(3)}\}$$

The value of a division into classes will become evident later ; and the number of distinct classes is a particularly important characteristic of a group.

**1.6. Invariant Subgroups. The Factor Group**

It is possible to define conjugate *subgroups* in much the same way as conjugate elements. If $H = \{A_1, A_2, \ldots, A_h\}$ is a subgroup of G and we denote by $RHR^{-1}$ the *collection* of conjugates, with respect to $R$ of all the elements of H then $RHR^{-1}$ is the subgroup conjugate to H with respect to $R$. The collection *is* a subgroup because its elements $(RA_1R^{-1}, RA_2R^{-1}, \ldots, RA_hR^{-1})$ evidently have the same multiplication table as H: thus if $A_iA_j = A_k$ it is identically true that $(RA_iR^{-1})(RA_jR^{-1}) = (RA_kR^{-1})$. A particularly important case arises when $RHR^{-1} = H$ (all $R$ in G) : H is then said to be an *invariant* or *normal* subgroup.

EXAMPLE 1.12. In the case of $C_{3v}$ only one of the four subgroups (example 1.11) is invariant, namely $\{E, C_3, \bar{C}_3\}$. The subgroup $\{E, \sigma^{(1)}\}$, for example, cannot be invariant because $C_3\sigma^{(1)}\bar{C}_3 = \sigma^{(2)}$ which is not contained in the subgroup.

When a group G possesses an invariant subgroup H, the collection consisting of H and its distinct cosets

$$H, \quad HR_2, \quad HR_3, \quad \ldots, \quad HR_n \quad (\text{with } R_1 = E)$$

can be regarded as a new group of order $n$ in which the subgroup and each coset plays the part of a *single element*, provided an appropriate law of combination is defined. For this purpose, the product of two collections is defined as the collection of all *distinct* results when all pairs of elements, one from each collection, are combined in the usual way. When collections are manipulated in this way, as single elements with a definite law of combination, they are referred to as *complexes*. It is not difficult to show that the system of complexes (subgroup and cosets) then forms a group. This new group is called the *factor group* of G and is denoted (purely formally) by G/H.

EXAMPLE 1.13. $C_{3v}$ has an invariant subgroup $H = \{E, C_3, \bar{C}_3\}$ and a right coset $H\sigma^{(1)} = \{\sigma^{(1)}, \sigma^{(2)}, \sigma^{(3)}\}$. If we combine an element of H and an element of $H\sigma^{(1)}$ there are nine possible results; but in fact, the only *different* answers are $\sigma^{(1)}$, $\sigma^{(2)}$, and $\sigma^{(3)}$—each arising three times. The product of the two collections is thus $\{\sigma^{(1)}, \sigma^{(2)}, \sigma^{(3)}\}$ and we can write, with this new law of combination, $\{H\}\{H\sigma^{(1)}\} = \{H\sigma^{(1)}\}$ (the braces being inserted to emphasize that the " elements " are themselves collections).

It should be noted that the factor group has the same multiplication table as the unit element and the distinct coset generators: in the example, these are $\{E, \sigma^{(1)}\}$. It follows that if we take any one element from each complex and combine them, *not distinguishing* elements which differ by a factor taken from the subgroup, we get a group with the same structure as the factor group. In this sense, the factor group results when we choose not to distinguish different elements within the subgroup and within each coset. This result is quite general.

### 1.7. Homomorphisms and Isomorphisms

In representation theory we are concerned with various collections of quantities which satisfy the same multiplication table as a given group, and which are therefore similar in structure. Such a *similarity of structure* is termed a *homomorphism.*

EXAMPLE 1.14. Consider, for illustration, the group $C_{3v}$ and let us associate with $E$ and the rotations the quantity $+1$, and with the reflections the quantity $-1$, thus:

| $E$ | $C_3$ | $\bar{C}_3$ | $\sigma^{(1)}$ | $\sigma^{(2)}$ | $\sigma^{(3)}$ |
|---|---|---|---|---|---|
| 1 | 1 | 1 | $-1$ | $-1$ | $-1$ |

Then the associated quantities do satisfy the same multiplication table. For example $C_3\sigma^{(1)} = \sigma^{(3)}$ and the associated quantities are similarly related $1(-1) = -1$. We note that the two distinct quantities which we have associated with the elements of the group themselves form a group of order 2. What we are doing is to make an association between the elements of two different groups, thus:

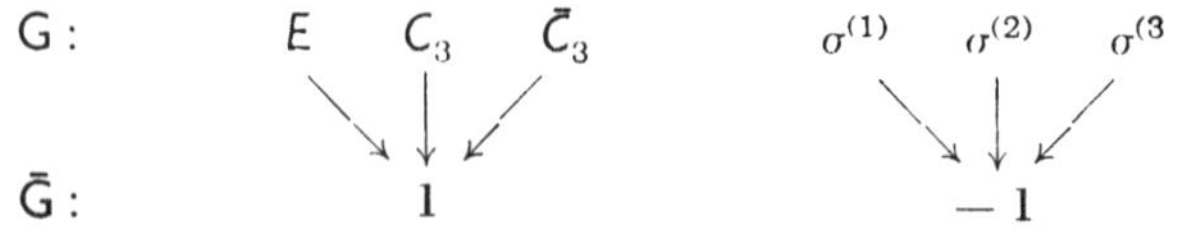

In the above example one group, G, is said to be " mapped " on to another, $\bar{G}$, each element of G having an " image " in $\bar{G}$; and the mapping is described as " three-to-one " because three elements of G are associated with each one of $\bar{G}$. Since in this case we associate the same symbol with a whole set of elements such a mapping fails to distinguish different elements within each set: and this is the kind of association already met, in example 1.13 *et seq.*, in defining the factor group. In fact, a group can always be mapped on to its factor group (in which the sets of elements are combined as complexes), or upon the identity $E$ and a complete set of coset generators (combined as single elements).

Example 1.15. The group $C_{3v}$ may be mapped on to its factor group $\{H, H\sigma^{(1)}\}$, where $H = \{E, C_3, \bar{C}_3\}$. Thus

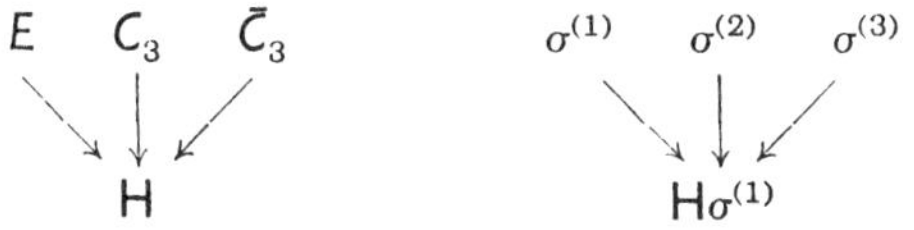


and when, for example, $C_3\sigma^{(1)} = \sigma^{(3)}$, the complexes associated with $C_3$, $\sigma^{(1)}$ and $\sigma^{(3)}$ (respectively) combine the same way:

$$\{H\}\,\{H\sigma^{(1)}\} = \{H\sigma^{(1)}\}$$

In the same way, the first three elements may be mapped on to $E$ and the second three on to $\sigma^{(1)}$ (identity and coset generator)

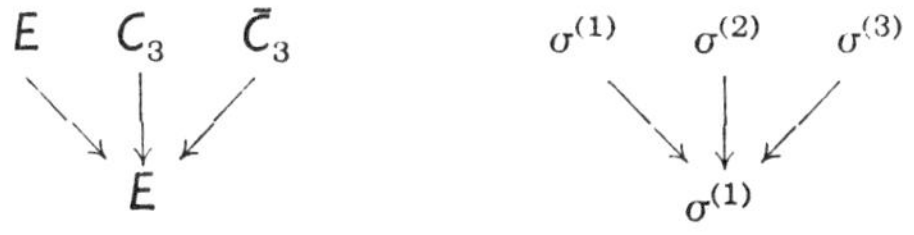


and when $C_3\sigma^{(1)} = \sigma^{(3)}$, the associated symbols combine in the same way: $E\sigma^{(1)} = \sigma^{(1)}$. We could also use any member of each set (e.g. $C_3$ and $\sigma^{(2)}$) in place of $E$ and $\sigma^{(1)}$ if we used the equality to mean " identical except for factors drawn from the subgroup". Thus, on replacing $\{E, \sigma^{(1)}\}$ by $\{C_3, \sigma^{(2)}\}$, $C_3\sigma^{(1)} = \sigma^{(3)}$ gives for the associated elements $C_3\sigma^{(2)} = \sigma^{(1)} = \sigma^{(2)}(\bar{C}_3)$. The associated elements thus combine the same way if we discard the subgroup element in brackets.

In general, the two groups G and $\bar{G}$ involved in a homomorphism are not truly *identical* in structure, containing different numbers of elements. On the other hand, had the association been one-to-one instead of three-to-one, the groups $\mathsf{G}$ and $\bar{\mathsf{G}}$ would have been *identical in structure* or " *isomorphic* ".‡

EXAMPLE 1.16. The abstract group in example 1.8 and the group $\{1, -1, i, -i\}$ are isomorphic under the one-to-one association

$$\begin{array}{cccc} A & B & C & D \\ \downarrow & \downarrow & \downarrow & \downarrow \\ 1 & -1 & i & -i \end{array}$$

After these preliminaries it remains only to give a formal definition of a homomorphic mapping. We suppose the groups are $\mathsf{G} = \{A, B, \ldots, R, \ldots\}$ and $\bar{\mathsf{G}} = \{\bar{A}, \bar{B}, \ldots, \bar{R}, \ldots\}$, and that $\{A', B', \ldots, R', \ldots\}$ is the collection of images of $A, B, \ldots, R, \ldots$. Then $A', B', \ldots, R', \ldots$ are elements of $\bar{\mathsf{G}}$, but each element may appear *more than once* in the collection $\{A', B', \ldots, R', \ldots\}$. Thus, in example 1.14 $\{A', B', \ldots, R', \ldots\} = \{1, 1, 1, -1, -1, -1\}$ while $\bar{\mathsf{G}} = \{1, -1\}$. Now the two collections $\{A, B, \ldots, R, \ldots\}$ and $\{A', B', \ldots, R', \ldots\}$ satisfy the same multiplication table provided $RS = T$ in the first implies that the associated elements in the second satisfy $R'S' = T'$. Since $T'$ is the image of $T$ (i.e. of $RS$) we may write $T' = (RS)'$ and require $(RS)' = R'S'$ : the image of the product must be the product of the images. A definition may now be formulated:

> A mapping of $\mathsf{G}$ on to $\bar{\mathsf{G}}$ is called *homomorphic* when the condition $(RS)' = R'S'$, $R'$, $S'$ being images in $\bar{\mathsf{G}}$, is fulfilled for all pairs of elements $R$ and $S$ of the group $\mathsf{G}$. If, in addition, the association $R \rightarrow R'$ is unique or one-to-one, so that each element of $\mathsf{G}$ has a different image in $\bar{\mathsf{G}}$, the mapping is *isomorphic*. (1.7.1)

It is of course at once evident that the distinct elements of $\{A', B', \ldots, R', \ldots\}$ must form a group. We now pass to an important illustration of these ideas.

### 1.8. Elementary Concept of a Representation

We have observed in the case of $\mathsf{G} = \mathsf{C}_{3v}$ that the association

| $E$ | $C_3$ | $\bar{C}_3$ | $\sigma^{(1)}$ | $\sigma^{(2)}$ | $\sigma^{(3)}$ |
|---|---|---|---|---|---|
| 1 | 1 | 1 | $-1$ | $-1$ | $-1$ |

‡ Sometimes an alternative terminology is used : isomorphic → " simply isomorphic ", homomorphic → " multiply isomorphic ".

introduces a collection of symbols which (under their own stated law of combination) satisfies the same multiplication table as the group G; and that the *distinct* elements of the new collection themselves form a group $\bar{G} = \{1, -1\}$. In technical language, we have set up a three-to-one homomorphic mapping of the group G on to the group $\bar{G}$. There is however an even more trivial association, $R \to 1$ (all $R$ in G), for obviously the collection of $g$ 1's under ordinary multiplication must satisfy the same multiplication table as *any* group G of order $g$. Every element of G has the *same* image, and the group $\bar{G}$ degenerates into the single element 1. This association defines a six-to-one mapping of $G = C_{3v}$ on to $\bar{G} = 1$. Both the mappings considered are examples of *representations* of the group $C_{3v}$, a term which we define more fully after considering one more example.

EXAMPLE 1.17. Let us consider the set of matrices ‡

$$\boldsymbol{M}_1 = \begin{pmatrix} 1 & 0 \\ 0 & 1 \end{pmatrix} \quad \boldsymbol{M}_2 = \begin{pmatrix} -\frac{1}{2} & -\frac{1}{2}\sqrt{3} \\ \frac{1}{2}\sqrt{3} & -\frac{1}{2} \end{pmatrix} \quad \boldsymbol{M}_3 = \begin{pmatrix} -\frac{1}{2} & \frac{1}{2}\sqrt{3} \\ -\frac{1}{2}\sqrt{3} & -\frac{1}{2} \end{pmatrix}$$

$$\boldsymbol{M}_4 = \begin{pmatrix} 1 & 0 \\ 0 & -1 \end{pmatrix} \quad \boldsymbol{M}_5 = \begin{pmatrix} -\frac{1}{2} & -\frac{1}{2}\sqrt{3} \\ -\frac{1}{2}\sqrt{3} & \frac{1}{2} \end{pmatrix} \quad \boldsymbol{M}_6 = \begin{pmatrix} -\frac{1}{2} & \frac{1}{2}\sqrt{3} \\ \frac{1}{2}\sqrt{3} & \frac{1}{2} \end{pmatrix}$$

Then, taking matrix multiplication as the law of combination for this collection, we can verify that the matrices have the same multiplication table as $C_{3v}$ provided we make the association

| $E$ | $C_3$ | $\bar{C}_3$ | $\sigma^{(1)}$ | $\sigma^{(2)}$ | $\sigma^{(3)}$ |
|---|---|---|---|---|---|
| $\boldsymbol{M}_1$ | $\boldsymbol{M}_2$ | $\boldsymbol{M}_3$ | $\boldsymbol{M}_4$ | $\boldsymbol{M}_5$ | $\boldsymbol{M}_6$ |

Thus $\sigma^{(1)}C_3 = \sigma^{(2)}$ and the images of the three elements, namely $\boldsymbol{M}_4$, $\boldsymbol{M}_2$ and $\boldsymbol{M}_5$, respectively, are related in precisely the same way, $\boldsymbol{M}_4\boldsymbol{M}_2 = \boldsymbol{M}_5$.

Although we can always agree to associate certain matrices with the elements of a group, there will be no point in doing this unless the matrices reflect the structure of the original group; a homomorphism is a very special kind of association. If, in the above example, we tried to associate the unit matrix $\boldsymbol{M}_1$ with $C_3$ (i.e. if we interchanged $\boldsymbol{M}_1$ and $\boldsymbol{M}_2$) the images would no longer satisfy the same multiplication table as $C_{3v}$ and the mapping would not be homomorphic. We note

‡ Throughout this book matrices are represented by bold letters. It is assumed that the reader is familiar with the elementary properties of matrices (see, for example, G. G. Hall, *Matrices and Tensors*, Pergamon Press, Oxford, 1963).

in the present case that the association is one-to-one since the six matrices are all different : the mapping is therefore isomorphic and the whole collection $\{\boldsymbol{M}_1, \boldsymbol{M}_2, \ldots, \boldsymbol{M}_6\}$ constitutes the second group $\mathsf{G}$.

We now define a representation :

> If we can find $g$ matrices $\{\boldsymbol{A}, \boldsymbol{B}, \ldots, \boldsymbol{R}, \ldots\}$ and associate them with the $g$ elements of a group $\mathsf{G} = \{A, B, \ldots, R, \ldots\}$ in such a way that when $RS = T$ then $\boldsymbol{RS} = \boldsymbol{T}$ for all pairs of elements $(R, S)$, the matrices are said to form a *representation* of $\mathsf{G}$. (1.8.1)

All the collections we have considered in this section are therefore representations of the group $\mathsf{C}_{3v}$ though the first two involve only $1 \times 1$ matrices (i.e. single numbers). The third representation, in terms of $2 \times 2$ matrices, is said to be *faithful* because of its one-to-one nature. No indication of the origin of this matrix group was given : the theory of representations of finite groups, taken up in detail in the later chapters, is concerned with determining the number and type of certain special representations (the so-called irreducible representations) and in simple cases with the explicit construction of the matrices.

In conclusion, we remark that the association of a matrix with a group element is often indicated by a " functional " notation. Instead of using corresponding bold letters $\boldsymbol{A}, \boldsymbol{B}, \ldots, \boldsymbol{R}, \ldots$ for the matrices associated with $A, B, \ldots, R, \ldots$ we may use a *common* letter (usually‡ $\boldsymbol{D}$ or $\boldsymbol{\Gamma}$) indicating the matrix associated with a particular element $R$ by enclosing the $R$ in brackets thus $\boldsymbol{D}(R)$ (" function of $R$ "). With this notation $\boldsymbol{A} = \boldsymbol{D}(A)$, $\boldsymbol{B} = \boldsymbol{D}(B)$, $\ldots$, $\boldsymbol{R} = \boldsymbol{D}(R)$, $\ldots$ and we speak of a " representation D by matrices $\boldsymbol{D}(R)$ ".

### 1.9. The Direct Product

Suppose we have two quite independent groups $\mathsf{A} = \{A_1, A_2, \ldots, A_m\}$ and $\mathsf{B} = \{B_1, B_2, \ldots, B_n\}$ and consider the set of elements $\{A_iB_j\}$. If we define a law of combination for the " products " by agreeing that factors of different type (one $A$, one $B$) commute while those of the same type (both $A$, or both $B$) are combined as in the separate groups, then the set of $mn$ products forms a new group. The new group, of order $mn$, is called the *direct product* of $\mathsf{A}$ and $\mathsf{B}$ and is denoted by

‡ $\boldsymbol{D}$ from the German " darstellung " (representation).

$\mathsf{A} \times \mathsf{B}$. It is easy to show that the group properties will be satisfied: for example,

$$(A_iB_j)(A_kB_l) = A_iA_kB_jB_l = A_pB_q$$

where $A_p$ and $B_q$ are elements of $\mathsf{A}$ and $\mathsf{B}$. Hence the product of two elements of the direct product group is another element of the direct product. It is in fact not necessary that the two groups shall be completely independent entities, so long as their elements commute with each other.

The properties of a direct product are clearly determined completely by those of the separate groups and it is therefore advantageous, wherever possible, to regard a group as a direct product of simpler groups. Examples often occur among the symmetry groups.

EXAMPLE 1.18. Let us consider again the symmetrical lamina of Fig. 1.2 (a-f) but admit a new symmetry operation $\sigma_h$ which reflects a point on the upper surface of the lamina to one directly beneath it on the lower surface. Then $E$ and $\sigma_h$ form a group ($\mathsf{A}$) in which $\sigma_h{}^2 = E$ and the operations of $\mathsf{A}$ commute with those of $\mathsf{C}_{3v}$ (which we momentarily call $\mathsf{B}$): it does not matter whether we move the $+$ to the lower surface and then rotate it, or rotate the cross and then move it to the lower surface. The group $\mathsf{A} \times \mathsf{B}$ then contains $2 \times 6 = 12$ operations:

$$\mathsf{A} \times \mathsf{B} = \{E, C_3, \bar{C}_3, \sigma^{(1)}, \sigma^{(2)}, \sigma^{(3)}, \sigma_h, \sigma_h C_3, \sigma_h \bar{C}_3, \sigma_h \sigma^{(1)}, \sigma_h \sigma^{(2)}, \sigma_h \sigma^{(3)}\}$$

and these yield 12 equivalent points, the six $+$'s on top of the lamina being accompanied by six beneath.

## 1.10. The Algebra of a Group

In conclusion we introduce certain concepts which really belong to modern algebra but which are nevertheless useful in later chapters.

In section 1.2 we observed that the real numbers formed a group in two senses (i) under addition and (ii) under multiplication. In the second case every number, *except zero*, was admitted, each having a unique inverse. When 0 is included we obtain a *field*, the field of real numbers, in which addition and multiplication are both defined and are associative and distributive:

$$(\alpha + \beta) + \gamma = \alpha + (\beta + \gamma), \qquad (\alpha\beta)\gamma = \alpha(\beta\gamma),$$
$$\alpha(\beta + \gamma) = \alpha\,\beta + \alpha\gamma$$

But the number 0 evidently has the special property, under multiplication, $0\xi = \xi 0 = 0$ ($\xi$ any number) and the equation $\alpha\xi = 1$ has a unique solution ($\xi = \alpha^{-1}$) only for $\alpha \neq 0$.

The real numbers in fact form a *commutative field*, since both laws of combination are commutative. Although the elements of this particular collection possess the group property we do not usually refer to them as " group elements " but simply as " numbers ". The *complex* numbers also form a commutative field ; and by " number " in general we shall understand an element of this field.

For some purposes it is useful to define quantities and laws of combination, involving both group elements proper (e.g. the symmetry operations in Fig. 1.2) and numbers. Let us write

$$\left.\begin{aligned} X &= x_1 R_1 + x_2 R_2 + \ldots + x_g R_g \\ Y &= y_1 R_1 + y_2 R_2 + \ldots + y_g R_g \\ &\text{etc.} \end{aligned}\right\} \quad (1.10.1)$$

in which $\{R_i\} = \mathsf{G}$ and $x_1, x_2, \ldots y_1, y_2, \ldots$ are numerical coefficients. These quantities are said to be elements of the *group algebra* associated with $\mathsf{G}$. The algebra is defined by specifying the conventions with which its elements are to be combined : let us adopt distributive and associative laws under both addition and multiplication, treating numerical coefficients commutatively. The sum of $X$ and $Y$ is then

$$Z = X + Y = (x_1 + y_1)R_1 + (x_2 + y_2)R_2 + \ldots$$

—another element of the algebra, with coefficients specified by

$$Z = X + Y: \qquad z_k = x_k + y_k \qquad (1.10.2)$$

The product law is obtained similarly :

$$Z = XY = (x_1 R_1 + x_2 R_2 + \ldots)(y_1 R_1 + y_2 R_2 + \ldots) = \sum_{i,j} x_i y_j R_i R_j$$

Since $R_i R_j$ is a member of the group, the product is also an element of the group algebra. The coefficient $z_k$ is the sum of products $x_i y_j$ for all those products $R_i R_j$ which are equal to $R_k$ and the product law may therefore be written

$$Z = XY: \qquad z_k = \sum_{\substack{i,j \\ (R_i R_j = R_k)}} x_i y_j \qquad (1.10.3)$$

The terms appearing in the sum are selected by referring to the multiplication table.

EXAMPLE 1.19. Two elements of the algebra of the group $C_{3v}$ are

$$X = E + 2C_3$$
$$Y = E + 3\sigma^{(1)}$$

Their sum is

$$Z = X + Y = 2E + 2C_3 + 3\sigma^{(1)}$$

Their product is

$$Z' = XY = (E + 2C_3)(E + 3\sigma^{(1)}) = E + 3\sigma^{(1)} + 2C_3 + 6\sigma^{(3)}$$

where $C_3\sigma^{(1)}$ has been reduced using the multiplication table.

It is clear that the group algebra, like the group itself, possesses the property of *closure* : no matter how many symbols are combined, in conformity with the agreed axioms, the result will be simply another element of the algebra of the form (1.10.1). The group elements‡ are referred to as the *basal units* of the group algebra, and their number is the " order " of the algebra. They are analogous to the unit vectors of elementary vector analysis, and just as there is a freedom of choice in selecting unit vectors there is a freedom in choosing basal units. If the algebra is of order $g$, *any* $g$ independent‡‡ linear combinations of the group elements may be used as basal units. When the basal units are not simply the elements of the group, the multiplication law is less simple : (1.10.3) must be replaced by

$$Z = XY: \qquad z_k = \sum_{i,j} A^k_{ij} x_i y_j \qquad (1.10.4)$$

where $A^k_{ij}$ is the coefficient of the $k$th basal unit in the product of the $i$th and $j$th units. The definitions of sum and product according to (1.10.2) and (1.10.4) then characterize a linear algebra in general (not just the group algebra) : the three-index symbols, $A^k_{ij}$, are known in general as the multiplication constants of the algebra. In the older literature such an algebra is often referred to as a system of " hyper-complex numbers ".

## BIBLIOGRAPHY

ALEXANDROFF, P. S., *An Introduction to the Theory of Groups*, Blackie 1959.
LEDERMAN, W., *Introduction to the Theory of Finite Groups*, Oliver & Boyd, 1957.

‡ Note the double usage of the term " element ".
‡‡ The general notion of " independence " will be clarified in the next chapter.

CHAPTER 2

# LATTICES AND VECTOR SPACES

## 2.1. Lattices. One Dimension

Let us take a *unit vector*, $a$, and try to generate from it a collection of elements forming a group under vector addition. The vector $a$ is here interpreted in the geometrical sense as a *displacement* (Fig. 2.1(a)) and the law of combination of displacements, defined as sequential performance, is adopted as the law of vector addition. In dealing with vector *addition* we indicate the mode of combination explicitly by the $+$ sign, writing, for example, $a+a$ instead of $aa$ and calling the

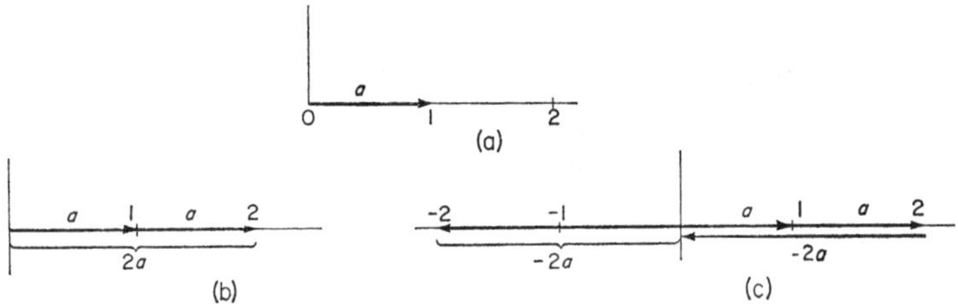


FIG. 2.1. Combination of displacements (one-dimensional). From the vector in (a) we derive the vector $2a$ in (b) by sequential performance. The vector $-2a$ in (c) is defined by its property of annulling $2a$, combination giving zero displacement.

result $2a$ instead of $a^2$. The observations of section 1.4 are easily translated into this "addition terminology". $n$-Fold repetition of the displacement $a$ is equivalent to a single displacement, denoted by $na$, which would carry a point from the origin to point $n$ (Fig. 2.1(b)): all such displacements must be elements of the group. The unit of the group must be *zero* displacement denoted by $\mathbf{0}$: $\mathbf{0}+na = na+\mathbf{0} = na$. And the inverse of $na$ must be the vector, which we call $-na$, which *returns* a point at $n$ to the origin (so that $na$ followed by $-na$ is equivalent to zero displacement): $na+(-na) = \mathbf{0}$. It is then apparent that (with a slight extension of notation: $\mathbf{0} = 0a$, $a = 1a$) all the displacements $\{na\}$ which carry a point from the origin to the points $\{n\}$ form a group in which the law of combination is

$$ma + na = (m+n)a \quad \text{(integral } m, n\text{, positive, negative or zero)} \qquad (2.1.1)$$

All the group elements may be expressed in terms of one *generator* $a$ and its inverse (i.e. as sums‡ of $a$'s or $-a$'s) and the group is accordingly a *cyclic group* in the sense of example 1.11. The group is also Abelian since the law of combination is commutative; and is infinite since there is a one-to-one correspondence between its elements and the integers $\{n\}$. In the language of section 1.7 there is an isomorphic mapping $na \rightarrow n$ (all $n$) of the group of displacements upon the group of integers (under addition).

Although the nature of the infinite cyclic group under addition is most readily grasped from a concrete geometrical realization, it should be noted that the group could be defined quite formally, without appeal to a geometrical picture, as in the case of the integers (example 1.4). Thus, from the group property, $a$, $2a$, $3a$, ..., $na$, ... must all be members of the collection. There is as yet no unit $x$ with the property $na + x = na$ (all positive $n$), so we introduce one, calling it $0$ (the *zero vector*). We must then define inverses ($x_n$, say) such that $na + x_n = 0$ (all positive $n$). If we denote the inverse of $a$ itself by $-a$, then since $2a + 2(-a) = a + a + (-a) + (-a) = a + (-a) = 0$ it follows that the inverse of $2a$, which we denote by $-2a$ is $2(-a)$. Generally the inverse of $na$ is $-na = n(-a)$. The resultant collection

$$\{\ldots,\quad -na,\quad \ldots,\quad -2a,\quad -a,\quad 0,\quad a,\quad 2a,\quad \ldots,\quad na,\quad \ldots\}$$

is then a group. Finally, we note that the terminology implies $0 = 0a$. For, in general,

$$-na + ma = \overbrace{(-a) + (-a) + \ldots}^{n\ \text{terms}} + \overbrace{a + a + \ldots}^{m\ \text{terms}} = (m - n)a,$$

where $(m - n)$ on the right-hand side is simply the number by which the $a$'s exceed the $-a$'s. The equivalence of $0$ and $0a$ follows when $m = n$; and similarly $a = 1a$, $-a = -1a$.‡‡ Finally, the group is $\{na\}$ where $n$ takes *all* integral values from $-\infty$ to $+\infty$ including zero, and the law of combination is readily seen to be $ma + na = (m + n)a$ for all integral values of $m$ and $n$ (positive, negative or zero).

Returning to the geometrical realization, we now place points at all the positions marked by the integers in Fig. 2.1 to obtain a *1-dimensional lattice* (Fig. 2.2) extending indefinitely in both directions. There are then two interpretations of the group we have been discussing. First, it is the infinite set of those displacements which carry a point *from the origin* to each one of the lattice points. When thought of in this way, as a means of identifying individual lattice points, the displacements are called *position vectors*. Secondly, the group may be regarded as the collection of *symmetry operations*—in this case translations (cf. the rotations and reflections of example 1.3)—which when applied to the whole point set bring it into self-coincidence.‡‡‡

‡ i.e. positive or negative " powers " if we used the " multiplication terminology " of section 1.4.

‡‡ It should be evident again that these considerations simply express in addition terminology the arguments in section 1.4 concerning the interpretation of " powers " of $a$ (multiplication terminology). In changing the terminology we merely replace $a^n$ by $na$, and the product of group elements by the sum.

‡‡‡ In chapter 3 we distinguish the two interpretations of a displacement vector by using a new symbol for the actual translational *operation*: the distinction is at present unimportant.

EXAMPLE 2.1. The lattice points in Fig. 2.2 may of course be replaced by copies of an *object* or " unit of pattern " at the origin, as in

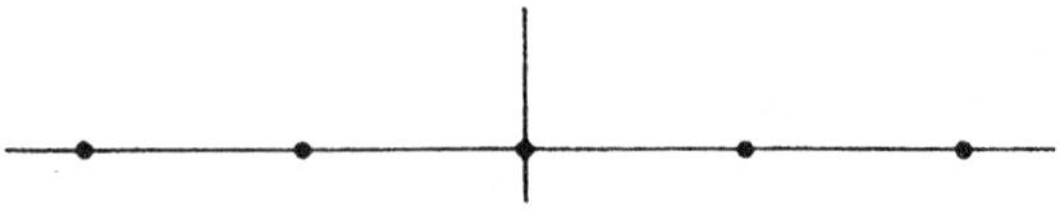

FIG. 2.2. One-dimensional lattice. The points are assumed to extend indefinitely in both directions.

Fig. 2.3(a) and the group of displacements is then the *translation group* of the pattern. It is possible, however, that the group of translations is only a subgroup of the full symmetry group. Figure 2.3(b) shows a pattern which clearly has a higher symmetry than (a): for there are

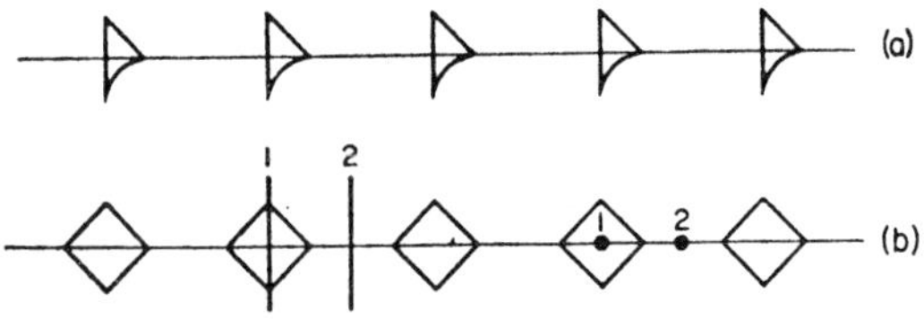


Fig. 2.3. Repeating patterns. (a) is brought into self-coincidence by translations alone, (b) by translations, rotations (about centres such as 1 and 2), and reflections (across lines such as 1 and 2 and the axis).

infinitely many centres (of types 1 and 2) about which the pattern may be *rotated* into self-coincidence, and infinitely many planes (also of types 1 and 2) through which the pattern may be *reflected* into self-coincidence.

As the examples show there are *two kinds* of operation which carry an infinite repeating pattern into self-coincidence. These are conveniently distinguished by noting what happens to the position vector of any point in the pattern. Certain symmetry operations leave one point in the pattern fixed and if this point is chosen as origin the position vectors of all other points suffer only rotations or reflections: such operations form a *point group*, one example of which has been discussed already (example 1.3). The infinite repeating patterns—which occur naturally in crystals—introduce *translational* symmetry operations under which every point suffers a common displacement, its position vector changing according to $\mathsf{r} \rightarrow \mathsf{r}' = \mathsf{r} + \mathsf{t}$ where $\mathsf{t}$ is the same for every point $\mathsf{r}$ ($\mathsf{t} = n\mathsf{a}$ in the one-dimensional case). The full symmetry group of such an object includes both translations and point group operations, and, by the group property, composite operations: it is then called a *space group*. Such groups will be discussed in detail in chapter 3.

## 2.2. Lattices. Two and Three Dimensions

There is no difficulty in generalizing the above considerations to two or three dimensions. In two dimensions we take two generators, $a_1$ and $a_2$ (displacements along any two different directions in space), and by repeated combination we are led to consider the set of all vectors $r = n_1a_1 + n_2a_2$ ($n_1$, $n_2$ integral). Again, this constitutes an infinite Abelian group under addition. The unit is the zero vector $0 = 0a_1 + 0a_2$; the inverse of $n_1a_1 + n_2a_2$ is $(-n_1a_1) + (-n_2a_2)$; and the law of combination is

$$(m_1a_1 + m_2a_2) + (n_1a_1 + n_2a_2) = (m_1 + n_1)a_1 + (m_2 + n_2)a_2 \tag{2.2.1}$$

Regarded as position vectors, these displacements lead from the

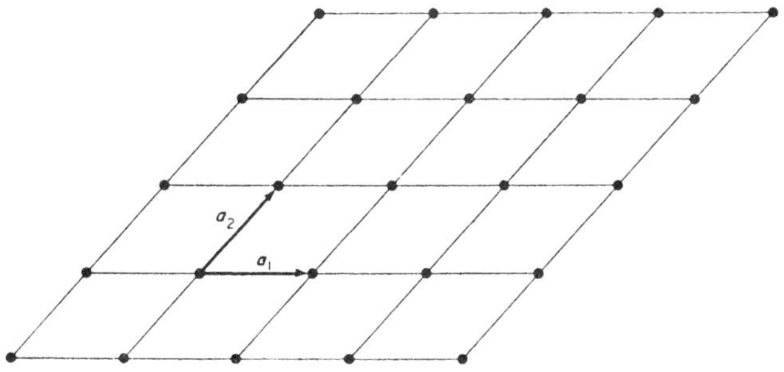


FIG. 2.4. Two-dimensional lattice. $a_1$ and $a_2$ are " primitive translations ".

origin to all the points of a two-dimensional lattice (Fig. 2.4): regarded as common translations of all the lattice points they form the translational symmetry group of the lattice.

EXAMPLE 2.2. The law of combination (2.2.1) expresses two important facts: these are exhibited geometrically in Fig. 2.5. The combination of displacements by sequential performance is both associative (Fig. 2.5(a)) and commutative (Fig. 2.5(b)). Each displacement is represented by a line segment pointing in a definite direction, but the *position* of the segment is irrelevant. Different points in space may suffer the *same* displacement (e.g. O → A and B → C in (b)).

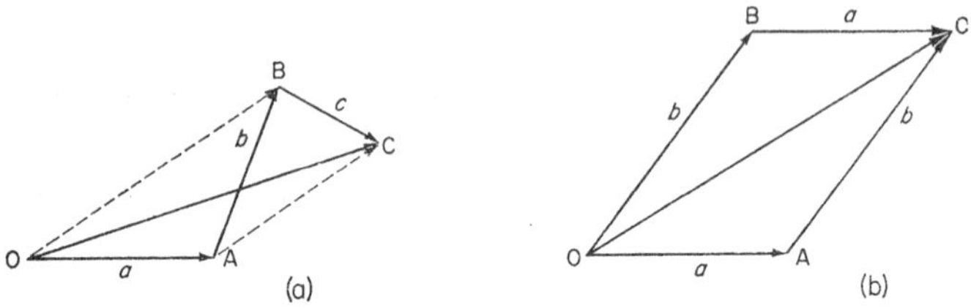


FIG. 2.5. Associative and commutative laws

(a) A displacement $\overrightarrow{OC}$, namely $c + b + a$, can be regarded equally well as $\overrightarrow{OB}$ followed by $\overrightarrow{BC}$ (i.e. as $c + (b + a)$) or as $\overrightarrow{OA}$ followed by $\overrightarrow{AC}$ (i.e. as $(c + b) + a$).

(b) A displacement $\overrightarrow{OC}$ can be regarded equally well as $\overrightarrow{OB}$ followed by $\overrightarrow{BC}$ (i.e. as $a + b$) or as $\overrightarrow{OA}$ followed by $\overrightarrow{AC}$ (i.e. as $b + a$).

Proceeding to three dimensions we take generators, $a_1$, $a_2$ and $a_3$, identified with basic displacements along three different directions in space, and form the set of all elements $r = n_1a_1 + n_2a_2 + n_3a_3$. The law of combination is

$$(m_1a_1 + m_2a_2 + m_3a_3) + (n_1a_1 + n_2a_2 + n_3a_3) = (m_1 + n_1)a_1 + (m_2 + n_2)a_2 + (m_3 + n_3)a_3 \quad (2.2.2)$$

and the unit and inverses may be defined exactly as before. The set of all such elements, regarded as position vectors, defines a *3-dimensional lattice* of points: regarded as common translations of all the lattice points, they form the translational symmetry group of the lattice.

We now see that the possible simple lattices in 1, 2 and 3 dimensions are completely determined by specifying 1, 2 or 3 *primitive translations*. The nature of the lattice and the possibility of its possessing additional point group symmetries (cf. Fig. 2.3(b)) will of course depend upon the behaviour of the whole system of lattice vectors under rotations and hence upon the lengths and inclinations of the primitive translations. Before such concepts can be properly understood it is necessary to introduce the essential ideas of the theory of vector spaces. This will also provide an opportunity of developing the basic mathematical equipment of group representation theory.

### 2.3. Vector Spaces

In discussing lattices we considered the group, G say, of all displacements, or " lattice vectors ", of the form $n_1a_1 + n_2a_2 + n_3a_3$ where $\{a_1, a_2, a_3\}$ was a system of generators and $n_1, n_2, n_3$ were integers. From the basic law of combination we can now formulate the following rules for combining lattice vectors ($r, s, \ldots$) and integers ($m, n, \ldots$):

(a) If $r$ is in G, then $mr$ is in G.
(b) $(m+n)r = mr + nr$ and $m(nr) = (mn)r$ (all $r$ in G).
(c) $mr = r$ (all $r$ in G) when, and only when, $m = 1$. (2.3.1)
$mr = 0$ (all $r$ in G) when, and only when, $m = 0$.
(d) $m(r+s) = mr + ms$ (all $r$, $s$ in G)

The properties of any lattice follow from these rules, which are accordingly sufficient to define a lattice " axiomatically ". Starting from the lattice concept, which they embody, we now pass to the " vector space " in which the lattice is replaced by the set of all points in space —a " point field ".

A *linear vector space* may be defined in terms of the lattice axioms (2.3.1), simply by relaxing the restriction that the only numbers which appear are integers. To see what this means geometrically it is sufficient to consider a 1-dimensional lattice.

EXAMPLE 2.3. We first ask for the interpretation of $(1/m)a$ and $(m/n)a$ ($m$, $n$ integers). From (b) of (2.3.1), $(1/n)a$ is that displacement which, when repeated $n$-fold, is equivalent to the generator $a$ itself. Geometrically (Fig. 2.6(a)), it is a step one $n$th as long as $a$. Similarly

FIG. 2.6. Non-integral multiples of a vector
(a) illustrates $3(\frac{1}{3}a) = a$.
(b) illustrates $3(\frac{2}{3}a) = 2a$.

$(m/n)a$ repeated $n$-fold is equivalent to the single lattice displacement $ma$. Geometrically (Fig. 2.6(b)), it is a step one $n$th as long as $ma$.

In this way, displacements of the form $(m/n)a$ are put into one-to-one correspondence with the rational numbers ; and when marked off from the origin their end points lie infinitely densely along the axis defined by $a$.

It is possible, however, to define displacements which *cannot* be written in the form $(m/n)a$ with integral $m$ and $n$: by an appeal to the idea of *continuity* we write these as $xa$ where $x$ is irrational, i.e. it can be *approached*, arbitrarily closely, from above and below, by numbers of the form $(m/n)$. Geometrically, we are simply *filling out* the space within each lattice " cell ", so that displacements are defined whose end points approach *all* points in the cell. This is, of course, equally true when we consider all three-dimensional displacements of the form $x_1a_1 + x_2a_2 + x_3a_3$ (where $x_1$, $x_2$, $x_3$ may be irrational) such displacements carrying a point from the origin to any point in three-dimensional space. Although this generalization is formally straightforward (we simply replace integers in the basic axioms by numbers) and geometrically transparent (instead of *lattice* points we now consider *all* points), the group elements are no longer expressible in terms of the generators in the usual way (section 2.1); there is a new operation by which one group element may be obtained from another, namely multiplication by *any* number (not merely the integers, which describe repetition of the basic displacement). A vector space is therefore often referred to as an Abelian group " *with an operator* "; and, for generality, the operator is allowed to be any *complex* number. To distinguish the numbers from the vectors we refer to them as scalars and indicate them by ordinary (i.e. lightface italic) type: we admit only one product of a scalar $x$ and a vector $r$, and it is immaterial whether we denote this by $xr$ or $rx$. The connection with the elementary vector algebra of three dimensions (see, for example, Hall, 1962) is now complete.

### 2.4. *n*-Dimensional Space. Basis Vectors

We must now consider more carefully the idea of *dimensionality* of a space. In " three-dimensional " space, or more briefly 3-space, every displacement can be represented in terms of three " basic vectors " $a_1$, $a_2$, $a_3$. Among *four* vectors, $r$, $a_1$, $a_2$, $a_3$, there must be a relationship

$$xr + x_1a_1 + x_2a_2 + x_3a_3 = 0$$

which, as may be argued from (2.3.1), is equivalent to

$$r = r_1a_1 + r_2a_2 + r_3a_3 \qquad (r_i = -x_i/x)$$

Such a relationship is termed a " *linear dependence* " among the four vectors. The vectors would be described as linearly *independent* if no such relationship existed (apart from the trivial one in which all coefficients were zero). The characteristic property of *three*-dimensional space is that any *four* vectors must be linearly dependent whilst fewer than four vectors *may* be linearly *in*dependent. Thus, two vectors not in the same direction are linearly independent; but if their directions are made to coincide, one becomes a scalar multiple of the other and their linear independence is lost. When the two vectors are linearly independent they are said to " span " a two-dimensional *space* or *manifold* (in this case a " plane "): here the manifold is simply the collection of all vectors which are linearly dependent on these two vectors (i.e. all vectors in the plane).

We shall often have occasion to consider sets of quantities which can be manipulated according to the rules (2.3.1) but which are not simply displacements in ordinary space. Nevertheless, it is convenient to refer to them as vectors, extending the terminology to include them and making full use of any geometrical analogies it suggests. In particular, we consider spaces of any finite number of dimensions, $n$, and extend the definition accordingly:

| | |
|---|---|
| A linear vector space in which it is impossible to find more than $n$ vectors with the property of linear independence is a space of $n$ dimensions. | (2.4.1) |

EXAMPLE 2.4. The five " 3d functions " which occur as solutions of the Schrödinger equation

$$\mathsf{H}\phi = E\phi$$

for the hydrogen atom (see, for example, Ref. 1), corresponding to the same value of the energy parameter ($E = E_{3d}$), are linearly independent: they cannot be related by $x_1\phi_1 + x_2\phi_2 + \ldots + x_5\phi_5 = 0$ for any choice of coefficients except $x_1 = x_2 = \ldots = x_5 = 0$. But because the differential equation is linear any combination of this form is still a solution, corresponding to the same energy value, $E = E_{3d}$. The five *functions* are then said to span a 5-dimensional space comprising all functions of the form $x_1\phi_1 + x_2\phi_2 + \ldots + x_5\phi_5$ (i.e. all solutions of the differential equation corresponding to $E = E_{3d}$). It is evident that addition of functions, and their multiplication by numbers, conform to the rules set out in (2.3.1). The space so defined is a " function space ".

In any space we can set up *basis vectors*, like the primitive translations which define a lattice. If we choose any $n$ linearly independent vectors $\mathbf{e}_1, \mathbf{e}_2, \ldots, \mathbf{e}_n$ it follows from the definition that *any* $(n+1)$th vector $\mathbf{r}$ must be linearly related to the e's:

$$x\mathbf{r} + x_1\mathbf{e}_1 + \ldots + x_n\mathbf{e}_n = \mathbf{0}$$

and hence that

$$\mathbf{r} = r_1\mathbf{e}_1 + r_2\mathbf{e}_2 + \ldots + r_n\mathbf{e}_n \qquad (r_i = -x_i/x) \quad (2.4.2)$$

The numbers $r_1, r_2, \ldots, r_n$ are called the *components* of $\mathbf{r}$ relative to the basis $\mathbf{e}_1, \mathbf{e}_2, \ldots, \mathbf{e}_n$. Geometrically, of course, the components may be regarded as parallel projections of a vector along the directions

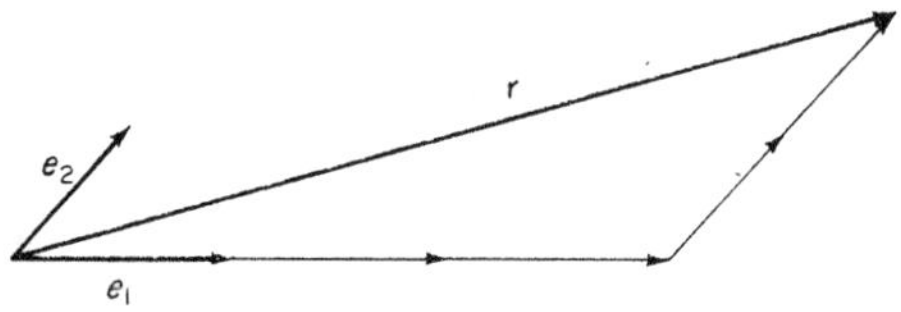


FIG. 2.7. Expression of a vector in terms of a basis: $\mathbf{r} = 3\mathbf{e}_1 + 2\mathbf{e}_2$.

specified by the basis vectors (Fig. 2.7). The laws for manipulating vectors (2.3.1) can then be expressed alternatively in terms of *number sets*. Thus

$$\mathbf{r} + \mathbf{s} = (r_1\mathbf{e}_1 + \ldots + r_n\mathbf{e}_n) + (s_1\mathbf{e}_1 + \ldots + s_n\mathbf{e}_n) = (r_1 + s_1)\mathbf{e}_1 + \ldots + (r_n + s_n)\mathbf{e}_n$$

and if the result is denoted by $\mathbf{t}$ with components $t_1, t_2, \ldots, t_n$ then

$$\mathbf{r} + \mathbf{s} = \mathbf{t} \qquad \text{implies} \qquad r_i + s_i = t_i \qquad (i = 1, 2, \ldots, n)$$

Vector addition means scalar addition of the corresponding components. It follows that the laws of combination could be written purely in terms of " number sets ". For example,

$$(r_1\ r_2 \ldots r_n) + (s_1\ s_2 \ldots s_n) = (r_1 + s_1 \quad r_2 + s_2 \ldots r_n + s_n)$$
$$x(r_1\ r_2 \ldots r_n) = (xr_1\ xr_2 \ldots xr_n)$$

The number sets in fact form a new group—of (1-row) matrices‡ under matrix addition—which is isomorphic with the group of vectors under vector addition: $\mathbf{r} \rightarrow (r_1\ r_2 \ldots r_n)$. Sometimes no distinction is made between a vector and the number set which represents it in a par-

‡ The sets may equally well be written as column matrices. This is, in fact, the standard practice in dealing with vector components (see section 2.5).

ticular basis, and the axioms are formulated entirely in terms of number sets. We prefer not to adopt this approach, however, because in physics and chemistry we like to denote a uniquely defined entity (such as an electric field) by a single symbol, whereas the number set (of field components) merely reflects one *mode of description*, which may be changed at will by changing the basis.

## 2.5. Components and Basis Changes

We now ask what happens to the components of a vector if we change from one basis to another. If instead of $\mathsf{e}_1, \mathsf{e}_2, \ldots, \mathsf{e}_n$ we choose a new basis $\bar{\mathsf{e}}_1, \bar{\mathsf{e}}_2, \ldots, \bar{\mathsf{e}}_n$, whose vectors are (by (2.4.2)) linear combinations of $\mathsf{e}_1, \mathsf{e}_2, \ldots, \mathsf{e}_n$, we may write $\mathsf{r} = \bar{r}_1\bar{\mathsf{e}}_1 + \ldots + \bar{r}_n\bar{\mathsf{e}}_n$ and must find the new components $\bar{r}_1, \bar{r}_2, \ldots, \bar{r}_n$. First we adopt a simple matrix notation. It is customary to write the components of a vector as a column (i.e. an $n \times 1$ matrix) and with this convention the expression

$$\mathsf{r} = \mathsf{e}_1 r_1 + \mathsf{e}_2 r_2 + \ldots \mathsf{e}_n r_n = \sum_{i=1}^{n} \mathsf{e}_i r_i$$

(no distinction being made between $\mathsf{e}_i r_i$ and $r_i \mathsf{e}_i$) may be written

$$\mathsf{r} = (\mathsf{e}_1\ \mathsf{e}_2 \ldots \mathsf{e}_n) \begin{pmatrix} r_1 \\ r_2 \\ | \\ r_n \end{pmatrix} = \mathbf{e}\boldsymbol{r} \tag{2.5.1}$$

provided the basis vectors are collected into a *row* matrix $\mathbf{e} = (\mathsf{e}_1\ \mathsf{e}_2 \ldots \mathsf{e}_n)$. Henceforth, throughout this book, bold face is used for *matrix* quantities and lightface (with subscripts) for individual elements. We continue to use distinctive characters (gill sans‡) to distinguish vectors, operations, etc. from scalars. The form of a matrix (e.g. row or column) is invariably evident from the context. Thus the vector $\mathsf{r} = \mathbf{e}\boldsymbol{r}$ is a linear combination of basis vectors, and $\mathbf{e}\boldsymbol{r}$ is consequently a row–column product. *Vector components always appear in columns, sets of vectors in rows.*

A new set of basis vectors may now be written in terms of the old according to (2.5.1):

$$\bar{\mathsf{e}}_i = \mathsf{e}_1 R_{1i} + \mathsf{e}_2 R_{2i} + \ldots + \mathsf{e}_n R_{ni} = \sum_{j=1}^{n} \mathsf{e}_j R_{ji} \quad (i = 1, 2, \ldots, n), \tag{2.5.2}$$

‡ In some applications (to function spaces) Greek letters are used.

and with a matrix notation parallel to (2.8),

$$(\bar{\mathbf{e}}_1\ \bar{\mathbf{e}}_2 \ldots \bar{\mathbf{e}}_n) = (\mathbf{e}_1\ \mathbf{e}_2 \ldots \mathbf{e}_n) \begin{pmatrix} R_{11} & R_{12} \ldots R_{1n} \\ R_{21} & R_{22} \ldots R_{2n} \\ \ldots & \ldots \\ R_{n1} & R_{n2} \ldots R_{nn} \end{pmatrix}$$

or, more briefly,

$$\boxed{\bar{\mathbf{e}} = \mathbf{e}\boldsymbol{R}} \tag{2.5.3}$$

The *i*th *column* of the square matrix $\boldsymbol{R}$ is just the set of components of the new basis vector $\bar{\mathbf{e}}_i$ in terms of the e's.

EXAMPLE 2.5. Figure 2.8 illustrates a 3-dimensional situation in which the basis vectors are three perpendicular unit vectors and

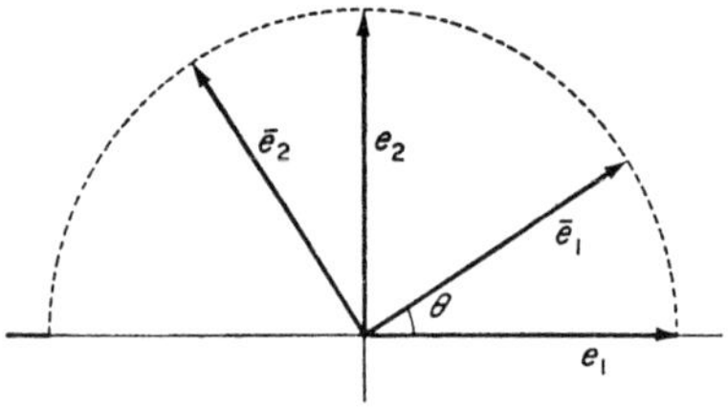


FIG. 2.8. Change of basis. The third basis vector ($\mathbf{e}_3$) is normal to the plane of the paper. The new basis is obtained by rotation about $\mathbf{e}_3$.

$(\bar{\mathbf{e}}_1\ \bar{\mathbf{e}}_2\ \bar{\mathbf{e}}_3)$ is obtained by rotating $(\mathbf{e}_1\ \mathbf{e}_2\ \mathbf{e}_3)$ through an angle $\theta$ about $\mathbf{e}_3$. In this case

$$\begin{aligned} \bar{\mathbf{e}}_1 &= \cos\theta\ \mathbf{e}_1 + \sin\theta\ \mathbf{e}_2 \\ \bar{\mathbf{e}}_2 &= -\sin\theta\ \mathbf{e}_1 + \cos\theta\ \mathbf{e}_2 \\ \bar{\mathbf{e}}_3 &= \mathbf{e}_3 \end{aligned}$$

Hence

$$(\bar{\mathbf{e}}_1\ \bar{\mathbf{e}}_2\ \bar{\mathbf{e}}_3) = (\mathbf{e}_1\ \mathbf{e}_2\ \mathbf{e}_3) \begin{pmatrix} \cos\theta & -\sin\theta & 0 \\ \sin\theta & \cos\theta & 0 \\ 0 & 0 & 1 \end{pmatrix}$$

and $\boldsymbol{R}$ in (2.5.3) is simply the square matrix on the right.

Now because, in general, the $n$ vectors $\bar{\mathbf{e}}_1, \bar{\mathbf{e}}_2, \ldots, \bar{\mathbf{e}}_n$ must be linearly independent (in order to span the same space as $\mathbf{e}_1, \mathbf{e}_2, \ldots, \mathbf{e}_n$), *they* provide a basis in terms of which $\mathbf{e}_1, \mathbf{e}_2, \ldots, \mathbf{e}_n$ could be expressed, and we could write

$$\mathbf{e} = \bar{\mathbf{e}}\boldsymbol{T} \tag{2.5.4}$$

If a matrix $\boldsymbol{T}$ with this property exists it is simply the inverse, under matrix multiplication, of $\boldsymbol{R}$: for if we multiply (2.5.3) from the right by $\boldsymbol{T}$ and compare the result with (2.5.4) it is necessary that $\boldsymbol{RT} = \mathbf{1}_n$ where $\mathbf{1}_n$ is the $n \times n$ unit matrix. This identifies $\boldsymbol{T}$ with the *inverse matrix* $\boldsymbol{R}^{-1}$ and the two bases are therefore related by

$$\bar{\mathbf{e}} = \mathbf{e}\boldsymbol{R}, \qquad \mathbf{e} = \bar{\mathbf{e}}\boldsymbol{R}^{-1} \tag{2.5.5}$$

We now turn to the components of an arbitrary vector $\mathsf{r}$. From (2.5.5) we may write

$$\mathsf{r} = \mathbf{e}\boldsymbol{r} = \bar{\mathbf{e}}\boldsymbol{R}^{-1}\boldsymbol{r} = \bar{\mathbf{e}}\bar{\boldsymbol{r}}, \text{ say.}$$

The new components $\bar{r}_1, \bar{r}_2, \ldots, \bar{r}_n$ (collected in $\bar{\boldsymbol{r}}$) are therefore determined by

$$\bar{\boldsymbol{r}} = \boldsymbol{R}^{-1}\boldsymbol{r}$$

and the change of basis is completely described by the equations

$$\mathbf{e} \to \bar{\mathbf{e}} = \mathbf{e}\boldsymbol{R}, \qquad \boldsymbol{r} \to \bar{\boldsymbol{r}} = \boldsymbol{R}^{-1}\boldsymbol{r} \tag{2.5.6}$$

These matrix equations may be written alternatively as

$$\mathsf{e}_i \to \bar{\mathsf{e}}_i = \sum_{j=1}^{n} \mathsf{e}_j R_{ji}, \quad r_i \to \bar{r}_i = \sum_{j=1}^{n} R^{-1}{}_{ij} r_j \quad (i = 1, 2, \ldots, n) \tag{2.5.7}$$

The basis vectors and the components of an arbitrary fixed vector are said to transform *contragrediently*; and the invariance of $\mathsf{r}$, which is the same entity irrespective of basis (cf. p. 31), is expressed by the fact that on changing the basis

$$\mathsf{r} \to \bar{\mathbf{e}}\bar{\boldsymbol{r}} = \mathbf{e}\boldsymbol{R}\boldsymbol{R}^{-1}\boldsymbol{r} = \mathbf{e}\boldsymbol{r} = \mathsf{r} \tag{2.5.8}$$

## 2.6. Mappings and Similarity Transformations

We have already considered mappings in which an image $\mathsf{R}'$ is associated with every element $\mathsf{R}$ of a group $\mathsf{G}$. We shall now consider mappings of a vector space upon itself by introducing operations which send every element (vector) over into *another* element (" rotated " vector). Mappings of this kind form the basis of representation theory.

We study the transformation of an arbitrary vector $\mathsf{r}$ which, instead of being fixed, is related in a definite way to the basis—so that when the basis is rotated the vector $\mathsf{r}$ is carried with it. Now the relationship of vector to basis is determined by its components: so we wish to find the vector $\mathsf{r}'$ which has the same components in the new basis as $\mathsf{r}$ had in the old. This vector we call the image of $\mathsf{r}$: evidently, when $\mathbf{e} \to \bar{\mathbf{e}} = \mathbf{e}\boldsymbol{R}$

$$\mathsf{r} \to \mathsf{r}' = \bar{\mathbf{e}}\boldsymbol{r} = \mathbf{e}\boldsymbol{R}\boldsymbol{r} = \mathbf{e}\boldsymbol{r}'$$

*relative to the original basis.* The components of $\mathsf{r}'$ are therefore

$$\boldsymbol{r}' = \boldsymbol{R}\boldsymbol{r} \tag{2.6.1}$$

This result should be contrasted with that in (2.5.6) in which the new column $\bar{\boldsymbol{r}}$ contained the components of a *fixed* vector *relative to a new basis.* In order to distinguish different interpretations of rather similar equations, we have used the bar to denote new components of a fixed vector (basis rotated) and a prime to denote the new components of a rotated vector (basis fixed). The matrix equation (2.6.1) shows how *all* vectors are sent into their images under a common rotation. It applies in particular to the basis vectors which may be regarded as images of the original set (i.e. $\bar{\mathsf{e}}_j = \mathsf{e}_j'$). Since the components of the $j$th basis vector are $(\mathsf{e}_j)_j = 1$, $(\mathsf{e}_j)_i = 0$ $(i \neq j)$ it is evident that the components of its image are $(\mathsf{e}_j')_i = \sum_{k=1}^{n} R_{ik}(\mathsf{e}_j)_k = R_{ij}$ $(i = 1, 2, \ldots, n)$; and $\mathsf{e}_j'$ therefore has as its components the numbers in the $j$th column of $\boldsymbol{R}$—a result consistent with (2.5.3) *et seq.*

It is convenient to regard a mapping $\mathsf{r} \rightarrow \mathsf{r}'$ as the result of an *operation* $\mathsf{R}$ which sends every vector $\mathsf{r}$ into an image $\mathsf{r}'$. In general, $\mathsf{R}$ would be described geometrically as a combination of a rotation and a stretch. With this notation

$$\mathsf{r}' = \mathsf{R}\mathsf{r} = \mathsf{R}\left(\sum_{i=1}^{n} \mathsf{e}_i r_i\right) \tag{2.6.2}$$

On the other hand $\mathsf{r}' = \bar{\mathbf{e}}\boldsymbol{r}$ gives (remembering $\bar{\mathsf{e}}_i = \mathsf{e}_i' = \mathsf{R}\mathsf{e}_i$ is the image of $\mathsf{e}_i$)

$$\mathsf{r}' = \sum_{i=1}^{n} \bar{\mathsf{e}}_i r_i = \sum_{i=1}^{n} \mathsf{R}\mathsf{e}_i r_i$$

Comparison shows that the mapping is distributive:

$$\mathsf{R}(\mathsf{e}_1 r_1 + \mathsf{e}_2 r_2 + \ldots) = \mathsf{R}\mathsf{e}_1 r_1 + \mathsf{R}\mathsf{e}_2 r_2 + \ldots$$

On writing the equation in matrix form we must therefore take

$$\mathsf{R}(\mathsf{e}_1\ \mathsf{e}_2\ \ldots\ \mathsf{e}_n) = (\mathsf{R}\mathsf{e}_1\ \mathsf{R}\mathsf{e}_2\ \ldots\ \mathsf{R}\mathsf{e}_n)$$

Equation (2.5.3), which relates the rotated basis to the original, then takes the form

$$\boxed{\mathsf{R}\mathbf{e} = \mathbf{e}\boldsymbol{R}} \tag{2.6.3}$$

This is a central result of the theory of mappings in vector space. It makes a one-to-one association between a mapping $\mathsf{R}$ in a vector space, and a matrix $\boldsymbol{R}$—the actual matrix depending on the choice of basis.

The effect of $R$ upon *any* vector is completely determined once we know its effect (2.6.3) upon $n$ independent basis vectors. When a whole set of mappings, $\{A, B, \ldots, R, \ldots\}$, is being considered, each will have its own matrix, and we may use the corresponding boldface letters to denote the various matrices:

$$A \to \boldsymbol{A}, \quad B \to \boldsymbol{B}, \quad \ldots, \quad R \to \boldsymbol{R}, \quad \ldots$$

This is the kind of association already met in section 1.7, where $A, B, \ldots, R, \ldots$ were group elements and $\boldsymbol{A}, \boldsymbol{B}, \ldots, \boldsymbol{R}, \ldots$ matrices: we shall show presently that the mappings do form a group and the matrices a representation. We also recall the alternative notation (section 1.8) in which the matrix on the right of (2.6.3) is denoted always by $\boldsymbol{D}$, say, and a "functional" notation is used to indicate that the form of $\boldsymbol{D}$ depends on the mapping considered. The particular matrix associated with $R$ is then $\boldsymbol{D}(R)$ and (2.6.3) may be written alternatively as

$$R\mathbf{e} = \mathbf{e}\boldsymbol{D}(R) \tag{2.6.4}$$

We now illustrate how the matrices are determined, along the lines of example 2.5.

EXAMPLE 2.6. Let us return to the point group $C_{3v}$ of symmetry operations which bring the equilateral triangle (Fig. 2.9) into self-coincidence. All such operations leave the centroid of the triangle

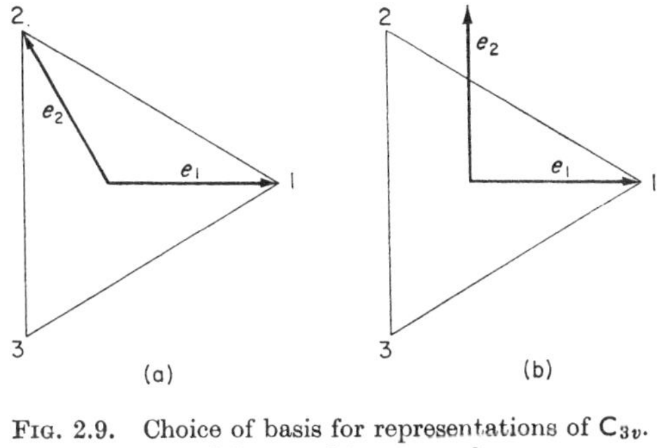


FIG. 2.9. Choice of basis for representations of $C_{3v}$. (a) oblique, (b) rectangular.

fixed. If we embed two vectors $\mathbf{e}_1$ and $\mathbf{e}_2$ in the lamina—the position vectors of two vertices relative to the centroid as origin—any symmetry operation can be described alternatively as a mapping $R$ in which the two vectors are sent into images $\mathbf{e}_1' = R\mathbf{e}_1$, $\mathbf{e}_2' = R\mathbf{e}_2$. By expressing the images in terms of the original vectors we obtain an explicit form of (2.6.4) and can therefore identify the $2 \times 2$ matrices which describe the group of mappings. From Fig. 2.9(a) we have, for example:

$$C_3\mathbf{e}_1 = \mathbf{e}_2 \text{ and } C_3\mathbf{e}_2 = -\mathbf{e}_1 - \mathbf{e}_2.$$

Hence $C_3(\mathbf{e}_1\ \mathbf{e}_2) = (\mathbf{e}_1\ \mathbf{e}_2)\begin{pmatrix} 0 & -1 \\ 1 & -1 \end{pmatrix}$ and $\boldsymbol{D}(C_3) = \begin{pmatrix} 0 & -1 \\ 1 & -1 \end{pmatrix}$

Similarly $\sigma^{(1)}\mathbf{e}_1 = \mathbf{e}_1$ and $\sigma^{(1)}\mathbf{e}_2 = -\mathbf{e}_1 - \mathbf{e}_2$.

Hence $\sigma^{(1)}(\mathbf{e}_1\ \mathbf{e}_2) = (\mathbf{e}_1\ \mathbf{e}_2)\begin{pmatrix} 1 & -1 \\ 0 & -1 \end{pmatrix}$ and $\boldsymbol{D}(\sigma^{(1)}) = \begin{pmatrix} 1 & -1 \\ 0 & -1 \end{pmatrix}$

The whole set of matrices is easily determined and it may then be verified that for every product of symmetry operations, $RS = T$, there is a corresponding matrix identity, $\boldsymbol{D}(R)\boldsymbol{D}(S) = \boldsymbol{D}(T)$. The matrices as defined above therefore form a representation of the group $\mathrm{C}_{3v}$ according to the definition in section 1.7.

Example 2.7. We consider the same problem but with a different choice of basis vectors (Fig. 2.9(b)). In this case we obtain, for example,

$$C_3\mathbf{e}_1 = -\tfrac{1}{2}\mathbf{e}_1 + \tfrac{1}{2}\sqrt{3}\mathbf{e}_2 \text{ and } C_3\mathbf{e}_2 = -\tfrac{1}{2}\sqrt{3}\,\mathbf{e}_1 - \tfrac{1}{2}\mathbf{e}_2.$$

Hence $C_3(\mathbf{e}_1\ \mathbf{e}_2) = (\mathbf{e}_1\ \mathbf{e}_2)\begin{pmatrix} -\frac{1}{2} & -\frac{1}{2}\sqrt{3} \\ \frac{1}{2}\sqrt{3} & -\frac{1}{2} \end{pmatrix}$ and $\boldsymbol{D}(C_3) = \begin{pmatrix} -\frac{1}{2} & -\frac{1}{2}\sqrt{3} \\ \frac{1}{2}\sqrt{3} & -\frac{1}{2} \end{pmatrix}$

When all operations have been examined it is found that the resultant set of matrices is that quoted, without explanation, in example 1.17. We have therefore found a pair of vectors which " carry " this particular representation.

The last two examples show how the introduction of vector spaces can provide a powerful method of constructing matrix representations. The representation spaces may generally be of $n$ dimensions and the basis vectors may be replaced by entities, such as the functions in example 2.4, which have a less direct geometrical significance. It is necessary only that every operation is accompanied by an associated linear transformation of these entities among themselves. This possibility is illustrated by a further example.

Example 2.8. Figure 2.10 indicates two degenerate normal modes of vibration of a circular membrane: these are described by amplitude functions which give the displacement ($z$) as a function of position

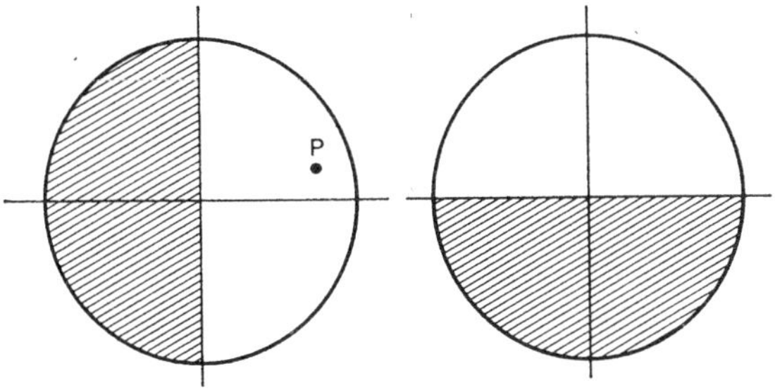


FIG. 2.10. Degenerate modes of vibration of a circular membrane. The shaded areas move down as the unshaded areas move up.

$(x, y)$. The two functions are of the form

$$\phi_1(x, y) = xf(x^2 + y^2), \qquad \phi_2(x, y) = yf(x^2 + y^2)$$

If the whole system is now rotated through an angle $\theta$, so that point P goes to point P′ in Fig. 2.11, the first system of displacements will

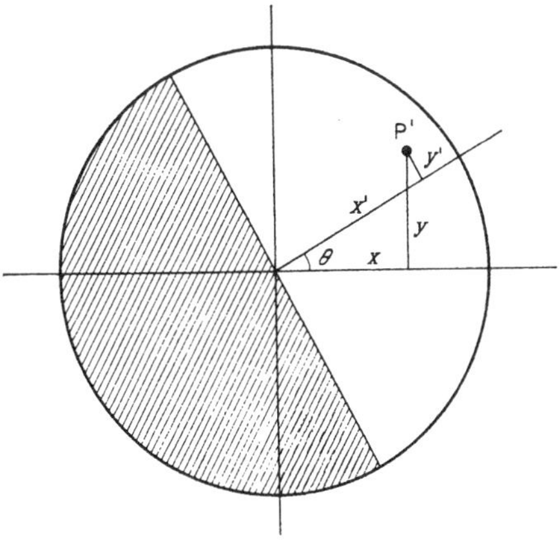


FIG. 2.11. " Rotated function " derived from the first mode in Fig. 2.10. The point P in Fig. 2.10, together with its displacement, is carried by rotation into the point P′.

be rotated into a new set, described by some function $\phi_1'(x, y)$, which may be defined as the original function of new coordinates, $x'$ and $y'$. Thus

$$\phi_1'(x, y) = \phi_1(x', y') = (x \cos\theta + y \sin\theta) f(x'^2 + y'^2) = \cos\theta\, \phi_1(x, y) + \sin\theta\, \phi_2(x, y)$$

since $x'^2 + y'^2 = x^2 + y^2$.

It may be said that the rotation of the system has "induced" a transformation of the functions $\phi_1(x, y)$ and $\phi_2(x, y)$ describing two particular modes of vibration. In fact

$$\phi_1(x, y) \to \cos\theta\, \phi_1(x, y) + \sin\theta\, \phi_2(x, y) = \mathsf{R}\phi_1(x, y)$$
$$\phi_2(x, y) \to -\sin\theta\, \phi_1(x, y) + \cos\theta\, \phi_2(x, y) = \mathsf{R}\phi_2(x, y)$$

where $\mathsf{R}\phi_1$ and $\mathsf{R}\phi_2$ are new function symbols (replacing $\phi_1'$ and $\phi_2'$) denoting the "rotated functions".

The functions play a rôle exactly analogous to that of the basis vectors $\mathbf{e}_1$ and $\mathbf{e}_2$ in example 2.5 and we may write

$$\mathsf{R}(\phi_1\ \phi_2) = (\phi_1\ \phi_2)\begin{pmatrix}\cos\theta & -\sin\theta \\ \sin\theta & \cos\theta\end{pmatrix}$$

Whenever the condition of a physical system is described by a function of position, defined with respect to axes embedded in the system and initially coinciding with fixed "external" axes, rotation will produce a function which has a new form when referred to the fixed axes. The rotation is said to *induce* a transformation in the function or, if the function is regarded as a vector in a function space, a rotation in function space.

Rotations in function space are of particular importance in quantum mechanics where the theory of group representations finds some of its most fruitful applications. In such applications, however, it is necessary to admit *complex* functions and this must be borne in mind when the theory is developed further. First, we examine more generally the possibilities suggested by the last three examples.

### 2.7. Representations. Equivalence

Examples 2.6–2.8 illustrate how we can set up a one-to-one association between mappings in an $n$-space and the $n \times n$ matrices which describe the behaviour of a given set of $n$ linearly independent vectors. It is evident that the mappings themselves form a group‡ and the examples suggest that the matrices provide a representation of the group. To show that this is indeed so, we consider sequential performance of two operations, first $\mathsf{S}$ and then $\mathsf{R}$ and show that if $\mathsf{RS} = \mathsf{T}$ then the associated matrices satisfy the corresponding identity

‡ Sequential performance is an associative law of combination. Two successive mappings lead to another vector of the same space and are therefore equivalent to some single mapping. There is a unit or identity mapping—that which leaves every vector undisturbed. Every mapping has an inverse—that which sends an image back into the original vector.

$\boldsymbol{RS} = \boldsymbol{T}$. Let $r$ be an arbitrary vector: then, from (2.6.3),

$$RSr = RS\mathbf{er} = R\mathbf{e}\boldsymbol{Sr} = \mathbf{e}\boldsymbol{RSr}$$

But if $RS = T$ the associated matrix $\boldsymbol{T}$ is defined by

$$RSr = Tr = T\mathbf{er} = \mathbf{e}\boldsymbol{Tr}$$

From the linear independence of the basis vectors, it follows that $\boldsymbol{RSr} = \boldsymbol{Tr}$ and since $\boldsymbol{r}$ is an arbitrary column this means $\boldsymbol{RS} = \boldsymbol{T}$. Hence, if matrices are associated with mappings according to (2.6.3) they form a group representation. We note, in particular, that the $n \times n$ unit matrix $\mathbf{1}_n$ is associated with the identity mapping and that the inverse of a matrix is associated with the inverse of the mapping it represents:

$$E \to \mathbf{1}_n, \qquad R^{-1} \to \boldsymbol{R}^{-1}$$

Also the matrices associated with distinct mappings must themselves be distinct. The matrix group defined in this way, as the set of matrices isomorphic with all the mappings of an $n$-dimensional vector space, is called the " full linear group in $n$ dimensions ".

In general, we shall say that a set of basis vectors " carries " a particular representation of the group of mappings. This means that when an arbitrary vector $r$ is sent into its image $r'$

(i) the basis vectors are changed to

$$\mathbf{e} \to \mathbf{e}' = R\mathbf{e} = \mathbf{e}\boldsymbol{D}(R) \tag{2.7.1}$$

(ii) the arbitrary vector with components $\boldsymbol{r}$ is changed to one with components $\boldsymbol{r}'$

$$\boldsymbol{r} \to \boldsymbol{r}' = \boldsymbol{D}(R)\boldsymbol{r} \tag{2.7.2}$$

Whenever it is more convenient to work in terms of components, the matrices are defined by the second relationship.

So far we have been concerned with the representation of the group of mappings in a given vector space, a representation which is essentially one-to-one since there is a unique matrix for each mapping. We have seen, however, that for a given abstract group, we can find representations in various different spaces: for $C_{3v}$ we already have (see section 1.8) two 1-dimensional representations and one 2-dimensional. In general, if we have a group $G = \{A, B, \ldots, R, \ldots\}$ we may find an $n:1$ homomorphism in which a mapping which we may now denote by $R_{(\iota)}$, in some vector space $V_\iota$, is associated with $R$. Correspondingly, $R$ will be associated with the matrix $\boldsymbol{D}_\iota(R)$ which describes the mapping in this particular space:

$$R \to R_{(\iota)} \to \boldsymbol{D}_\iota(R)$$

The subscript $i$ (or some other notational modification) which distinguishes the mapping in vector space $V_i$ from the given group element $R$ is often discarded, as, for instance, in example 2.8, where the significance of $R$ (as a transformation in function space), is clear from the " vectors " on which it works. Different representations are indicated simply by the subscript on the $\boldsymbol{D}$. The representations of $C_{3v}$ which we have already obtained are thus conveniently collected as :

| $R$ | $E$ | $C_3$ | $\bar{C}_3$ | $\sigma^{(1)}$ | $\sigma^{(2)}$ | $\sigma^{(3)}$ |
|---|---|---|---|---|---|---|
| $\boldsymbol{D}_1(R)$ | 1 | 1 | 1 | 1 | 1 | 1 |
| $\boldsymbol{D}_2(R)$ | 1 | 1 | 1 | $-1$ | $-1$ | $-1$ |
| $\boldsymbol{D}_3(R)$ | $\begin{pmatrix} 1 & 0 \\ 0 & 1 \end{pmatrix}$ | $\begin{pmatrix} -\frac{1}{2} & -\frac{1}{2}\sqrt{3} \\ \frac{1}{2}\sqrt{3} & -\frac{1}{2} \end{pmatrix}$ | $\begin{pmatrix} -\frac{1}{2} & \frac{1}{2}\sqrt{3} \\ -\frac{1}{2}\sqrt{3} & -\frac{1}{2} \end{pmatrix}$ | $\begin{pmatrix} 1 & 0 \\ 0 & -1 \end{pmatrix}$ | $\begin{pmatrix} -\frac{1}{2} & -\frac{1}{2}\sqrt{3} \\ -\frac{1}{2}\sqrt{3} & \frac{1}{2} \end{pmatrix}$ | $\begin{pmatrix} -\frac{1}{2} & \frac{1}{2}\sqrt{3} \\ \frac{1}{2}\sqrt{3} & \frac{1}{2} \end{pmatrix}$ |

(2.7.3)

The matrices $\boldsymbol{D}_1(R)$ refer to mappings in a 1-dimensional representation space $V_1$ ; matrices $\boldsymbol{D}_2(R)$ to corresponding mappings in another 1-dimensional space, $V_2$‡; and matrices $\boldsymbol{D}_3(R)$ to mappings in a 2-dimensional space $V_3$. Finally, we note from examples 2.6 and 2.7 that the matrices which describe a single set of mappings (in this case 2-dimensional rotations) are completely changed if we make a new choice of basis. Since the basis is arbitrary, it appears that an infinite number of alternative representations can be obtained simply by changing the basis but that these will not differ in any fundamental way. We now show how the matrices in two such representations are related. As in section 2.5, where we considered the effect of basis change upon the components of a fixed vector, we suppose $\mathbf{e} \rightarrow \bar{\mathbf{e}} = \mathbf{e}\boldsymbol{T}$ and write (according to (2.6.3))

$$\left.\begin{array}{lll} \text{(a)} & R\mathbf{e} = \mathbf{e}\boldsymbol{R} & \text{(original basis)} \\ \text{(b)} & R\bar{\mathbf{e}} = \bar{\mathbf{e}}\bar{\boldsymbol{R}} & \text{(new basis)} \end{array}\right\} \quad (2.7.4)$$

We wish to know how $\bar{\boldsymbol{R}}$ is related to $\boldsymbol{R}$. It follows at once that

$$R\bar{\mathbf{e}} = R\mathbf{e}\boldsymbol{T} = \mathbf{e}\boldsymbol{R}\boldsymbol{T} = \bar{\mathbf{e}}\boldsymbol{T}^{-1}\boldsymbol{R}\boldsymbol{T}$$

and hence, by comparison with (2.7.4b)

$$\bar{\boldsymbol{R}} = \boldsymbol{T}^{-1}\boldsymbol{R}\boldsymbol{T} \quad \text{or} \quad \bar{\boldsymbol{D}}(R) = \boldsymbol{T}^{-1}\boldsymbol{D}(R)\boldsymbol{T} \qquad (2.7.5)$$

‡ We cannot choose any vector in ordinary space so that it behaves in this way under rotations and reflections. This emphasizes that the space which carries a group representation may be purely formal.

This compares with the transformation law for vector components $\boldsymbol{r} \to \bar{\boldsymbol{r}} = \boldsymbol{T}^{-1}\boldsymbol{r}$. Naturally, because the new matrix group must also be isomorphic with the same set of mappings, $\boldsymbol{RS} = \boldsymbol{P}$ must imply $\bar{\boldsymbol{R}}\bar{\boldsymbol{S}} = \bar{\boldsymbol{P}}$. Inspection shows that the new matrices do indeed possess this property. Representations D and $\bar{\mathrm{D}}$ which differ in this way are said to be related by a *similarity transformation* and are described as *equivalent*.

EXAMPLE 2.9. The representations of example 2.6 and 2.7, corresponding only to different choice of basis vectors ($\mathbf{e}$ and $\bar{\mathbf{e}}$ say), should be equivalent. From Figs. 2.9(a) and 2.9(b) it follows that $\mathrm{e}_1 = \bar{\mathrm{e}}_1$, $\mathrm{e}_2 = -\frac{1}{2}\bar{\mathrm{e}}_1 + \frac{1}{2}\sqrt{3}\ \bar{\mathrm{e}}_2$. Hence $\bar{\mathrm{e}}_2 = \frac{1}{3}\sqrt{3}\ \mathrm{e}_1 + \frac{2}{3}\sqrt{3}\mathrm{e}_2$, and

$$(\bar{\mathrm{e}}_1\ \bar{\mathrm{e}}_2) = (\mathrm{e}_1\ \mathrm{e}_2)\begin{pmatrix} 1 & \frac{1}{3}\sqrt{3} \\ 0 & \frac{2}{3}\sqrt{3} \end{pmatrix}, \quad (\mathrm{e}_1\ \mathrm{e}_2) = (\bar{\mathrm{e}}_1\ \bar{\mathrm{e}}_2)\begin{pmatrix} 1 & -\frac{1}{2} \\ 0 & \frac{1}{2}\sqrt{3} \end{pmatrix}.$$

The two matrices are $\boldsymbol{T}$ and $\boldsymbol{T}^{-1}$ respectively. From (2.7.5) we then obtain, for example,

$$\bar{\boldsymbol{D}}(C_3) = \boldsymbol{T}^{-1}\begin{pmatrix} 0 & -1 \\ 1 & -1 \end{pmatrix}\boldsymbol{T} = \begin{pmatrix} -\frac{1}{2} & -\frac{1}{2}\sqrt{3} \\ \frac{1}{2}\sqrt{3} & -\frac{1}{2} \end{pmatrix}$$

which is the matrix found in example 2.7.

## ‡2.8. Length and Angle. The Metric

The length of a vector, or the distance between two points whose position vectors are given, has not yet been defined even in real 3-space. It might be said that the vector $n\mathbf{r}$ is " $n$ times as long as $\mathbf{r}$ " and, indeed, this is the convention adopted in measuring distances along a *given* straight line. But to compare the lengths of vectors which point in different directions it is necessary to consider in some detail the ideas of length and angle. A space in which lengths and angles are defined is said to possess a *metric*. This simply means that there is some rule for comparing the lengths of *any* two vectors, and hence of assigning a length to *every* vector by comparison with a unit.

We recall the properties of the scalar product of elementary vector theory. For any two vectors $\mathbf{r}$, $\mathbf{s}$ we can define a single number (the product of their lengths times the cosine of the angle between them), written $\mathbf{r}.\mathbf{s}$ and called the " scalar product " of $\mathbf{r}$ and $\mathbf{s}$. It is then evident geometrically that

‡ The rest of this chapter is not essential for an understanding of much of what follows and need be referred to only briefly on first reading ; it is important for later applications, particularly in quantum mechanics.

$$\left.\begin{array}{lll}\text{(i)} & r.s = s.r & \text{(scalar product symmetrical in the two vectors)} \\ \text{(ii)} & r.(s+t) = r.s + r.t & \text{(distributive law satisfied)} \\ \text{(iii)} & r.(as) = a(r.s) & \text{(scalar product proportional to length of each vector)} \\ \text{(iv)} & r.r > 0, \text{ all } r \neq 0 & \text{(length}^2\text{ positive, zero for the zero vector only)}\end{array}\right\} \quad (2.8.1)$$

Here lengths and angles are defined using elementary Euclidean geometry, the length being taken as a positive quantity (the positive root in (iv)). It follows from these properties that if two vectors are expressed in terms of basis vectors, $r = \sum_{i=1}^{3} r_i \mathbf{e}_i$, $s = \sum_{j=1}^{3} s_j \mathbf{e}_j$, then

$$r.s = \sum_{i,j=1}^{3} r_i s_j (\mathbf{e}_i . \mathbf{e}_j) \qquad (2.8.2)$$

The scalar product is thus a " bilinear form " in the vector components, with coefficients which are basis vector scalar products.

Let us now try to generalize these ideas to vectors with any number of components, real or complex, using (iv) to *define* the length of a vector as the square root of the scalar product with itself. First we enquire whether the original definition will serve if we admit linear combinations of the e's (which for the moment we may regard as oblique basis vectors in ordinary space) with *complex* coefficients. This may be desirable for formal manipulations, even though such combinations have no simple physical interpretation, as the following example shows:

EXAMPLE 2.10. The so-called " spherical " basis vectors are related to the unit vectors of example 2.5 by

$$\mathbf{e}_{+1} = -(\mathbf{e}_1 + i\mathbf{e}_2)/\sqrt{2}, \quad \mathbf{e}_0 = \mathbf{e}_3, \quad \mathbf{e}_{-1} = (\mathbf{e}_1 - i\mathbf{e}_2)/\sqrt{2}$$

These have simple transformation properties when the system is rotated about the $\mathbf{e}_3$-axis. Thus, from example 2.5,

$$\begin{aligned}\mathbf{e}_{+1} &\to -(\cos\theta\, \mathbf{e}_1 + \sin\theta\, \mathbf{e}_2 - i\sin\theta\, \mathbf{e}_1 + i\cos\theta\, \mathbf{e}_2)/\sqrt{2} \\ &= -(e^{-i\theta}\mathbf{e}_1 + ie^{-i\theta}\mathbf{e}_2)\sqrt{2} \\ &= e^{-i\theta}\mathbf{e}_{+1}\end{aligned}$$

Similarly, $\mathbf{e}_{-1} \to e^{i\theta}\mathbf{e}_{-1}$. The spherical basis vectors are therefore simply multiplied by complex " phase factors " when the system is rotated.

Now, if we used (2.8.2) directly, the squared length of a vector would be

$$r \cdot r = r_1^2(e_1 \cdot e_1) + r_2^2(e_2 \cdot e_2) + r_3^2(e_3 \cdot e_3)$$
$$+ 2r_1r_2(e_1 \cdot e_2) + 2r_1r_3(e_1 \cdot e_3) + 2r_2r_3(e_2 \cdot e_3)$$

But the concept of length would then break down because this would be a complex number and $r \cdot r$ would no longer be analogous to the quantity defined in the elementary theory: it would, for instance, be meaningless to postulate $r \cdot r > 0$. This difficulty can be overcome by introducing, along with every vector $r$ a vector $r^*$ (sometimes called the "dual" or "adjoint" of $r$) whose components are the complex conjugates of those of $r$ and defining the scalar product of $r$ with itself as $r^* \cdot r$. Thus, from (2.8.2.),

$$r^* \cdot r = r_1^* r_1(e_1 \cdot e_1) + r_2^* r_2(e_2 \cdot e_2) + r_3^* r_3(e_3 \cdot e_3)$$
$$+ r_1^* r_2(e_1 \cdot e_2) + r_1^* r_3(e_1 \cdot e_3) + r_2^* r_3(e_2 \cdot e_3)$$
$$+ r_2^* r_1(e_2 \cdot e_1) + r_3^* r_1(e_3 \cdot e_1) + r_3^* r_2(e_3 \cdot e_2)$$

This is an essentially *real* quantity (assuming still that the basis vector scalar products are real) since each non-real term is accompanied by its complex conjugate. The corresponding result for two different vectors, $r$ and $s$, could be written

$$r^* \cdot s = \sum_{i,j=1}^{3} r_i^* s_j (e_i \cdot e_j) \tag{2.8.3}$$

Now this result exhibits a symmetry property differing from that for real vectors: by rearrangement we obtain (with $(e_j \cdot e_i) = (e_i \cdot e_j)$)

$$r^* \cdot s = \sum_{i,j=1}^{3} (s_i^* r_j)^* (e_i \cdot e_j) = (s^* \cdot r)^*$$

The scalar product in this wider sense is thus a *complex number* associated with each *ordered* pair of vectors; if we reverse the order, a scalar product is replaced by its complex conjugate. This observation allows us to discard the assumption of real e's. For $e_i \cdot e_j$ is the special case of $e_i^* \cdot e_j$ in which the vectors are real; and we need only replace $e_i \cdot e_j$ in (2.8.3) by $e_i^* \cdot e_j$ to obtain a completely general and internally consistent definition of the scalar product. This simply means replacing each $e_i$ by $e_i^*$ in the definition of the dual of $r$. In an $n$-dimensional space the dual of

$$r = r_1 e_1 + r_2 e_2 + \ldots + r_n e_n$$

would thus be

$$r^* = r_1^* e_1^* + r_2^* e_2^* + \ldots + r_n^* e_n^*$$

and the scalar product of $r$ and $s$ (in that order) would be

$$r^* s = \sum_{i,j=1}^{n} r_i^* s_j (e_i^* e_j) = \sum_{i,j=1}^{n} r_i^* M_{ij} s_j$$

where the dot in the product has been dropped since the star on the left-hand member is sufficient to indicate the scalar product.‡ The scalar product defined in this way is described as *Hermitian* (after the French mathematician Hermite) and the space in which it occurs is a "Hermitian vector space". The matrix $\boldsymbol{M}$ whose elements $M_{ij}$ are the basis vector scalar products, is called the *metrical matrix* since it completely determines the metrical properties of the space. Its elements may be regarded as given numbers characteristic of the space and the basis which spans it. The scalar product may then be written in the convenient matrix form

$$r^* s = \boldsymbol{r}^\dagger \boldsymbol{M} \boldsymbol{s} \tag{2.8.4}$$

where $\boldsymbol{r}$, $\boldsymbol{s}$ are the usual columns of components and the dagger applied to a matrix means transposition accompanied by "starring" of the elements. For an arbitrary matrix, $\boldsymbol{X}^\dagger$ is variously referred to as the "Hermitian transpose", the "adjoint" or the "associate" of $\boldsymbol{X}$: we shall employ the first of these terms. When applied to a single vector, the star and dagger are of course equivalent: for consistency of appearance we use the star. When applied to a matrix product the dagger also reverses the sequence of the factors:

$$(\boldsymbol{AB})^\dagger = \boldsymbol{B}^\dagger \boldsymbol{A}^\dagger$$

The use of the dagger in (2.8.4) is then completely consistent, the same result following by direct matrix multiplication. For, on taking the Hermitian transpose, $r^* = (\mathbf{e}\boldsymbol{r})^\dagger = \boldsymbol{r}^\dagger \mathbf{e}^\dagger$ and hence

$$r^* s = \boldsymbol{r}^\dagger \mathbf{e}^\dagger \mathbf{e} \boldsymbol{s} = \boldsymbol{r}^\dagger \boldsymbol{M} \boldsymbol{s}$$

where

$$\boldsymbol{M} = \mathbf{e}^\dagger \mathbf{e} = \begin{pmatrix} e_1^* \\ e_2^* \\ | \\ e_n^* \end{pmatrix} (e_1 \; e_2 \; \ldots \; e_n) = \begin{pmatrix} e_1^* e_1 & e_1^* e_2 & \ldots \\ e_2^* e_1 & e_2^* e_2 & \ldots \\ \ldots & \ldots & \ldots \end{pmatrix} \tag{2.8.5}$$

The simplest and most familiar metric space is ordinary 3-space with rectangular axes:

Example 2.11. In elementary vector algebra $e_1$, $e_2$ and $e_3$ are three perpendicular vectors of unit length: this means $e_i^* e_i = 1$ ($i = 1, 2, 3$),

‡ Other notations for this scalar product are $(r|s)$, and $(r, s)$. See also p. 50.

$\mathsf{e}_i^*\mathsf{e}_j = 0$ $(i, j = 1, 2, 3;\ i \neq j)$ where the stars are merely formal since the basis is real. This is usually written $\mathsf{e}_i^*\mathsf{e}_j = \delta_{ij}$ where $\delta_{ij}$, the "Kronecker delta", is unity for $i = j$ but zero otherwise. The components of a vector are then "rectangular Cartesian" components, and if we take

$$\mathsf{r}_1 = x_1\mathsf{e}_1 + y_1\mathsf{e}_2 + z_1\mathsf{e}_3$$

$$\mathsf{r}_2 = x_2\mathsf{e}_1 + y_2\mathsf{e}_2 + z_2\mathsf{e}_3$$

the scalar product is

$$\mathsf{r}_1^*\mathsf{r}_2 = (x_1\ y_1\ z_1)\begin{pmatrix}1 & 0 & 0\\ 0 & 1 & 0\\ 0 & 0 & 1\end{pmatrix}\begin{pmatrix}x_2\\ y_2\\ z_2\end{pmatrix} = x_1x_2 + y_1y_2 + z_1z_2$$

(all components being real). The length of a vector $\mathsf{r}$ is $\sqrt{(\mathsf{r}^*\mathsf{r})}$ or

$$|\mathsf{r}| = \sqrt{(\mathsf{r}^*\mathsf{r})} = \sqrt{(x^2 + y^2 + z^2)}$$

And in terms of scalar products the *angle* between $\mathsf{r}_1$ and $\mathsf{r}_2$ is given by

$$\cos\theta_{12} = \frac{\mathsf{r}_1^*\mathsf{r}_2}{|\mathsf{r}_1|\,|\mathsf{r}_2|} = \frac{x_1x_2 + y_1y_2 + z_1z_2}{\sqrt{\{(x_1^2 + y_1^2 + z_1^2)(x_2^2 + y_2^2 + z_2^2)\}}}$$

Since the scalar product has been defined for a Hermitian vector space in general by (2.8.4), the length $|\mathsf{r}|$ of a vector $\mathsf{r}$, and the angle between two vectors $\mathsf{r}_1$ and $\mathsf{r}_2$, may be defined exactly as in 3-space (example 2.11) by

$$|\mathsf{r}| = (\mathsf{r}^*\mathsf{r})^{\frac{1}{2}} = (\boldsymbol{r}^\dagger\boldsymbol{M}\boldsymbol{r})^{\frac{1}{2}} \qquad (2.8.6)$$

$$\cos\theta_{12} = \frac{\mathsf{r}_1^*\mathsf{r}_2}{|\mathsf{r}_1|\,|\mathsf{r}_2|} = \frac{\boldsymbol{r}_1^\dagger\boldsymbol{M}\boldsymbol{r}_2}{\{(\boldsymbol{r}_1^\dagger\boldsymbol{M}\boldsymbol{r}_1)(\boldsymbol{r}_2^\dagger\boldsymbol{M}\boldsymbol{r}_2)\}^{\frac{1}{2}}} \qquad (2.8.7)$$

$\theta_{12}$ is of course only formally an "angle", being in general a complex number, but the terminology is useful. In particular, two vectors are said to be perpendicular or *orthogonal* if $\mathsf{r}_1^*\mathsf{r}_2 = 0$ and this implies $\theta_{12} = \pi/2$. If the vectors of a basis are mutually orthogonal and of unit length

$$\mathsf{e}_i^*\mathsf{e}_j = \delta_{ij} \qquad (2.8.8)$$

—which means (cf. example 2.11) that the metric is the $n$-dimensional unit matrix—the basis is said to be *orthonormal* or *unitary*. A rectangular Cartesian basis results when all quantities are *real*. With oblique basis vectors (which may, for instance, be more appropriate in describing a crystal) $\boldsymbol{M}$ has off-diagonal elements and the scalar product expression is a general quadratic form. One useful consequence of orthonormality is that every component of a vector can be expressed as a scalar product: for

$$\mathbf{e}_k^*\mathbf{r} = \mathbf{e}_k^* \sum_i \mathbf{e}_i r_i = \sum_i \delta_{ki} r_i = r_k$$

i.e.

$$r_k = \mathbf{e}_k^*\mathbf{r} \tag{2.8.9}$$

EXAMPLE 2.12. The "spherical basis" of example 2.10 and its dual are

$$\mathbf{e}_{+1} = -(\mathbf{e}_1 + i\mathbf{e}_2)/\sqrt{2}, \quad \mathbf{e}_0 = \mathbf{e}_3, \quad \mathbf{e}_{-1} = (\mathbf{e}_1 - i\mathbf{e}_2)/\sqrt{2}$$

$$\mathbf{e}_{+1}^* = -(\mathbf{e}_1^* - i\mathbf{e}_2^*)/\sqrt{2}, \quad \mathbf{e}_0^* = \mathbf{e}_3^*, \quad \mathbf{e}_{-1}^* = (\mathbf{e}_1^* + i\mathbf{e}_2^*)/\sqrt{2}$$

The elements of $\boldsymbol{M}$ are then

$$M_{+1,+1} = \mathbf{e}_{+1}^*\mathbf{e}_{+1} = \tfrac{1}{2}(\mathbf{e}_1^* - i\mathbf{e}_2^*)(\mathbf{e}_1 + i\mathbf{e}_2) = 1,$$

etc. and we obtain once again $\boldsymbol{M} = \mathbf{1}_3$: the basis is unitary, though no longer Cartesian.

EXAMPLE 2.13. Finally, let us take as the basis vectors $\mathbf{e}_1, \mathbf{e}_2, \ldots$ a set of *functions* of a variable $\theta$ defined over some interval $a \leqslant \theta \leqslant b$ and of "*integrable square*": $\mathbf{e}_k = e_k(\theta)$. This means that

$$\int_a^b |e_k(\theta)|^2 \, d\theta = \text{finite value (all } k)$$

A linear combination of the basis functions is then another function, defined over the same interval and also of integrable square. The axioms (2.3.1) are evidently valid and the functions span a certain space (see also example 2.4). A familiar set is

$$e_k(\theta) = \frac{\exp(ik\theta)}{\sqrt{(2\pi)}} \quad (-\pi \leqslant \theta \leqslant \pi) \quad k = 0, \pm 1, \pm 2, \ldots$$

The general vector in the space which they span is

$$\mathbf{f} = f(\theta) = \sum_{k=-\infty}^{+\infty} f_k e_k(\theta)$$

and the expression in terms of these functions corresponds to the *Fourier resolution* of an arbitrary function $f(\theta)$, defined in the interval $(-\pi, \pi)$. The requirement of an integrable square suggests that we define a metric by

$$\mathbf{e}_j^*\mathbf{e}_k = \int_{-\pi}^{\pi} e_j^*(\theta) e_k(\theta) \, d\theta$$

which is finite for $j = k$ and also satisfies the symmetry condition $\mathsf{e}_j^*\mathsf{e}_k = (\mathsf{e}_k^*\mathsf{e}_j)^*$. With this definition

$$\mathsf{e}_k^*\mathsf{e}_k = \frac{1}{2\pi}\int_{-\pi}^{\pi} \exp(-ik\theta)\exp(ik\theta)\, d\theta = 1$$

and the basis vectors are of *unit length*. It also follows that for $j \neq k$ $\mathsf{e}_j^*\mathsf{e}_k = 0$ and the basis is therefore orthonormal.

The definition of a scalar product is of course arbitrary but is naturally made on grounds of practical utility. Thus, in example 2.13, the expression for the vector components (2.8.9) becomes

$$f_k = \mathsf{e}_k^* f = \frac{1}{\sqrt{(2\pi)}}\int_{-\pi}^{\pi} \exp(-ik\theta) f(\theta)\, d\theta$$

which is the usual Fourier coefficient expression. In other words, the Fourier coefficients play the part of vector components in a unitary space.

In conclusion, we note that the axiomatic definition of the scalar product (2.8.1) now takes the following amended form. With every ordered pair of vectors $r$, $s$ we associate a complex number $r^*s$ called the scalar product, with the properties

$$\left.\begin{array}{ll} \text{(i)} & r^*s = (s^*r)^* \\ \text{(ii)} & r^*(s+t) = r^*s + r^*t \\ \text{(iii)} & r^*(cs) = c(r^*s) \\ \text{(iv)} & r^*r > 0 \ (\text{all } r \neq 0) \end{array}\right\} \quad (2.8.10)$$

Had these axioms been adopted as a starting point, they could have been used to infer the expression (2.8.4) in terms of a set of basis vectors. We wished to show, however, *why* these axioms form the natural generalization of those which appear in elementary vector theory. Our choice was guided largely by the property embodied in (iv): every non-zero vector has a length (the positive root of $r^*r$) which is a real positive quantity. A space with this property is said to have a *positive definite* metric—a term which is also used to describe the corresponding metrical matrix $\boldsymbol{M}$, for which $\boldsymbol{r}^\dagger\boldsymbol{M}\boldsymbol{r} > 0$ (all $\boldsymbol{r} \neq 0$).

**2.9. Unitary Transformations**

The formal expression for the metric (2.8.5) leads immediately to its transformation law under change of basis. If we introduce a new basis, $\bar{\mathbf{e}} = \mathbf{e}\boldsymbol{T}$, the metrical matrix $\bar{\boldsymbol{M}}$ for the new set is

$$\bar{\boldsymbol{M}} = \bar{\mathbf{e}}^\dagger\bar{\mathbf{e}} = \boldsymbol{T}^\dagger\mathbf{e}^\dagger\mathbf{e}\boldsymbol{T}$$

or

$$\bar{\boldsymbol{M}} = \boldsymbol{T}^\dagger\boldsymbol{M}\boldsymbol{T} \tag{2.9.1}$$

The scalar product is then clearly invariant against basis change—as it must be :

$$\bar{\boldsymbol{r}}^\dagger\bar{\boldsymbol{M}}\bar{\boldsymbol{s}} = \boldsymbol{r}^\dagger\boldsymbol{T}^{-1\dagger}(\boldsymbol{T}^\dagger\boldsymbol{M}\boldsymbol{T})\boldsymbol{T}^{-1}\boldsymbol{s} = \boldsymbol{r}^\dagger\boldsymbol{M}\boldsymbol{s}$$

This, of course, is independent of whether the basis vectors are orthogonal or oblique. But unitary bases, with $\boldsymbol{M} = \mathbf{1}_n$, are usually most convenient, and we are consequently interested in basis changes which preserve this property. If $\boldsymbol{T}$ leads from one unitary basis to another we must have

$$\bar{\boldsymbol{M}} = \boldsymbol{T}^\dagger\mathbf{1}_n\boldsymbol{T} = \mathbf{1}_n \qquad \text{or} \qquad \boldsymbol{T}^\dagger\boldsymbol{T} = \mathbf{1}_n$$

Non-singular square matrices which satisfy this condition are described as "*unitary matrices*". Since $\boldsymbol{T}$ is non-singular a unique inverse exists and $\boldsymbol{T}^\dagger = \boldsymbol{T}^{-1}$. Consequently

$$\boldsymbol{T}^\dagger\boldsymbol{T} = \boldsymbol{T}\boldsymbol{T}^\dagger = \mathbf{1}_n \tag{2.9.2}$$

We shall frequently use the letter $\boldsymbol{U}$ to denote a unitary matrix.

EXAMPLE 2.14. The matrix which relates the spherical basis vectors of example 2.10 to the Cartesian basis, according to $(\mathbf{e}_{+1}\ \mathbf{e}_0\ \mathbf{e}_{-1}) = (\mathbf{e}_1\ \mathbf{e}_2\ \mathbf{e}_3)\boldsymbol{U}$, is

$$\boldsymbol{U} = \begin{pmatrix} -\frac{1}{2}\sqrt{2} & 0 & \frac{1}{2}\sqrt{2} \\ -\frac{1}{2}i\sqrt{2} & 0 & -\frac{1}{2}i\sqrt{2} \\ 0 & 1 & 0 \end{pmatrix}$$

from which we see $\boldsymbol{U}^\dagger\boldsymbol{U} = \mathbf{1}_3$. This also expresses the fact that any two columns, $\boldsymbol{u}_i$, $\boldsymbol{u}_j$ are orthonormal in the unitary sense $\boldsymbol{u}_i^\dagger\boldsymbol{u}_j = \delta_{ij}$. A similar result holds for the rows.

As always, $\boldsymbol{U}$ may also be associated with a mapping $\mathsf{U}$, according to (2.6.3), which sends an arbitrary vector $\mathsf{r}$ into its image $\mathsf{r}' = \mathsf{U}\mathsf{r}$. Mappings of this kind have the special property of preserving lengths and angles—for the scalar product between $\mathsf{r}' = \mathsf{U}\mathsf{r}$ and $\mathsf{s}' = \mathsf{U}\mathsf{s}$ is the same as that between $\mathsf{r}$ and $\mathsf{s}$. To show this we take a unitary basis and have, since $\boldsymbol{r}' = \boldsymbol{U}\boldsymbol{r}$, $\boldsymbol{s}' = \boldsymbol{U}\boldsymbol{s}$,

$$\mathsf{r}'^*\mathsf{s}' = \boldsymbol{r}'^\dagger\boldsymbol{s}' = \boldsymbol{r}^\dagger\boldsymbol{U}^\dagger\boldsymbol{U}\boldsymbol{s} = \boldsymbol{r}^\dagger\boldsymbol{s} = \mathsf{r}^*\mathsf{s}$$

Here we have assumed, by putting $\boldsymbol{M} = \mathbf{1}_n$, that a unitary basis *can be found.* It may be shown that provided the metric is positive definite (p. 47), so that $r^*r > 0$ (all $r \neq 0$), a unitary basis can always be found.

For completeness we indicate the proof:

Any Hermitian matrix $\boldsymbol{M}$ can be brought to diagonal form by a unitary transformation, i.e. there exists a $\boldsymbol{U}$ such that $\boldsymbol{U}^\dagger \boldsymbol{M} \boldsymbol{U} = \boldsymbol{D}$ (diagonal). $\boldsymbol{M}$ can therefore be expressed as $\boldsymbol{M} = \boldsymbol{U}\boldsymbol{D}\boldsymbol{U}^\dagger$. But from (2.9.1) $\boldsymbol{D}$ is the metric for some other basis and positive definiteness therefore requires that all its (diagonal) elements are real and positive. We can then define $\boldsymbol{D}^{\frac{1}{2}} = \boldsymbol{D}^{\dagger\frac{1}{2}} = \text{diag}\,(D_{11}^{\frac{1}{2}}\ D_{22}^{\frac{1}{2}}\ldots)$ and introduce $\boldsymbol{V} = \boldsymbol{U}\boldsymbol{D}^{\frac{1}{2}}$. In this case $\boldsymbol{M} = \boldsymbol{V}\boldsymbol{V}^\dagger$ and, since different bases must be related by *non-singular* transformation matrices, the inverses exist and give $\boldsymbol{V}^{-1}\boldsymbol{M}\boldsymbol{V}^{-1\dagger} = \boldsymbol{T}^\dagger\boldsymbol{M}\boldsymbol{T} = \boldsymbol{1}_n$ where $\boldsymbol{T} = \boldsymbol{V}^{-1\dagger}$. It follows that a basis transformation (2.9.1) can be found for which the new metrical matrix becomes the unit matrix. The existence of unitary bases is therefore ensured by the metric axioms.

### 2.10. Matrix Elements as Scalar Products

It was noted in section 2.8 that the introduction of a metric gave a convenient expression (2.8.9) for the $i$th component of a vector $r$ as a scalar product: with a unitary basis $r_i = \mathsf{e}_i^* r$. It follows similarly that the elements of the matrix describing a mapping are metrically determined; and again it is advantageous to use a unitary basis. From (2.6.3) we have $R\mathbf{e} = \mathbf{e}\boldsymbol{R}$ and on multiplying by $\mathbf{e}^\dagger$

$$\mathbf{e}^\dagger R\mathbf{e} = \mathbf{e}^\dagger\mathbf{e}\boldsymbol{R} = \boldsymbol{M}\boldsymbol{R} \qquad (2.10.1)$$

If $\boldsymbol{M} = \boldsymbol{1}_n$ this expresses $\boldsymbol{R}$ as an array of scalar products

$$\boldsymbol{R} = \mathbf{e}^\dagger R\mathbf{e} = \begin{pmatrix} \mathsf{e}_1^* \\ \mathsf{e}_2^* \\ | \end{pmatrix} (R\mathsf{e}_1\ R\mathsf{e}_2\ \ldots) = \begin{pmatrix} \mathsf{e}_1^* R\mathsf{e}_1 & \mathsf{e}_1^* R\mathsf{e}_2 & \ldots \\ \mathsf{e}_2^* R\mathsf{e}_1 & \mathsf{e}_2^* R\mathsf{e}_2 & \ldots \\ \ldots & \ldots & \ldots \end{pmatrix} \qquad (2.10.2)$$

The $ij$-element of $\boldsymbol{R}$ is

$$R_{ij} = \mathsf{e}_i^* R\mathsf{e}_j \qquad (2.10.3)$$

—the scalar product of $\mathsf{e}_i$ and the image $R\mathsf{e}_j$ of $\mathsf{e}_j$. It follows that if we have a space with a metric it is not necessary to determine $R_{ij}$ by expressing $R\mathsf{e}_j$ in terms of $\mathsf{e}_1$, $\mathsf{e}_2$, ... and picking out the $i$th component: we can simply evaluate the scalar product $\mathsf{e}_i^* R\mathsf{e}_j$. This possibility should be verified in the case of example 2.7. The following example is less trivial.

Example 2.15. Any operation on the function $f(\theta)$ leading to a new function $f'(\theta)$ may be regarded as a mapping in a function space (cf. example 2.13). Thus, with $f = f(\theta)$, $f' = f'(\theta)$ and $D = \mathrm{d}/\mathrm{d}\theta$, we may have

$$f' = Df$$

When $f(\theta)$ is expanded over a complete orthonormal set of functions $e_1(\theta)$, $e_2(\theta)$, ... we write $f = \mathbf{e}f$ and the expansion coefficients for $f'$ are determined by $\boldsymbol{f}' = \boldsymbol{D}\boldsymbol{f}$, for $f' = \mathsf{D}\mathbf{e}f = \mathbf{e}\boldsymbol{D}f = \mathbf{e}f'$. But instead of regarding $D_{ij}$ as the coefficient of $e_i(\theta)$ in the Fourier expansion of $\mathsf{D}e_j(\theta)$—according to $\mathsf{D}e_j(\theta) = \sum_i e_i(\theta)D_{ij}$—we may simply write

$$D_{ij} = \mathbf{e}_i^*\mathsf{D}\mathbf{e}_j = \int_{-\pi}^{\pi} e_i^*(\theta)\,\frac{\mathrm{d}}{\mathrm{d}\theta}\,e_j(\theta)\,\mathrm{d}\theta$$

from our definition of the scalar product of two functions. The matrix $\boldsymbol{D}$ is called the *matrix representative* of the operator $\mathsf{D}$, relative to the basis of functions $\{e_i(\theta)\}$ while $D_{ij}$, defined as in (2.10.3) is the *matrix element* of the operator between functions $e_i(\theta)$ and $e_j(\theta)$. Since $\mathsf{D}$ is, from our present point of view, simply a mapping, it follows from section 2.7 that the matrix associated with a sequence of operators $\mathsf{RS}$ ($\mathsf{S}$ acting first) is the matrix product $\boldsymbol{RS}$. There is then a homomorphism between a group of operators and the associated group of matrices.

The only generalization we have to accept, in dealing with function spaces, is the admission of an *infinite* number of dimensions: for an arbitrary function can generally be represented only in terms of an *infinite* set of orthogonal functions (e.g. $f(\theta)$ in terms of $\{e_k(\theta)\}$ in example 2.13). This raises no formal difficulty, since the number of dimensions is not mentioned in the basic axioms (2.3.1) and (2.8.10). But it is clear that the basis should be *complete* (i.e. that the expansion may be made with unlimited accuracy by admitting more and more functions). Representations of this kind in terms of *infinite matrices*, are characteristic of quantum mechanics and are fully discussed elsewhere (Ref. 2). We shall be concerned more particularly, however, with *finite* sets of functions, conforming to certain special requirements (e.g. the functions in example 2.8 which are solutions of a differential equation, both corresponding to the same given vibrational frequency) and with symmetry operators rather than with general differential operators.

Finally, it is worth making the connection with the quantum mechanical notation due to Dirac (Ref. 2). Vectors $\mathbf{e}_1, \mathbf{e}_2, \ldots, \mathbf{e}_n$ are then written as $|1\rangle, |2\rangle, \ldots, |n\rangle$ and called "kets" while their duals, $\mathbf{e}_1^*, \mathbf{e}_2^*, \ldots, \mathbf{e}_n^*$ are written $\langle 1|, \langle 2|, \ldots, \langle n|$ and called "bras". A scalar product $\mathbf{e}_i^*\mathbf{e}_j$ then becomes a "bra(c)ket" (or "bra-line-ket")

$\langle i|j\rangle$. An arbitrary "ket vector" is

$$\mathsf{v} = v_1\mathsf{e}_1 + v_2\mathsf{e}_2 + \ldots + v_n\mathsf{e}_n$$

or

$$|\rangle = v_1|1\rangle + v_2|2\rangle + \ldots + v_n|n\rangle$$

When the basis vectors are orthonormal ($\langle i|j\rangle = \delta_{ij}$), $v_i$ may be expressed by (2.8.9) as the scalar product $\langle i|\rangle$, and we may write

$$|\rangle = \sum_i |i\rangle\langle i|\rangle \tag{2.10.4}$$

Again, *provided the basis is orthonormal*, the $ij$-element of the matrix which represents a mapping $\mathsf{R}$ is given by (2.10.3). Consequently vector components and matrix elements are transcribed as follows:

$$v_i = \langle i|\rangle, \qquad R_{ij} = \langle i|\mathsf{R}|j\rangle \tag{2.10.5}$$

When the basis vectors are labelled by a considerable number of indices (as in many quantum mechanical problems) the Dirac notation is particularly useful since it avoids multiple subscripts. It should be borne in mind, however, that whenever *oblique* basis vectors are used the notation breaks down completely. The matrices which give a representation of the operators are then determined (according to (2.10.1)) by

$$\left.\begin{aligned} &\boldsymbol{R} = \boldsymbol{M}^{-1}\boldsymbol{R}^D \\ \text{where}\quad &M_{ij} = \langle i|j\rangle, \qquad R^D_{ij} = \langle i|\mathsf{R}|j\rangle \end{aligned}\right\} \tag{2.10.6}$$

Operator equations are then no longer easily translated into Dirac language.

### 2.11. The Eigenvalue Problem

One problem which arises frequently is how to find a set of vectors which are not *rotated* by a given mapping, $\mathsf{F}$ say, but are merely changed in length. Any such vector must then satisfy the equation

$$\mathsf{F}\mathsf{v} = \lambda\mathsf{v} \tag{2.11.1}$$

Here $\lambda$ is a scalar multiplier and for any solution, $\mathsf{v} = \mathsf{v}_i$, takes some corresponding value, $\lambda = \lambda_i$. Equation (2.11.1) is called an *eigenvalue equation*, $\mathsf{v}_i$ and $\lambda_i$ being the $i$th *eigenvector* and *eigenvalue*. When we choose a basis $\mathbf{e}$ the equation takes the matrix form

$$\boldsymbol{F}\boldsymbol{v} = \lambda\boldsymbol{v} \tag{2.11.2}$$

and is equivalent to a set of simultaneous equations

$$\left.\begin{aligned} (F_{11}-\lambda)v_1 + F_{12}v_2 + \ldots + F_{1n}v_n &= 0 \\ F_{21}v_1 + (F_{22}-\lambda)v_2 + \ldots + F_{2n}v_n &= 0 \\ \ldots\ldots\ldots\ldots\ldots\ldots\ldots\ldots\ldots\ldots \\ F_{n1}v_1 + F_{n2}v_2 + \ldots + (F_{nn}-\lambda)v_n &= 0 \end{aligned}\right\} \tag{2.11.3}$$

It is known from elementary algebra that these equations are consistent only for those values of $\lambda$ which satisfy

$$\begin{vmatrix} (F_{11}-\lambda) & F_{12} & \cdots & F_{1n} \\ F_{21} & (F_{22}-\lambda) & \cdots & F_{2n} \\ \cdots & \cdots & \cdots & \cdots \\ F_{n1} & F_{n2} & \cdots & (F_{nn}-\lambda) \end{vmatrix} = 0 \qquad (2.11.4)$$

The eigenvalues are thus the roots of the $n$th degree polynomial which appears on expanding the determinant. On substituting each root, $\lambda = \lambda_i$, back into (2.11.3) we can solve for $v_1, v_2, \ldots, v_n$, obtaining say $v_{i1}, v_{i2}, \ldots, v_{in}$ which define a column $\boldsymbol{v}_i$ and hence the eigenvector $\mathsf{v}_i$. The column $\boldsymbol{v}_i$ is also loosely called an "eigenvector" of the matrix $\boldsymbol{F}$.

In most applications $\mathsf{F}$ has the *Hermitian symmetry* $\mathsf{e}_i^*\mathsf{F}\mathsf{e}_j = (\mathsf{e}_j^*\mathsf{F}\mathsf{e}_i)^*$ and accordingly $\boldsymbol{F}^\dagger = \boldsymbol{F}$. The following results are then well known (see, for example, Hall, 1963):

(i) There are just $n$ *real* eigenvalues

(ii) Eigenvectors with different eigenvalues are orthogonal in the unitary sense: $\mathsf{v}_i^*\mathsf{v}_j = \boldsymbol{v}_i^\dagger \boldsymbol{v}_j = 0 \quad (\lambda_i \neq \lambda_j)$

(iii) Any linear combination of eigenvectors with the *same* eigenvalue is also an eigenvector: if there are $p$ such independent eigenvectors, then we can find $p$ *orthogonal* combinations with the same eigenvalue. Consequently a full set of $n$ independent orthogonal eigenvectors can always be found.

Since any $\mathsf{v}_i$ may be multiplied by a constant and will still satisfy (2.11.1) it is convenient to normalize the solutions, and in this case

$$\mathsf{v}_i^*\mathsf{v}_j = \boldsymbol{v}_i^\dagger \boldsymbol{v}_j = \delta_{ij} \qquad (2.11.5)$$

If now we collect the columns, $\boldsymbol{v}_i$, into a square matrix

$$\boldsymbol{V} = \left( \boldsymbol{v}_1 \,\middle|\, \boldsymbol{v}_2 \,\middle|\, \cdots \,\middle|\, \boldsymbol{v}_n \right) \qquad (2.11.6)$$

it is clear that $\boldsymbol{V}$ is unitary,

$$\boldsymbol{V}^\dagger \boldsymbol{V} = \boldsymbol{V}\boldsymbol{V}^\dagger = \mathbf{1}_n \qquad (2.11.7)$$

and follows easily that

$$\boldsymbol{V}^\dagger \boldsymbol{F} \boldsymbol{V} = \text{diag}\,(\lambda_1, \lambda_2, \ldots, \lambda_n) \qquad (2.11.8)$$

In other words the solutions of the eigenvalue problem define a unitary transformation which brings $\boldsymbol{F}$ to *diagonal form*.

## REFERENCES

1. PAULING, L., and WILSON, E. B., *Introduction to Quantum Mechanics* (Chap. V), McGraw-Hill, 1935.
2. DIRAC, P. A. M., *The Principles of Quantum Mechanics* (4th Edition, p. 16), Oxford University Press, 1958.

## BIBLIOGRAPHY

BIRKHOFF, G., and McLANE, S., *A Survey of Modern Algebra*, Macmillan, 1941. (A very readable account of many aspects of modern algebra, including some of the topics discussed in this chapter and in chapter 1.)

HALL, G. G., *Matrices and Tensors* (Topic 1, vol. 4 of this Encyclopedia), Pergamon Press, 1963. (This book contains an account of elementary vector algebra and also of all the matrix theory required in the present volume.)

MURNAGHAN, F. D., *The Theory of Group Representations*, Johns Hopkins Press, 1938. (Chapter 1 contains a rather more advanced treatment of the elements of abstract group theory and of linear vector spaces.)

WEYL, H., *Symmetry*, Princeton University Press, 1952. (These lectures provide a beautiful introduction to the subject matter of the present volume. The third lecture, on *Ornamental Symmetry* is particularly relevant to this chapter.)

CHAPTER 3

# POINT AND SPACE GROUPS

## 3.1. Symmetry Operations as Orthogonal Transformations

In chapter 1 we discussed the symmetry operations which brought a triangular lamina into self-coincidence (example 1.3). Such operations all leave one point of the figure unmoved and are said to form a *point group*. A very large number of familiar objects exhibit some type of point group symmetry (Fig. 3.1). On the other hand, we have also met symmetry operations which do *not* leave any point of an object

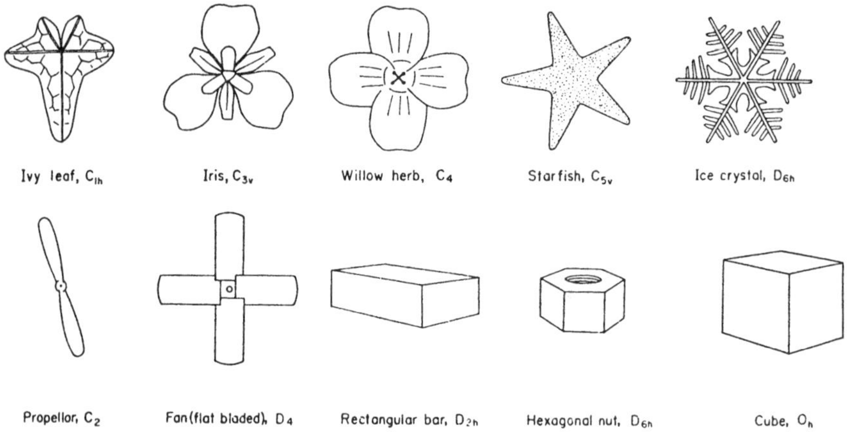


FIG. 3.1.

fixed, namely, the *translations* which bring an infinite repeating pattern into self-coincidence (example 2.1). Both rotations and translations may be described by indicating their effect on the position vector of a general point in the pattern, the origin conveniently being the point about which the rotations are defined. The rotations send the point $\mathsf{r}$ into its image $\mathsf{r}' = \mathsf{R}\mathsf{r}$. If we choose a basis, in which $\mathsf{r}$ and $\mathsf{r}'$ have columns of components $\boldsymbol{r}$ and $\boldsymbol{r}'$, the mapping $\mathsf{R}$ is represented by a matrix $\boldsymbol{R}$ and we have, for any point group operation

$$\mathsf{r}' = \mathsf{R}\mathsf{r}, \quad \boldsymbol{r}' = \boldsymbol{R}\boldsymbol{r}$$

Translation has a different effect. In a translation $\mathsf{t}$ the point with position vector $r$ is sent into that with $r' = r + t$, columns of components being related similarly:

$$r' = r + t, \quad \boldsymbol{r}' = \boldsymbol{r} + \boldsymbol{t}$$

The two types of operation may be combined. A point group operation $R$ (which we shall henceforth refer to as a "rotation"), followed by a translation $\mathsf{t}$ produces an image $r' = Rr + t$. It is convenient to use a notation due to Seitz (Ref. 1), writing

$$r' = Rr + t = (R \mid t)r \tag{3.1.1}$$

or in matrix form

$$\boldsymbol{r}' = \boldsymbol{R}\boldsymbol{r} + \boldsymbol{t} = (\boldsymbol{R} \mid \boldsymbol{t})\boldsymbol{r} \tag{3.1.2}$$

Since there is a one-to-one correspondence between the two modes of description it is immaterial which is used: a statement about vectors and mappings can be changed into one about columns (of components) and matrices simply by a change of typeface. We recall that, given any basis $\mathbf{e} = (\mathrm{e}_1\ \mathrm{e}_2\ \mathrm{e}_3)$ the matrix $\boldsymbol{R}$ associated with rotation $R$ is defined by (2.6.3), namely

$$R\mathbf{e} = \mathbf{e}\boldsymbol{R} = \mathbf{e}\boldsymbol{D}(R)$$

$\boldsymbol{D}(R)$ being used as an alternative to $\boldsymbol{R}$ when we wish to stress that the matrices provide a representation D. With the notation (3.1.1) a translation without rotation would be $(E \mid t)$ and a rotation without translation would be $(R \mid 0)$: thus, $(R \mid t)$ may be regarded as a composite operation

$$(R \mid t) = (E \mid t)(R \mid 0) \quad \text{(rotation followed by translation)} \tag{3.1.3}$$

It should be noted that the same operations in the reverse order produce a different image,

$$(R \mid 0)(E \mid t)r = (R \mid 0)(r + t) = Rr + Rt = (R \mid t')r \quad (t' = Rt),$$

but that the resultant operation is nevertheless of the form (3.1.1). Finally, the law of combination of two general operations is readily seen to be

$$(R \mid t)(S \mid t') = (RS \mid t + Rt') \tag{3.1.4}$$

and it follows that the inverse of an operation is defined by

$$(R \mid t)^{-1} = (R^{-1} \mid -R^{-1}t) \tag{3.1.5}$$

We shall be concerned particularly with the matrices, $\boldsymbol{R} = \boldsymbol{D}(R)$, which provide a representation of the rotations $R$. The properties of such matrices, which have already been constructed for the group of the equilateral triangle (example 2.7) must now be examined in more detail. All symmetry operations have one characteristic feature: they describe transformations which preserve lengths and angles—for otherwise the image of the pattern would not "fit" the original pattern. It is clear that the translational part of a symmetry operation always satisfies this requirement, for under the operation $(E \mid t)$

$$r \to r' = r + t, \quad s \to s' = s + t$$

and the vector separation of any two points remains unchanged:

$$r' - s' = (r + t) - (s + t) = r - s$$

Consequently, we need only consider the restrictions imposed on the rotation matrices, $\boldsymbol{R}$.

Mappings which preserve lengths and angles have been discussed in section 2.9 for a rather general $n$-space, in which case they are described by *unitary* matrices. In the present case $n = 3$ and the basis may be chosen so that all vector components are real: this makes $\boldsymbol{R}$ a $3 \times 3$ unitary matrix with *real* elements. Such a matrix is said to be *real orthogonal*, and since $\boldsymbol{R}^\dagger = \tilde{\boldsymbol{R}}$ its distinctive property is, from (2.9.2), $\boldsymbol{R}\tilde{\boldsymbol{R}} = \mathbf{1}_3$ or

$$\tilde{\boldsymbol{R}} = \boldsymbol{R}^{-1} \tag{3.1.6}$$

We shall now show that such matrices are of two types—one associated with pure rotations, the other with operations which involve a reflection. First, it is clear that, since $|\tilde{\boldsymbol{R}}| = |\boldsymbol{R}|$ and $|\boldsymbol{AB}| = |\boldsymbol{A}||\boldsymbol{B}|$ (from the elementary theory of determinants),

$$|\boldsymbol{R}|^2 = 1, \qquad |\boldsymbol{R}| = \pm 1 \tag{3.1.7}$$

The upper and lower signs, respectively, characterize the two types of operation referred to: when $|\boldsymbol{R}| = +1$, $R$ is said to describe a *proper rotation* and when $|\boldsymbol{R}| = -1$, an *improper rotation.* It is immediately verified that the matrices of example 2.6, listed in (2.7.3), fall into these two classes. Now it is known from section 2.11 that $\boldsymbol{R}$ has eigenvalues which occur as the roots of a cubic equation with real coefficients, and that a unitary matrix $\boldsymbol{U}$ may be found such that‡

$$\boldsymbol{U}^\dagger \boldsymbol{R} \boldsymbol{U} = \hat{\boldsymbol{R}} = \text{diag}\ (\lambda_1\ \lambda_2\ \lambda_3) \tag{3.1.8}$$

Geometrically, this means that with a suitable new basis, $\boldsymbol{u} = \boldsymbol{e}\boldsymbol{U}$, the matrix describing the mapping $R$ will become diagonal. Generally however, $\boldsymbol{U}$, and hence $\hat{\boldsymbol{R}}$, may contain complex elements. This means simply that a formally more convenient basis is obtained by combining the original basis vectors with *complex* coefficients (cf.

‡ We shall use diag $(d_1\ d_2\ d_3 \ldots)$ to mean the matrix with $d_1, d_2, d_3, \ldots$ along the diagonal and zeros elsewhere.

example 2.10). Now the Hermitian conjugate of (3.1.8) is

$$\boldsymbol{U}^{\dagger}\boldsymbol{R}^{\dagger}\boldsymbol{U} = \hat{\boldsymbol{R}}^{\dagger} = \text{diag}\,(\lambda_1^* \,\lambda_2^* \,\lambda_3^*) \tag{3.1.9}$$

If we multiply together (3.1.8) and (3.1.9) and require $\boldsymbol{R}^{\dagger}\boldsymbol{R} = \boldsymbol{U}^{\dagger}\boldsymbol{U} = \mathbf{1}_3$ we obtain

$$\lambda_1\lambda_1^* = \lambda_2\lambda_2^* = \lambda_3\lambda_3^* = 1$$

Besides being unimodular, the $\lambda$'s are roots of a cubic equation and any complex root must be accompanied by its conjugate: the eigenvalues must therefore be

$$\lambda_1 = \mathrm{e}^{i\theta}, \qquad \lambda_2 = \mathrm{e}^{-i\theta}, \qquad \lambda_3 = \pm 1 \tag{3.1.10}$$

(in which the order along the diagonal may, of course, be chosen arbitrarily). Clearly the alternative signs of $\lambda_3$ correspond to the proper and improper rotations with $|\boldsymbol{R}| = |\hat{\boldsymbol{R}}| = +1$ and $-1$ respectively.

In order to obtain a *real* basis, a further transformation of basis vectors is necessary; this will associate with $R$ a new matrix, obtained from $\hat{\boldsymbol{R}}$ by a further unitary transformation, with a simple geometrical interpretation. Thus, if we choose

$$\boldsymbol{V} = \frac{1}{\sqrt{2}}\begin{pmatrix} 1 & i & 0 \\ 1 & -i & 0 \\ 0 & 0 & 1 \end{pmatrix} \tag{3.1.11}$$

we obtain

$$R \to \bar{\boldsymbol{R}} = \boldsymbol{V}^{\dagger}\hat{\boldsymbol{R}}\boldsymbol{V} = \begin{pmatrix} \cos\theta & -\sin\theta & 0 \\ \sin\theta & \cos\theta & 0 \\ 0 & 0 & \pm 1 \end{pmatrix} \tag{3.1.12}$$

If we denote the corresponding basis by $\mathsf{v} = \boldsymbol{u}\boldsymbol{V} = \mathbf{e}\boldsymbol{U}\boldsymbol{V}$, the effect of the rotation $R$ is given by $R\mathsf{v} = \mathsf{v}\bar{\boldsymbol{R}}$ or

$$\begin{aligned} \mathsf{v}_1 \to \mathsf{v}_1' &= \cos\theta\, \mathsf{v}_1 + \sin\theta\, \mathsf{v}_2 \\ \mathsf{v}_2 \to \mathsf{v}_2' &= -\sin\theta\, \mathsf{v}_1 + \cos\theta\, \mathsf{v}_2 \\ \mathsf{v}_3 \to \mathsf{v}_3' &= \pm \mathsf{v}_3 \end{aligned}$$

On taking the upper sign, $R$ is seen to describe a positive rotation through angle $\theta$ about the $\mathsf{v}_3$ axis (cf. example 2.5); with the lower sign this rotation is accompanied by reflection across the plane, normal to the axis, containing the other two basis vectors. We note also that $\mathsf{v}_3$ is an *eigenvector* of the rotation $R$ (eigenvalue $\pm 1$), while in the special case of a pure reflection ($\theta = 0$) $\mathsf{v}_1$ and $\mathsf{v}_2$ are also eigenvectors (eigenvalues $+1$). Thus, any operation which preserves lengths and angles and leaves one point unmoved can be regarded as a rotation (proper or improper) about a unique axis through that point. In other words, the only operations which a point group can contain are either rotations about an axis or rotations accompanied by reflections across a plane normal to the axis. In particular, the cases $\theta = 0$ and $\theta = \pi$ correspond to the simple operations of a *pure reflection* ($\sigma$) and *an inversion* ($i$), respectively, with matrices

$$\boldsymbol{D}(\sigma) = \begin{pmatrix} 1 & 0 & 0 \\ 0 & 1 & 0 \\ 0 & 0 & -1 \end{pmatrix}, \qquad \boldsymbol{D}(i) = \begin{pmatrix} -1 & 0 & 0 \\ 0 & -1 & 0 \\ 0 & 0 & -1 \end{pmatrix}$$

Any improper rotation can then be regarded as the product (clearly commutative) of a proper rotation and either a reflection or an inversion: for

$$\begin{pmatrix} \cos\theta & -\sin\theta & 0 \\ \sin\theta & \cos\theta & 0 \\ 0 & 0 & -1 \end{pmatrix} = \begin{pmatrix} 1 & 0 & 0 \\ 0 & 1 & 0 \\ 0 & 0 & -1 \end{pmatrix} \begin{pmatrix} \cos\theta & -\sin\theta & 0 \\ \sin\theta & \cos\theta & 0 \\ 0 & 0 & 1 \end{pmatrix} =$$

$$= \begin{pmatrix} -1 & 0 & 0 \\ 0 & -1 & 0 \\ 0 & 0 & -1 \end{pmatrix} \begin{pmatrix} \cos(\theta+\pi) & -\sin(\theta+\pi) & 0 \\ \sin(\theta+\pi) & \cos(\theta+\pi) & 0 \\ 0 & 0 & 1 \end{pmatrix} \qquad (3.1.13)$$

We note, for future reference, that the *trace* of a rotation matrix is independent of choice of axes. For, making use of (2.7.5),

$$\text{tr}\,\bar{\boldsymbol{R}} = \text{tr}\,\boldsymbol{T}^{-1}\boldsymbol{R}\boldsymbol{T} = \text{tr}\,\boldsymbol{R}\boldsymbol{T}\boldsymbol{T}^{-1} = \text{tr}\,\boldsymbol{R}$$

The trace is, in fact, dependent only on the *rotation angle*, (3.1.12) giving

$$\text{tr}\,\boldsymbol{R} = \pm 1 + 2\cos\theta \qquad (3.1.14)$$

($\pm 1$ according as the rotation is proper or improper). Finally, we note that in the finite groups the rotation angles are restricted to the values $k(2\pi/n)$ where $k$ and $n$ are *integers*; $n$ is called the "order" of the corresponding axis of symmetry. This follows immediately because if $R$ is a symmetry operation then so are $R^2$, $R^3$, . . . and unless there is some integer $n$ for which $R^n = E$ the group of rotations will be infinite. In the finite case it is usual to denote the generator $R$ by $C_n$: then $C_n^{\,n}$ corresponds to a rotation through $n\,\theta = 2\pi$ and $C_n^{\,k}$ to rotation through $k\theta = k(2\pi/n)$. The rotations $E$, $C_n$, . . . , $C_n^{\,n-1}$ then form a cyclic group $C_n$ (possibly a *sub*-group of the full symmetry group) whose order is equal to the order of the axis.

We shall now discuss first the point groups, in which only the rotations occur, and then the *space groups* in which translations also occur and the symmetry operations are generally of the composite form $(R\,|\,t)$.

### 3.2. The Axial Point Groups

The simpler types of point group symmetry are characterized by a *principal axis* whose order is higher than that of any other axis. Such an axis is usually called a *symmetry element*‡: another typical symmetry element is a *plane of symmetry*, such that points on one side of the plane are mirror images of those on the other. In illustrations the principal axis is usually shown vertical and taken as the $z$ axis. Operations on a 3-dimensional figure are then conveniently described, as in Fig. 1.2, by indicating their effect upon the horizontal projections (usually shown in the plane of the paper) of some reference point. Any point is sent, by the symmetry operations, into a set of images or " equivalent points ": an image *above* the paper is indicated by a cross, at the foot of the perpendicular on to the paper, while a circle is used to denote a similar image at the same distance *beneath* the paper. The set of equivalent points produced by the operations of $\mathsf{C}_{3v}$ has already been given in Fig. 1.2. The pure rotation groups, $\mathsf{C}_1, \mathsf{C}_2, \ldots, \mathsf{C}_6$, give the projection diagrams in the first row of Fig. 3.2, which also shows regular plane figures which are carried into self-coincidence. Symmetry axes of higher order occur in some molecules but are uncommon. In developing the subject we use the Schoenflies notation (which is nearly always used in molecular theory and molecular spectroscopy—e.g. Herzberg (Ref. 2) ).

Perhaps the simplest way of specifying the point groups and of investigating their structure is to take a set of generators and then to find all symmetry operations by exhaustive combination, recording the results in a projection diagram. In addition to rotations about a principal axis, with a single generator $\mathsf{C}_n$ we must consider *reflections* and rotations (of order less than $n$) *about other axes*. A little consideration shows that the operations available are severely limited in number and type, simply by the fact that a symmetry operation must send a symmetry element into a new symmetry element. Thus (Fig. 3.3) a second rotation axis inclined at an arbitrary angle to the principal axis would imply the existence of another axis, of order equal to that of the principal axis, in some quite different direction ; this would contradict the assumption of a single principal axis unless the second axis were (i) perpendicular to the principal axis and (ii) of order 2 (in which case the new symmetry operations send the principal axis into itself). Now if there is *one* two-fold axis normal to the principal

‡ The term should not be confused with the element of a symmetry group.

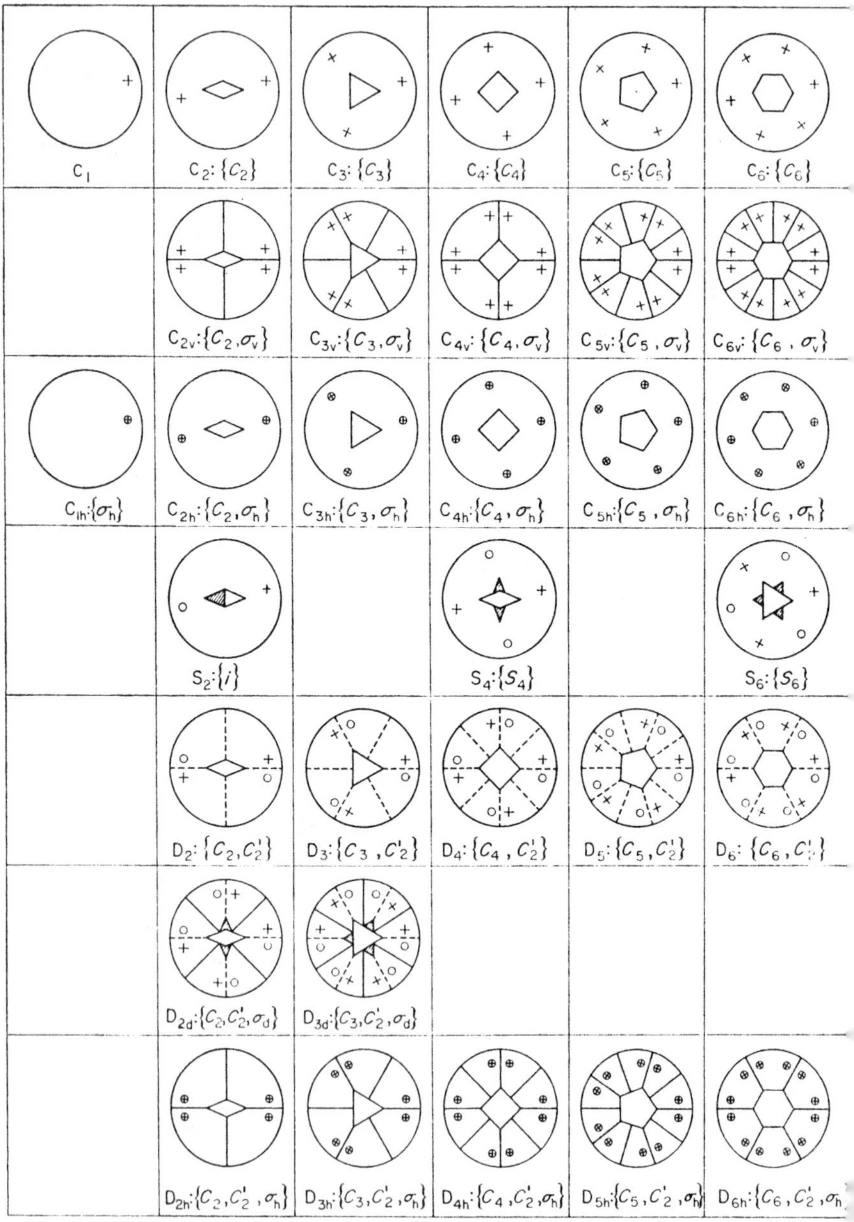


FIG. 3.2. Projection diagrams for axial point groups.

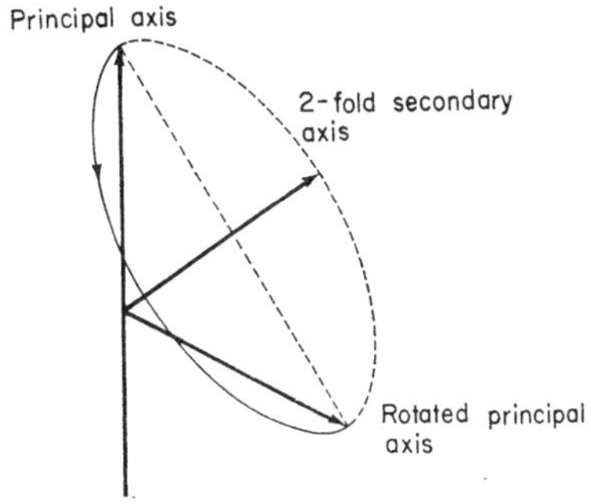


FIG. 3.3. Generation of a new symmetry element by a symmetry operation. Rotation about the 2-fold axis, yields a new principal axis.

$n$-fold axis, there must be $n$, since each operation $C_n$ will generate another. Similar reasoning shows that the only reflections we need to consider are (i) across planes containing the principal axis (bisecting the angles between secondary axes, if any), and (ii) across a plane normal to the principal axis. These distinct types of operation, and the projection diagrams with which they are associated are illustrated in Fig. 3.4: they serve to characterize all the point groups. The last

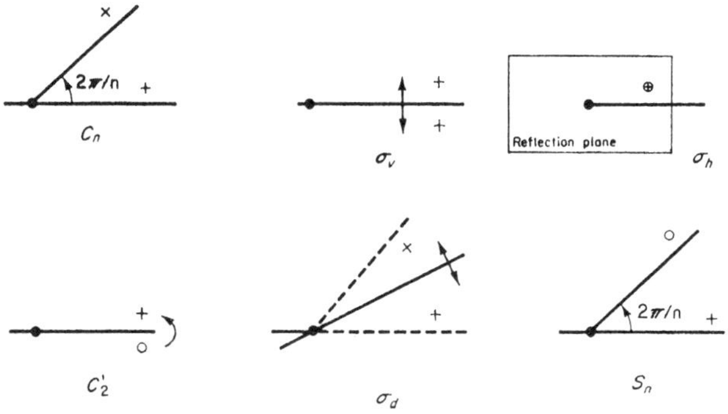


FIG. 3.4. Types of axial group operation.

$C_n$ = positive rotation through $2\pi/n$
$\sigma_v$ = reflection across vertical plane (used for family which includes $xz$ plane).
$\sigma_h$ = reflection across horizontal plane (plane of paper)
$C_2'$ = rotation through $\pi$ about axis normal to principal axis
$\sigma_d$ = reflection across " dihedral " plane, containing principal axis and bisecting angle between adjacent 2-fold axes (also used in $C_{4v}$ and $C_{6v}$ for planes of the family which does *not* include the $xz$ plane)
$S_n$ = improper positive rotation through $2\pi/n$ ($S_n = \sigma_h C_n = i\bar{C}_n$). (The bar is used to indicate a *negative* rotation. $S_2$ is usually denoted by $i$.)

operation in the figure, the improper rotation $S_n$, is described by the matrix (3.1.13) with $\theta = 2\pi/n$ and is thus equivalent to $\sigma_h C_n$; but since it is not necessary that $\sigma_h$ and $C_n$ appear *separately* in the group, $S_n$ must be listed as a possible independent generator. In particular, $S_2$ is usually denoted by $i$ since it represents an *inversion* in which every point is sent into an image across the centre of symmetry. When we wish to give names to all the individual elements of an axial point group we shall do so in accordance with Fig. 3.5: the $n$-fold

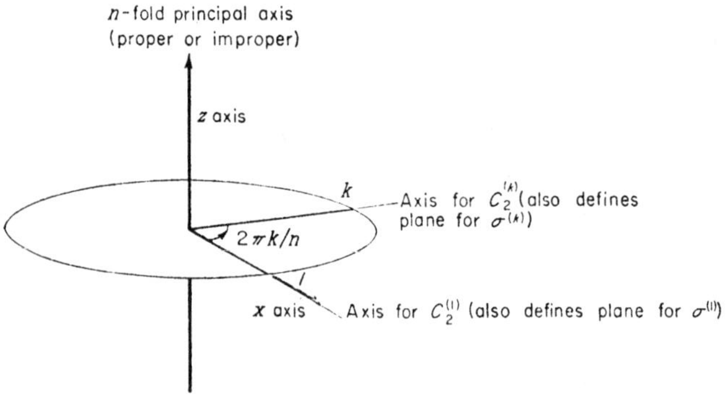


FIG. 3.5. Nomenclature for axial group elements.

axis of symmetry (proper or improper) is taken as $z$ axis, while secondary axes or reflection planes are numbered with superscripts 1, 2, . . . , $n$ increasing in the positive direction and with 1 defining the $x$ axis. A plane or axis mid-way between two others ($i$ and $j$) then gives operations with an ($ij$) superscript. When this notation is used the $v$ or $d$ subscript is redundant and may be suppressed, though it is convenient when referring to a whole class of operations (e.g. the $\sigma_v$ operations $\sigma^{(1)}$, $\sigma^{(2)}$, etc.).

Systems of generators, and the sets of images resulting from exhaustive repetition and combination of all operations, are indicated in Fig. 3.2 for all the axial symmetry groups up to $n = 6$. Presence of an $S_n$ operation is indicated by a black polygon beneath the white polygon representing $C_{n/2}$—two such polygons, stuck together, exhibiting $S_n$ symmetry. Let us briefly consider one column in the figure, that for $n = 2$. In the first example (see Fig. 3.6(i)), the points 1, 2 are generated from 1 by application of $E$, $C_2$ respectively. In the second, the additional points 1′, 2′ follow using the second generator $\sigma^{(1)}$ ($\sigma_v$): they are obtained from 1 using $\sigma^{(1)}$ and $\sigma^{(1)}C_2$, respectively.

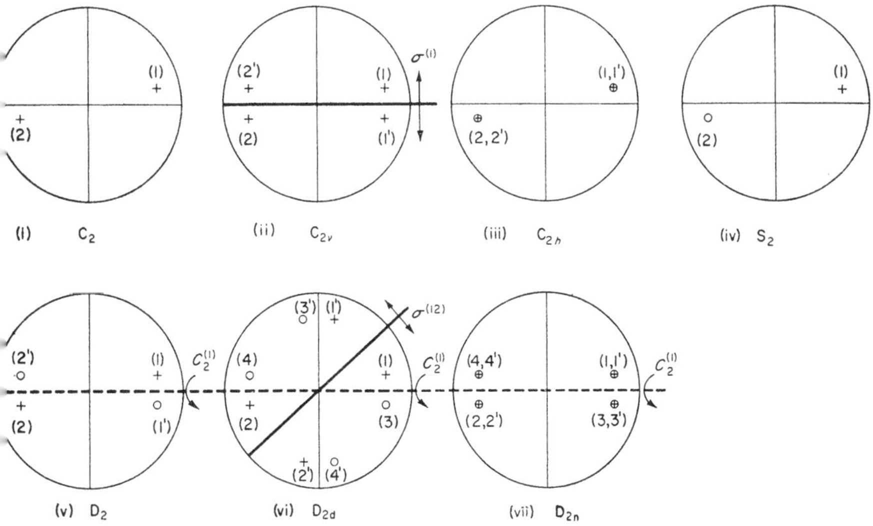


FIG. 3.6. Generation of equivalent points. Axes and planes are indicated only for the generators.

In the third example the two equivalent points *beneath* the plane are obtained using the generator $\sigma_h$ instead of $\sigma^{(1)}$. The nomenclature for these groups, namely $C_2$, $C_{2v}$ and $C_{2h}$, is self-explanatory. Molecules illustrating these three symmetry types are depicted in Fig. 3.7. In the fourth example, Fig. 3.6(iv), the group consists only of $\{E, i\}$: not many molecules have such low symmetry if they are symmetrical at all. The remaining groups with $n = 2$ contain $D_2$, with generators $C_2$ and $C_2^{(1)}$ (about an axis perpendicular to the principal axis). The choice of " principal " axis is, in this case ($n = 2$), somewhat arbitrary, since other axes of equal order must occur, and in referring to systems with three dissimilar 2-fold axes it is necessary to indicate clearly the axes chosen. In some cases, conventions may be adopted. Thus, for planar molecules (e.g. naphthalene) the principal axis is almost invariably taken perpendicular to the plane. Again, a subscript is used to distinguish the various " dihedral " groups: thus, $D_{2d}$ represents $D_2$ with an additional generator $\sigma_d$, while $D_{2h}$ contains $\sigma_h$ in place of $\sigma_d$. In the $D_2$ diagram, points 1 and 2 arise using $E$ and $C_2$; points 1′ and 2′ using $C_2^{(1)}$ and $C_2^{(1)}C_2$. The four additional points in the $D_{2h}$ diagram are obtained simply by a further reflection across the plane of the paper. In the case of $D_{2d}$ we meet the new operation

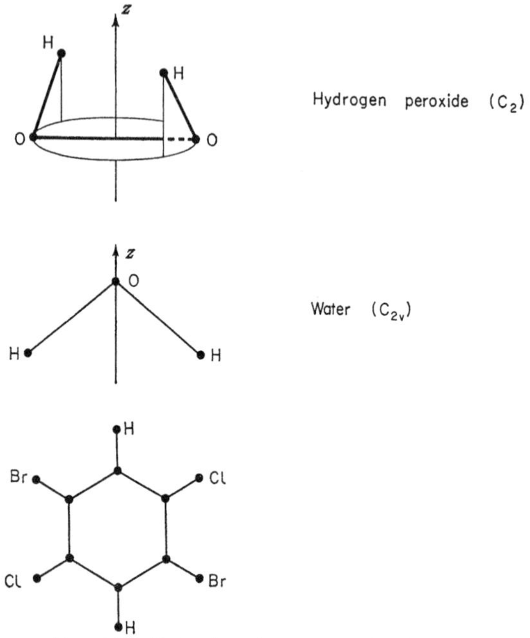


FIG. 3.7. Molecules of symmetry $C_2$, $C_{2v}$, $C_{2h}$.

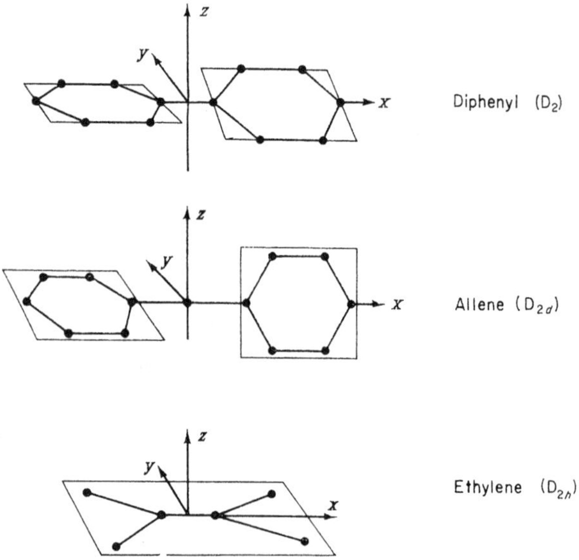


FIG. 3.8. Molecules of symmetry $D_2$, $D_{2d}$, $D_{2h}$. In diphenyl the rings are twisted out of plane by steric hindrance between hydrogens (not shown). In allene the rings are perpendicular and the x axis is an improper 4-fold axis.

$\sigma^{(12)}(=\sigma_d)$‡: if 1, 2, 3, 4 are the images produced by the operations of $\mathsf{D}_2$ then $\sigma^{(12)}$ sends them into 1′, 2′, 3′, 4′, respectively. Molecules illustrating the dihedral symmetry types are depicted in Fig. 3.8.

It is natural to enquire why we cannot get new groups (perhaps "$\mathsf{C}_{2vh}$", for example), by choosing other sets of generators, and also why certain groups—such as $\mathsf{D}_{4d}$—are not listed in Fig. 3.2. These questions are answered by the following examples.

EXAMPLE 3.1. Let us try to construct a new group $\mathsf{C}_{2vh}$ using the generators $\{C_2, \sigma^{(1)}, \sigma_h\}$—a set not so far considered. The distinct operations are then seen to be

$$\{E, C_2 \mid \sigma^{(1)}, \sigma^{(1)}C_2 \mid \sigma_h, \sigma_h C_2, \sigma_h \sigma^{(1)}, \sigma_h \sigma^{(1)} C_2\}$$

in which each subgroup (first $\mathsf{C}_2$, then $\mathsf{C}_{2v}$) is extended by forming its left coset with a new generator (first $\sigma^{(1)}$, then $\sigma_h$). Each operation is then seen to be equivalent to a single operation of one of the types shown in Fig. 3.4. Thus, if we label the directions as in Fig. 3.6(ii), the group becomes

$$\mathsf{C}_{2vh} = \{E, C_2 \mid \sigma^{(1)}, \sigma^{(12)} \mid \sigma_h, i, C_2^{(1)}, C_2^{(12)}\}$$

On the other hand, the generators of $\mathsf{D}_{2h}$ give

$$\begin{aligned}\mathsf{D}_{2h} &= \{E, C_2 \mid C_2^{(1)}, C_2^{(1)}C_2 \mid \sigma_h, \sigma_h C_2, \sigma_h C_2^{(1)}, \sigma_h C_2^{(1)} C_2\} \\ &= \{E, C_2 \mid C_2^{(1)}, C_2^{(12)} \mid \sigma_h, i, \sigma^{(1)}, \sigma^{(12)}\}\end{aligned}$$

It is evident that the specification of a new set of generators does not necessarily lead to a new group; the group we called $\mathsf{C}_{2vh}$ is identical with $\mathsf{D}_{2h}$ and hence two sets of generators may be entirely equivalent. When there is any ambiguity in the choice of generators we naturally adopt that which gives the simpler nomenclature.

The above example brings out another important point: the existence of certain typical symmetry operations may imply the existence of others. Thus when $\sigma_h$ and $\sigma_v$ are symmetry operations we must also admit a secondary 2-fold axis: in example 3.1, and generally, $\sigma_h \sigma^{(i)} = C_2^{(i)}$. A group containing both kinds of operation is therefore essentially of $\mathsf{D}$ type. It is for this reason also that some groups, such as $\mathsf{D}_{4d}$, do not appear in Fig. 3.2: they are identical with groups of higher symmetry which appear later in the table.

‡ The notation $\sigma^{(12)}$ is appropriate since there is evidently an *improper* 4-fold axis: $\sigma^{(12)}$ is across the plane midway between directions 1 and 2 (cf. Fig. 3.5).

EXAMPLE 3.2. Let us consider the projection diagram for $\mathsf{D}_4$, adding a reflection operation to produce $\mathsf{D}_{4d}$. Evidently, $C_2^{(1)}C_4 = C_2^{(12)}$—rotation about a two-fold axis midway between axes 1 and 2. If, therefore, we bisect the angle between these axes with a plane giving the reflection $\sigma^{(12)}$, the resultant group must contain $\sigma^{(12)}C_2^{(12)}$, which, by inspection, is equivalent to $\sigma_h$. Thus, in the case $n = 4$ we cannot have dihedral planes without also having $\sigma_h$ which is characteristic of the symmetry groups $\mathsf{D}_{nh}$ : in fact $\mathsf{D}_{4d} = \mathsf{D}_{4h}$.

EXAMPLE 3.3. As a final example, consider the non-appearance of $\mathsf{S}_3$. Repetition of the rotary-reflection $S_3$ evidently generates the set of equivalent points associated with $\mathsf{C}_{3h}$ and therefore, $\mathsf{S}_3 = \mathsf{C}_{3h}$. This may be seen algebraically on putting $S_3 = \sigma_h C_3$ (also remembering that the operations commute and $\sigma_h{}^2 = C_3{}^3 = E$). The powers of $S_3$ are

$$\{S_3, S_3{}^2(= C_3{}^2), S_3{}^3(= \sigma_h), S_3{}^4(= C_3), S_3{}^5(= \sigma_h C_3{}^2), S_3{}^6(= E)\}$$

and these are just the operations contained in $\mathsf{C}_{3h}$.

Considerations of this kind, for each $n$-fold principal axis, show that all distinct axial symmetry groups (up to $n = 6$) are included in Fig. 3.2. The classes into which the various operations fall are also readily obtained. Tables 3.1–3.7 show the structure of each group, in terms of its generators, the operations being grouped into classes. It should be noted that many of the point groups may be regarded as direct products of two simpler groups (see section 1.9). $\mathsf{D}_{4h}$, for example, contains all the elements of $\mathsf{D}_4$ plus the coset generated by $\sigma_h$, but since $E$ and $\sigma_h$ form a group whose operations commute with those of $\mathsf{D}_4$, we may write $\mathsf{D}_{4h} = \mathsf{C}_{1h} \times \mathsf{D}_4$.

TABLE 3.1

*The $\mathsf{C}_n$ groups : generator $\{C_n\}$*

| | | | | | | |
|---|---|---|---|---|---|---|
| $\mathsf{C}_1$ : | $E$ | | | | | |
| $\mathsf{C}_2$ : | $E$ | $C_2$ | | | | |
| $\mathsf{C}_3$ : | $E$ | $C_3$ | $\bar{C}_3$ | | | |
| $\mathsf{C}_4$ : | $E$ | $C_4$ | $C_2$ | $\bar{C}_4$ | | |
| $\mathsf{C}_5$ : | $E$ | $C_5$ | $C_5^2$ | $\bar{C}_5^2$ | $\bar{C}_5$ | |
| $\mathsf{C}_6$ : | $E$ | $C_6$ | $C_3$ | $C_2$ | $\bar{C}_3$ | $\bar{C}_6$ |

*Notes:* $C_n$ and $\bar{C}_n$ are rotations through $\pm 2\pi/n$. There is one generator, $C_n$, and each element is in a class by itself (vertical lines indicate the division into classes).

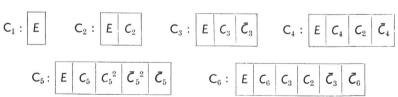

## Table 3.2

*The $C_{nv}$ groups : generators $\{C_n, \sigma^{(1)}\}$*

$C_{2v}$ :

| $E$ | $C_2$ |
|---|---|
| $\sigma^{(1)}$ | $\sigma^{(2)}$ |

$C_{3v}$ :

| $E$ | $C_3$ | $\bar{C}_3$ |
|---|---|---|
| $\sigma^{(1)}$ | $\sigma^{(2)}$ | $\sigma^{(3)}$ |

$C_{4v}$ :

| $E$ | $C_2$ | $C_4$ | $\bar{C}_4$ |
|---|---|---|---|
| $\sigma^{(1)}$ | $\sigma^{(2)}$ | $\sigma^{(23)}$ | $\sigma^{(12)}$ |

$C_{5v}$ :

| $E$ | $C_5$ | $\bar{C}_5$ | $C_5^2$ | $\bar{C}_5^2$ |
|---|---|---|---|---|
| $\sigma^{(1)}$ | $\sigma^{(3)}$ | $\sigma^{(4)}$ | $\sigma^{(5)}$ | $\sigma^{(2)}$ |

$C_{6v}$ :

| $E$ | $C_3$ | $\bar{C}_3$ | $C_2$ | $C_6$ | $\bar{C}_6$ |
|---|---|---|---|---|---|
| $\sigma^{(1)}$ | $\sigma^{(3)}$ | $\sigma^{(2)}$ | $\sigma^{(23)}$ | $\sigma^{(34)}$ | $\sigma^{(12)}$ |

*Notes:* The first row in each block contains the elements of the subgroup $C_n$. The second row contains the left coset (elements in the same order) with respect to the first element in the row. Vertical lines divide the group into its classes.

## Table 3.3

*The $C_{nh}$ groups : generators $\{C_n, \sigma_h\}$*

$C_{1h}$ :

| $E$ |
|---|
| $\sigma_h$ |

$C_{2h}$ :

| $E$ | $C_2$ |
|---|---|
| $\sigma_h$ | $i$ |

$C_{3h}$ :

| $E$ | $C_3$ | $\bar{C}_3$ |
|---|---|---|
| $\sigma_h$ | $S_3$ | $\bar{S}_3$ |

$C_{4h}$ :

| $E$ | $C_4$ | $C_2$ | $\bar{C}_4$ |
|---|---|---|---|
| $\sigma_h$ | $S_4$ | $i$ | $\bar{S}_4$ |

$C_{5h}$ :

| $E$ | $C_5$ | $C_5^2$ | $\bar{C}_5^2$ | $\bar{C}_5$ |
|---|---|---|---|---|
| $\sigma_h$ | $S_5$ | $S_5^2$ | $\bar{S}_5^2$ | $\bar{S}_5$ |

$C_{6h}$ :

| $E$ | $C_6$ | $C_3$ | $C_2$ | $\bar{C}_3$ | $\bar{C}_6$ |
|---|---|---|---|---|---|
| $\sigma_h$ | $S_6$ | $S_3$ | $i$ | $\bar{S}_3$ | $\bar{S}_6$ |

*Notes:* The same remarks apply as in Table 3.2. The second row is the left coset of the first with respect to the first element ($\sigma_h$). $S_n$ and $\bar{S}_n$ are improper rotations through $\pm 2\pi/n$ and $i(= S_2)$ is the inversion.

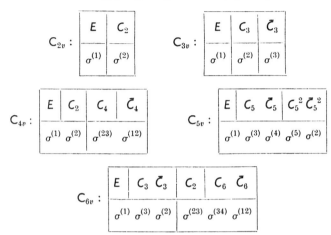

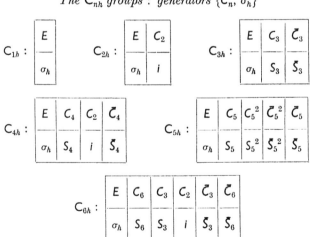

## Table 3.4

*The* $\mathsf{S}_n$ *groups : generator* $\{S_n\}$

$S_2$ :

| | |
|---|---|
| $E$ | $i$ |

$S_4$ :

| | | | |
|---|---|---|---|
| $E$ | $S_4$ | $C_2$ | $\bar{S}_4$ |

$S_6$ :

| | | | | | |
|---|---|---|---|---|---|
| $E$ | $S_6$ | $C_3$ | $i$ | $\bar{C}_3$ | $\bar{S}_6$ |

*Note:* There is one generator $S_n$ and each element is in a class by itself.

## Table 3.5

*The* $\mathsf{D}_n$ *groups : generators* $\{C_n, C_2^{(1)}\}$

$D_2$ :

| | |
|---|---|
| $E$ | $C_2$ |
| $C_2^{(1)}$ | $C_2^{(12)}$ |

$D_3$ :

| | |
|---|---|
| $E$ | $C_3$ $\bar{C}_3$ |
| $C_2^{(1)}$ $C_2^{(2)}$ $C_2^{(3)}$ | |

$D_4$ :

| | | | |
|---|---|---|---|
| $E$ | $C_2$ | $C_4$ | $\bar{C}_4$ |
| $C_2^{(1)}$ $C_2^{(2)}$ | | $C_2^{(23)}$ $C_2^{(12)}$ | |

$D_5$ :

| | | |
|---|---|---|
| $E$ | $C_5$ $\bar{C}_5$ | $C_5^2$ $\bar{C}_5^2$ |
| $C_2^{(1)}$ $C_2^{(3)}$ $C_2^{(4)}$ $C_2^{(5)}$ $C_2^{(2)}$ | | |

$D_6$ :

| | | | |
|---|---|---|---|
| $E$ | $C_3$ $\bar{C}_3$ | $C_2$ | $C_6$ $\bar{C}_6$ |
| $C_2^{(1)}$ $C_2^{(3)}$ $C_2^{(2)}$ | | $C_2^{(23)}$ $C_2^{(34)}$ $C_2^{(12)}$ | |

*Notes:* The same remarks apply as in Table 3.2. The superscript in brackets indicates the 2-fold axis in accordance with Fig. 3.5.

## Table 3.6

*The* $\mathsf{D}_{nd}$ *groups: generators* $\{C_n, C_2^{(1)}, \sigma^{(12)}\}$.

$D_{2d}$ :

| | | |
|---|---|---|
| $E$ | $C_2$ | $C_2^{(1)}$ $C_2^{(2)}$ |
| $\sigma^{(12)}$ $\sigma^{(23)}$ | | $S_4$ $\bar{S}_4$ |

$D_{3d}$ :

| | | | |
|---|---|---|---|
| $E$ | $C_3$ $\bar{C}_3$ | $C_2^{(1)}$ $C_2^{(2)}$ $C_2^{(3)}$ | |
| $\sigma^{(12)}$ $\sigma^{(34)}$ $\sigma^{(23)}$ | | $S_6$ $\bar{S}_6$ | $i$ |

*Notes :* The first row contains a $D_n$ subgroup (Table 3.5) which in turn contains a cyclic subgroup (a second vertical stroke indicating the end of the cyclic subgroup). The second row contains the left coset with respect to the new generator. Since $D_{nd}$ contains improper rotations of order $2n$ the numbering of the 2-fold axes in the $D_n$ subgroup is modified (Fig. 3.5)

$$\{C_2^{(1)}\ C_2^{(12)}\} \text{ in } D_2 \rightarrow \{C_2^{(1)}\ C_2^{(2)}\} \text{ in } D_{2d}$$

$$\{C_2^{(1)}\ C_2^{(2)}\ C_2^{(3)}\} \text{ in } D_3 \rightarrow \{C_2^{(1)}\ C_2^{(3)}\ C_2^{(2)}\} \text{ in } D_{3d}$$

Note also that alternative generators are

$$D_{2d} : \{S_4, C_2^{(1)}\} \qquad D_{3d} : \{S_6, C_2^{(1)}\}$$

TABLE 3.7

*The* $D_{nh}$ *groups: generators* $\{C_n, C_2^{(1)}, \sigma_h\}$.

$D_{2h}$:

| | | | |
|---|---|---|---|
| $E$ | $C_2$ | $C_2^{(1)}$ | $C_2^{(12)}$ |
| $\sigma_h$ | $i$ | $\sigma^{(1)}$ | $\sigma^{(12)}$ |

$D_{3h}$:

| | | | | | |
|---|---|---|---|---|---|
| $E$ | $C_3$ | $\bar{C}_3$ | $C_2^{(1)}$ | $C_2^{(2)}$ | $C_2^{(3)}$ |
| $\sigma_h$ | $S_3$ | $\bar{S}_3$ | $\sigma^{(1)}$ | $\sigma^{(2)}$ | $\sigma^{(3)}$ |

$D_{4h}$:

| | | | | | | | |
|---|---|---|---|---|---|---|---|
| $E$ | $C_2$ | $C_4$ | $\bar{C}_4$ | $C_2^{(1)}$ | $C_2^{(2)}$ | $C_2^{(23)}$ | $C_2^{(12)}$ |
| $\sigma_h$ | $i$ | $S_4$ | $\bar{S}_4$ | $\sigma^{(1)}$ | $\sigma^{(2)}$ | $\sigma^{(23)}$ | $\sigma^{(12)}$ |

$D_{5h}$:

| | | | | | | | | | |
|---|---|---|---|---|---|---|---|---|---|
| $E$ | $C_5$ | $\bar{C}_5$ | $C_5^2$ | $\bar{C}_5^2$ | $C_2^{(1)}$ | $C_2^{(3)}$ | $C_2^{(4)}$ | $C_2^{(5)}$ | $C_2^{(2)}$ |
| $\sigma_h$ | $S_3$ | $\bar{S}_5$ | $S_5C_5$ | $\bar{S}_5\bar{C}_5$ | $\sigma^{(1)}$ | $\sigma^{(3)}$ | $\sigma^{(4)}$ | $\sigma^{(5)}$ | $\sigma^{(2)}$ |

$D_{6h}$:

| | | | | | | | | | | | |
|---|---|---|---|---|---|---|---|---|---|---|---|
| $E$ | $C_3$ | $\bar{C}_3$ | $C_2$ | $C_6$ | $\bar{C}_6$ | $C_2^{(1)}$ | $C_2^{(3)}$ | $C_2^{(2)}$ | $C_2^{(23)}$ | $C_2^{(34)}$ | $C_2^{(12)}$ |
| $\sigma_h$ | $S_3$ | $\bar{S}_3$ | $i$ | $S_6$ | $\bar{S}_6$ | $\sigma^{(1)}$ | $\sigma^{(3)}$ | $\sigma^{(2)}$ | $\sigma^{(23)}$ | $\sigma^{(34)}$ | $\sigma^{(12)}$ |

*Notes:* The first row contains a $D_n$ subgroup (Table 3.5) which in turn contains a cyclic subgroup (preceding the double vertical stroke). The second row contains the left coset with respect to the new generator. In $D_{5h}$, $S_5C_5$ and $\bar{S}_5\bar{C}_5$ are improper rotations through $\pm 4\pi/5$.

### 3.3. The Tetrahedral and Octahedral Point Groups

In some systems of high symmetry it is not possible to single out a unique principal axis, there being several equivalent $n$-fold axes. The groups we shall discuss are derived from those of the regular tetrahedron and the regular octahedron under rotations. The pure rotation groups of these objects are denoted by T and O, respectively, subscripts being added (as in the axial groups) to indicate the presence of additional generators. The group of the regular octahedron is also the group of the cube, as may be seen from Fig. 3.9 where the octahedron is inscribed within a cube, its vertices being at the face centres of the cube. Both systems clearly have the same symmetry operations. Moreover, the group of the tetrahedron, T, must be a subgroup of O because the tetrahedron can, in turn, be inscribed in the cube (Fig. 3.10); but four cube vertices are then distinguished from the other

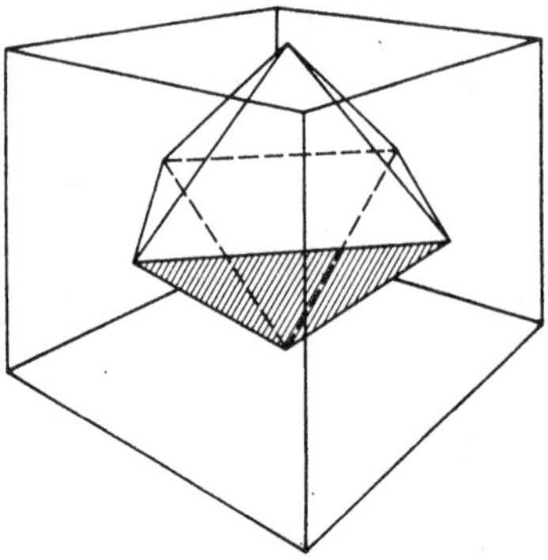

FIG. 3.9. Regular octahedron inscribed in a cube.

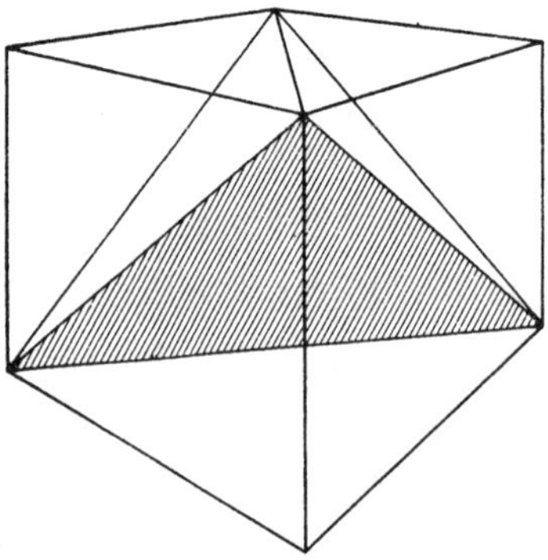

FIG. 3.10. Regular tetrahedron inscribed in a cube.

four and we are thus restricted to the subgroup of operations which do not mix the two types of cube vertex. Regular objects of still higher symmetry do exist (the icosohedron and dodecahedron), but their point groups are of little importance in molecular applications. Since the cubic groups contain up to 48 elements, some care is necessary in devising a simple but unambiguous notation for the operations. It is convenient to denote the Cartesian axes (with sense + or −) by $x$, $y$, $z$, $\bar{x}$, $\bar{y}$, $\bar{z}$. Symmetry axes passing through a cube face, edge, or corner are then labelled by giving one, two or three of these quantities (Fig. 3.11). Rotations through $(2\pi/n)$ about the various axes are then denoted by

$$C_n^{\alpha}, \quad C_n^{\alpha\beta}, \quad C_n^{\alpha\beta\gamma} \quad (\alpha, \beta, \gamma = x, y, z, \bar{x}, \bar{y}, \bar{z})$$

Since a plane is determined by giving its *normal*, which may be specified like any other axis, the reflection operations may be labelled similarly:

$$\sigma^{\alpha}, \quad \sigma^{\alpha\beta}, \quad \sigma^{\alpha\beta\gamma} \quad (\alpha, \beta, \gamma = x, y, z, \bar{x}, \bar{y}, \bar{z})$$

It should be noted that this notation differs somewhat from that used for the axial groups. Typical operations are indicated in Fig. 3.11.

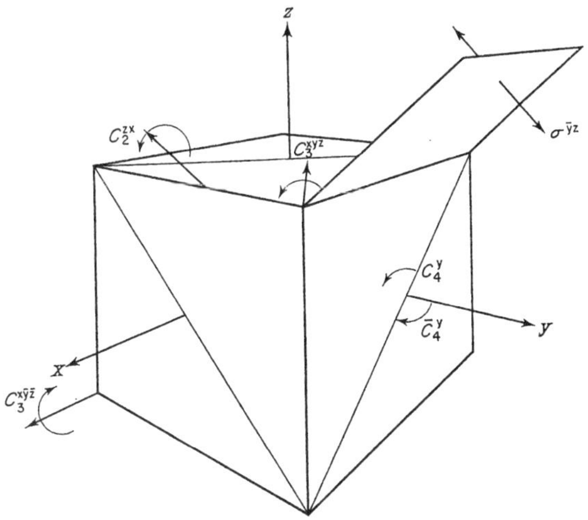


FIG. 3.11. Nomenclature for cubic group elements.

The important groups are as follows:

*The Group* T. This is the basic group, appearing as a subgroup in all the others. From Fig. 3.10 it is clear that $C_2^x$ is a symmetry operation, turning two face diagonals (front and rear faces) through $\pi$ about their midpoints, and permuting the other edges of the tetrahedron. $C_2^y$ has a similar effect; and since its axis is perpendicular to that of $C_2^x$ there must be a third axis—obviously $C_2^z$. But there is also a three-fold axis, namely the body diagonal which joins the nearest and farthest cube corners; and since the tetrahedron has four equivalent vertices there will be four equivalent axes. Here, therefore, we have an example of a 3-fold axis accompanied by 2-fold axes. The group is determined by two generators which (with the conventions adopted) may be chosen as $C_2^z$ and $C_3^{xyz}$. For $C_3^{xyz}$ evidently rotates the axis of $C_2^z$ into that of $C_2^x$, while a second application produces the axis of $C_2^y$; and similarly $C_2^z$ rotates the $xyz$ axis into another 3-fold axis in the $\bar{x}\bar{y}\bar{z}$ direction, while use of $C_2^x$ and $C_2^y$ yields the other 3-fold axes. Hence, T is generated by $C_2^z$ and $C_3^{xyz}$: altogether there are twelve distinct operations falling into four classes, these being conveniently displayed as in Table 3.8.

TABLE 3.8

*The group* T: *generators* $\{C_2^z, C_3^{xyz}\}$.

| | | | | |
|---|---|---|---|---|
| $E$ | $C_2^x$ | $C_2^y$ | $C_z^2$ | $\bar{C}_3^{xyz} = C_3^{xyz\,2}$ |
| $C_3^{xyz}$ | $C_3^{x\bar{y}\bar{z}}$ | $C_3^{\bar{x}y\bar{z}}$ | $C_3^{\bar{x}\bar{y}z}$ | $C_2^x = C_3^{xyz}\, C_2^z\, \bar{C}_3^{xyz}$ |
| $\bar{C}_3^{xyz}$ | $\bar{C}_3^{x\bar{y}\bar{z}}$ | $\bar{C}_3^{\bar{x}y\bar{z}}$ | $\bar{C}_3^{\bar{x}\bar{y}z}$ | $C_2^y = C_3^{xyz\,2}\, C_2^z\, \bar{C}_3^{xyz\,2}$ |

*Notes :* Elements in the first row and column are expressible in terms of the generators as indicated. The table entry in the $R$-row and $S$-column is $SRS^{-1}$ : each of the last two rows is therefore the class of its first element.

*The Group* $T_d$. If we add to T a reflection operation $\sigma^{y\bar{z}}$, a reflection across a plane whose normal is the $y\bar{z}$ axis, we can generate twelve new operations. Thus $C_2^z$ rotates the plane of $\sigma^{y\bar{z}}$ until it lies midway between the $\bar{y}$ and $z$ axes, with a corresponding reflection $\sigma^{yz}$. Since the $x$, $y$ and $z$ axes are equivalent, six planes arise in this way. A second type of improper operation is also evident. Thus, the rotary reflection $S_4^z$ has the effect indicated in Fig. 3.12 and can evidently be

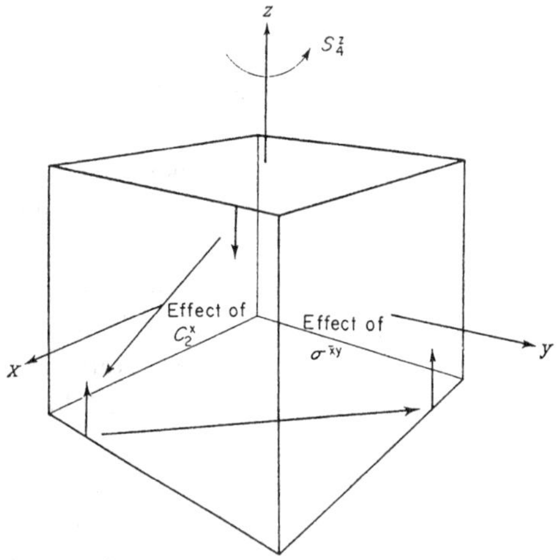


FIG. 3.12. Improper rotations in $T_d$. Rotation followed by reflection brings the short arrow from its initial to its final position : the result is the same as that of $S_4^z$ and hence $S_4^z = \sigma^{\bar{x}y}C_2^x$

expressed in terms of the operations already discussed: $S_4^z = \sigma^{\bar{x}y} C_2^x$. Also, since $S_4^{z\,2} = C_2^z$ it is no longer necessary to regard $C_2^z$ as a generator. *Two* generators determine the group if one of them is chosen as an improper 4-fold rotation. For reasons which will appear later, it is most convenient to adopt $\bar{S}_4^z$ and $C_3^{xyz}$. The twenty-four operations fall into five classes (Table 3.9). The tetrahedral molecule $CH_4$ (methane) and also many complex ions (e.g. $FeCl_4^-$) have $\mathsf{T}_d$ symmetry.

TABLE 3.9

*The group* $\mathsf{T}_d$ : *generators* $\{\bar{S}_4^z, C_3^{xyz}\}$

| | | | | | | | | | | | | |
|---|---|---|---|---|---|---|---|---|---|---|---|---|
| T | E | $C_2^x$ | $C_2^y$ | $C_2^z$ | $C_3^{xyz}$ | $C_3^{x\bar{y}\bar{z}}$ | $C_3^{\bar{x}y\bar{z}}$ | $C_3^{\bar{x}\bar{y}z}$ | $\bar{C}_3^{yxz}$ | $\bar{C}_3^{x\bar{y}\bar{z}}$ | $\bar{C}_3^{\bar{x}y\bar{z}}$ | $\bar{C}_3^{\bar{x}\bar{y}z}$ |
| $\bar{S}_4^z$ T | | $\sigma^{xy}$ | $\sigma^{x\bar{y}}$ | | $\sigma^{yz}$ | | | $\sigma^{y\bar{z}}$ | | $\sigma^{z\bar{x}}$ | $\sigma^{zx}$ | |
| | $\bar{S}_4^z$ | | | $S_4^z$ | | $\bar{S}_4^x$ | $S_4^x$ | | $S_4^y$ | | | $\bar{S}_4^y$ |

*Notes :* The first row contains the subgroup T : subsequent rows contain elements of its left coset with respect to the new generator, chosen as $\bar{S}_4^z$. Each class appears in a separate box.

*The Group* $\mathsf{T}_h$. This group is obtained by adding the generator $i$ instead of a reflection and again contains twenty-four operations (Table 3.10). It is not the group of a regular tetrahedral object but rather of a pair of interlocking tetrahedra (each giving its partner on inversion through the origin) known as a di-dodecahedron.

TABLE 3.10

*The group* $\mathsf{T}_h$ : *generators* $\{C_2^z, C_3^{xyz}, i\}$

| | | | | | | | | | | | | |
|---|---|---|---|---|---|---|---|---|---|---|---|---|
| T | E | $C_2^x$ | $C_2^y$ | $C_2^z$ | $C_3^{xyz}$ | $C_3^{x\bar{y}\bar{z}}$ | $C_3^{\bar{x}y\bar{z}}$ | $C_3^{\bar{x}\bar{y}z}$ | $\bar{C}_3^{xyz}$ | $\bar{C}_3^{x\bar{y}\bar{z}}$ | $\bar{C}_3^{\bar{x}y\bar{z}}$ | $\bar{C}_3^{\bar{x}\bar{y}z}$ |
| $i$T | $i$ | $\sigma^x$ | $\sigma^y$ | $\sigma^z$ | $\bar{S}_6^{xyz}$ | $\bar{S}_6^{x\bar{y}\bar{z}}$ | $\bar{S}_6^{\bar{x}y\bar{z}}$ | $\bar{S}_6^{\bar{x}\bar{y}z}$ | $S_6^{xyz}$ | $S_6^{x\bar{y}\bar{z}}$ | $S_6^{\bar{x}y\bar{z}}$ | $S_6^{\bar{x}\bar{y}\bar{z}}$ |

*Notes :* The first row contains the subgroup T, the second its left coset with respect to $i$. Each class appears in a separate box.

*The Group* O. The octahedral group may be obtained by replacing the 2-fold axes in T by 4-fold axes (Fig. 3.11) and noting that a new set of six 2-fold axes also arises. It is sufficient to take as generators one rotation of each type, for example $C_4^z$ and $C_3^{xyz}$. By using these symmetry operations on each other we can generate three 4-fold, four 3-fold and six 2-fold axes, giving twenty-four operations falling into five classes (Table 3.11). There is an isomorphism between O and $T_d$ in which $C_4^z \leftrightarrow \bar{S}_4^z$ : by taking these as corresponding generators we obtain an isomorphic association between elements at corresponding places in Tables 3.9 and 3.11.

### Table 3.11

*The group O: generators $\{C_4^z, C_3^{xyz}\}$.*

| | | | | | | | | | | | | |
|---|---|---|---|---|---|---|---|---|---|---|---|---|
| T | E | $C_2^x$ | $C_2^y$ | $C_2^z$ | $C_3^{xyz}$ | $C_3^{x\bar{y}\bar{z}}$ | $C_3^{\bar{x}y\bar{z}}$ | $C_3^{\bar{x}\bar{y}z}$ | $\bar{C}_3^{xyz}$ | $\bar{C}_3^{x\bar{y}\bar{z}}$ | $\bar{C}_3^{\bar{x}y\bar{z}}$ | $\bar{C}_3^{\bar{x}\bar{y}z}$ |
| $C_4^z$T | | $C_2^{xy}$ | $C_2^{x\bar{y}}$ | | $C_2^{yz}$ | | | $C_2^{y\bar{z}}$ | | $C_2^{z\bar{x}}$ | $C_2^{zx}$ | |
| $C_4^z$T | $C_4^z$ | | | $\bar{C}_4^z$ | | $C_4^x$ | $\bar{C}_4^x$ | | $\bar{C}_4^y$ | | | $C_4^y$ |

*Notes :* The first row contains the subgroup T ; subsequent rows its left coset with respect to the new generator $C_4^z$. Each class appears in a separate box. There is an isomorphic association between the elements of this group and those which appear in the corresponding places in Table 3.9.

### Table 3.12

*The group $O_h$: generators $\{C_4^z, C_3^{xyz}, i\}$.*

| | | | | | | | | | | | | |
|---|---|---|---|---|---|---|---|---|---|---|---|---|
| T | E | $C_2^x$ | $C_2^y$ | $C_2^z$ | $C_3^{x\bar{y}z}$ | $C_3^{x\bar{y}\bar{z}}$ | $C_3^{\bar{x}y\bar{z}}$ | $C_3^{\bar{x}\bar{y}z}$ | $\bar{C}_3^{xyz}$ | $\bar{C}_3^{x\bar{y}\bar{z}}$ | $\bar{C}_3^{\bar{x}y\bar{z}}$ | $\bar{C}_3^{\bar{x}\bar{y}z}$ |
| $i$T | $i$ | $\sigma^x$ | $\sigma^y$ | $\sigma^z$ | $\bar{S}_6^{xyz}$ | $\bar{S}_6^{x\bar{y}\bar{z}}$ | $\bar{S}_6^{\bar{x}y\bar{z}}$ | $\bar{S}_6^{\bar{x}\bar{y}z}$ | $S_6^{xyz}$ | $S_6^{x\bar{y}\bar{z}}$ | $S_6^{\bar{x}y\bar{z}}$ | $S_6^{\bar{x}\bar{y}z}$ |
| $C_4^z$T | | $C_2^{xy}$ | $C_2^{x\bar{y}}$ | | $C_2^{yz}$ | | | $C_2^{y\bar{z}}$ | | $C_2^{z\bar{x}}$ | $C_2^{zx}$ | |
| $C_4^z$T | $C_4^z$ | | | $\bar{C}_4^z$ | | $C_4^x$ | $\bar{C}_4^x$ | | $\bar{C}_4^y$ | | | $C_4^y$ |
| $C_4^z$ $i$T | | $\sigma^{xy}$ | $\sigma^{x\bar{y}}$ | | $\sigma^{yz}$ | | | $\sigma^{y\bar{z}}$ | | $\sigma^{z\bar{x}}$ | $\sigma^{zx}$ | |
| $C_4^z$ $i$T | $\bar{S}_4^z$ | | | $S_4^z$ | | $\bar{S}_4^x$ | $S_4^x$ | | $S_4^y$ | | | $\bar{S}_4^y$ |

*Notes :* The first two rows contain the subgroup $T_h$ (the first row itself being the subgroup T). The remaining rows contain the elements of the left coset of $T_h$ with respect to $C_4^z$. Each class appears in a separate box. The group is also the direct product $S_2 \times O$ : the rows T and $C_4^z$T together form O, while $i$T and $iC_4^z$T ($=C_4^z i$T) contain corresponding elements of the coset $i$O.

*The Group* $\mathsf{O}_h$. On adding to $\mathsf{O}$ the generator $i$ we obtain twenty-four new operations. It is easily seen that these may also be described in terms of reflection planes which are obviously an important feature of the symmetry of an actual cube: thus, from Fig. 3.11, we see $iC_2^z = \sigma^z$. The structure of the group is indicated in Table 3.12.

### 3.4. Compatibility of Symmetry Operations

The point groups above are the ones of most importance in molecular and solid state theory. We shall also show presently that only axes of order 1, 2, 3, 4, 6 are compatible with the translational symmetry of crystals. With this restriction, which we now adopt, all possibilities are exhausted by the groups considered. We have built up the groups step by step, adding to a group $\mathsf{H}$ a new generator $P$ and forming a new group $\mathsf{G}$ of which $\mathsf{H}$ is a subgroup.‡ This process comes to a natural end, and even the most complicated group, $\mathsf{O}_h$, has only three generators. Why is this so?

Let $\mathsf{H} = \{R_1, R_2, \ldots, R_h\}$ be a symmetry group and $P$ a new generator, $\mathsf{H}$ and its coset with $P$ forming a new group $\mathsf{G}$. Then the collection

$$\bar{\mathsf{H}} = \{PR_1P^{-1}, \quad PR_2P^{-1}, \quad \ldots, \quad PR_hP^{-1}\}$$

is also a subgroup of symmetry operations, *isomorphic* with $\mathsf{H}$. Thus, if $\mathsf{H}$ describes an axis of order $n$, $\bar{\mathsf{H}}$ describes another axis of order $n$ : This simply expresses the idea that symmetry operations applied to symmetry *elements* (e.g. axes or planes) must generate new symmetry elements of the same type. To justify this assertion we consider a rotation $R$,

$$\mathbf{r} \to \mathbf{r}' = R\mathbf{r}$$

and observe that, for any two vectors related to $\mathbf{r}$ and $\mathbf{r}'$ by a common rotation $P$, $P\mathbf{r}' = (PRP^{-1})P\mathbf{r}$ or $\bar{\mathbf{r}}' = \bar{R}\bar{\mathbf{r}}$ ($\bar{R} = PRP^{-1}$). The operator describing a mapping similar to $R$ but in which all vectors (and hence all geometrical objects—such as symmetry elements) have been shifted round by a common rotation $P$, is thus

$$\bar{R} = PRP^{-1} \tag{3.4.1}$$

Now if $\mathsf{G}$ is to be a *symmetry group* we must verify that $PR_i$ and $PR_iP^{-1}$ (all $i$) are admissible elements. This requirement places severe restrictions on the generator $P$. Thus, for example, although $PR_i$ is

‡ If $E, P_1, P_2 \ldots$ is itself a group $\mathsf{P}$, the totality of elements $\mathsf{H}$, $P_1\mathsf{H}$, $P_2\mathsf{H}$, . . . is called the "semi-direct" product of $\mathsf{P}$ and $\mathsf{H}$. We have therefore constructed the point groups as semi-direct products. Various theorems can be established for such products but in the interests of simplicity we have used only first principles.

necessarily a rotation it is *not* necessarily of order 1, 2, 3, 4 or 6 and may consequently be inadmissible as a *symmetry* operation. This is an important restriction: but an even more stringent condition was encountered in the axial groups (section 3.2). We postulated a *unique* axis, of order higher than that of any other axis: the group $\bar{\mathsf{H}}$, of rotations about a new axis, must then either coincide with $\mathsf{H}$ or contradict the assumption. The axial groups are therefore groups in which the rotations about a principal axis form an *invariant subgroup* (section 1.6):

$$\bar{\mathsf{H}} = P\mathsf{H}P^{-1} = \mathsf{H} \text{ (all } P \text{ not in } \mathsf{H}) \tag{3.4.2}$$

If $R_i$ is an element of $\mathsf{H}$ this means of course that

$$PR_iP^{-1} = R_j \tag{3.4.3}$$

—the transform of any principal axis rotation is another such rotation.

We now consider the matrix form of such restrictions. Suppose we have an $n$-fold axis (the $z$ axis), with generator $C_n^z$ and try to add an $m$-fold axis ($z'$) inclined at an angle $\theta$ (Fig. 3.13), with generator

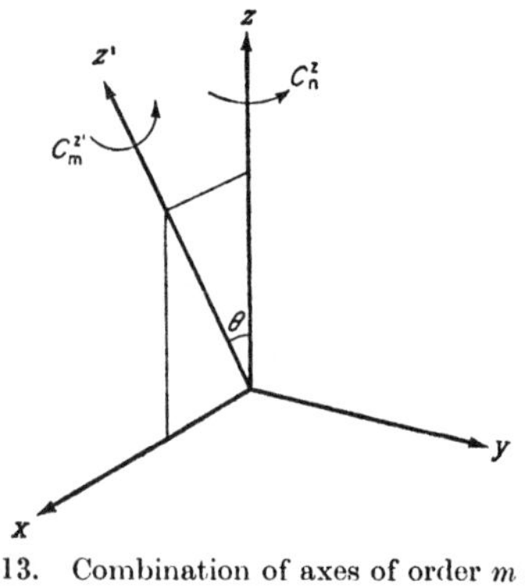


FIG. 3.13. Combination of axes of order $m$ and $n$.

$C_m^{z'}$. The matrix form of a rotation about the $z$ axis is given in (3.1.12). By appeal to (3.4.1), that of a rotation about the $z'$ axis may be written $\boldsymbol{D}(C_m^{z'}) = \boldsymbol{D}(C_\theta^y)\boldsymbol{D}(C_m^z)\boldsymbol{D}(C_{-\theta}^y)$ where

$$\boldsymbol{D}(C_m^z) = \begin{pmatrix} c_m & -s_m & 0 \\ s_m & c_m & 0 \\ 0 & 0 & 1 \end{pmatrix}, \qquad \boldsymbol{D}(C_{\pm\theta}^y) = \begin{pmatrix} c_\theta & 0 & \pm s_\theta \\ 0 & 1 & 0 \\ \pm s_\theta & 0 & c_\theta \end{pmatrix}$$

and $c_m = \cos 2\pi/m$, $c_\theta = \cos\theta$, etc. The result is

$$\boldsymbol{D}(C_m^{z'}) = \begin{pmatrix} c_\theta^2 c_m + s_\theta^2 & -c_\theta s_m & -c_\theta s_\theta(c_m - 1) \\ c_\theta s_m & c_m & -s_\theta c_m \\ -c_\theta s_\theta(c_m - 1) & s_\theta c_m & s_\theta^2 c_m + c_\theta^2 \end{pmatrix}$$

We now ask whether $C_m^{z'}C_n^z$ will be admissible as another symmetry operation. This can be so only if the resultant rotation, through $\phi$ say, generates a group of allowed order, $p = 1, 2, 3, 4$ or $6$. We may test this, without finding the rotation explicitly, by examining the *trace* of its matrix, for by (3.1.14) this is $1 + 2\cos\phi$ (irrespective of axial direction) and must therefore take one of the values $-1, 0, 1, 2,$ or $3$ corresponding to $\phi = k(2\pi/p)$ ($k$ integral). The trace is easily evaluated, and the condition becomes

$$c_\theta^2(c_nc_m - c_n - c_m + 1) - 2c_\theta s_m s_n + (c_mc_n + c_n + c_m)$$
$$= 1 + 2\cos 2\pi k/p = -1, 0, 1, 2, 3 \qquad (3.4.4)$$

This equation restricts the possible $\theta$ values for any choice of $m$, $n$ and $p$. It must be stressed, however, that the conditions which it expresses, though necessary, are *not sufficient*. A generator can be rejected if it fails to satisfy (3.4.4), but one which *does* satisfy must be tested further, using the closure property—every element generated must be contained in a *point group*. We consider a typical example.

EXAMPLE 3.4. At what angle can two 3-fold axes be inclined, and what third axis do they determine?

Here $c_m = c_n = -\frac{1}{2}$ and $s_m = s_n = \frac{1}{2}\sqrt{3}$. From (3.4.4)

$$\tfrac{1}{4}(9c_\theta^2 - 6c_\theta - 3) = -1, 0, 1, 2, 3$$

The last possibility is trivial, giving $\theta = 0$ or $\pi$ which corresponds to a single axis. The other possibilities are shown in the following table:

| | $-1$ | $0$ | $1$ | $2$ |
|---|---|---|---|---|
| $c_\theta$ | $\frac{1}{3}$ | $\frac{1}{3} \pm \frac{2}{3}$ | $\frac{1}{3} \pm \frac{2}{3}\sqrt{2}$ | $\frac{1}{3} \pm \frac{2}{3}\sqrt{3}$ |
| $\theta$ | 70° 32′ | 109° 28′ | 127° 33′ | 145° 14′ |
| $(p, k)$ for resultant operation $(C_p)^k$ | $\{(2, 1), (4, 2)\}$ | $\{(3, \pm 1), (6, \pm 2)\}$ | $(4, \pm 1)$ | $(6, \pm 1)$ |

The first two $\theta$ values correspond to axes inclined at the tetrahedral angle (cf. Fig. 3.10), the first possibility giving a 2- or 4-fold resultant axis and the second a 3- or 6-fold axis. The remaining values look less plausible. We therefore use the closure property, asserting that $C_m^{z'2} C_n^z$ must also be in the symmetry group. The trace of this operation follows from the left-hand side of (3.19) on halving the value of $m$ and gives the further condition $\frac{1}{4}(9c_\theta^2 + 6c_\theta - 3) = -1, 0, 1, 2$. But the $c_\theta$ values for the 4- and 6-fold axes in the last two columns do not satisfy this equation and are therefore, after all, inadmissible. The 6-fold axis in the second column can be ruled out in a similar way. The remaining possibilities are self-consistent: the 2-fold and 4-fold axes (first column) occur in the tetrahedron and cube, respectively, while the 3-fold axis (second column) appears in both.

Finally we note that all the combinations which occur in the *axial* groups may be determined alternatively by postulating a cyclic invariant subgroup. To do this we need only evaluate the matrix of the transformed rotation $C_m^{z'} C_n^z \bar{C}_m^{z'}$ and require this to be of the block form characteristic of the invariant subgroup: this gives $c_\theta = 0$, $\theta = \pi/2$ and with this reduction $s_m = 1$ and only *two* fold axes are allowed. These are the propositions on which the geometrical discussion was based. By systematically applying the considerations of this section, we therefore obtain an algebraic proof of the impossibility of finding further point groups.

### 3.5. Symmetry of Crystal Lattices

We now resume the discussion of composite operations $(R \mid t)$, introduced in section 3.1, which include translation as well as rotation. The operations with $t = 0$ fall into sets describing the various types of point group symmetry discussed in sections 3.2 and 3.3. Limitations upon the possible symmetry types arise, as was seen in section 3.4, from the need for compatibility among the operations admitted in each group: every combination of the generating operations must bring *some* regular polyhedron into self-coincidence in order to qualify as a symmetry operation. When operations with $t \neq 0$ are introduced we must again ask what basic operations are compatible: every combination of the basic rotations *and translations* must now bring *some lattice* into self-coincidence. We shall find that only 32 groups of rotational operations are compatible with the existence of a lattice, these determining the 32 " crystal classes ", and that only certain of the 14 distinct lattice types—the 14 " Bravais lattices "—can be admitted in each crystal class. As a result, it is possible to enumerate just 230 distinct space groups.

First, we note the general principle upon which the argument is based. The fundamental property of all repeating patterns is that the *translational* symmetry operations form an invariant subgroup. Other symmetry operations can then occur only if they are compatible with this requirement and this places severe restrictions on the permissible rotations and reflections. The situation is exactly similar to that in section 3.2: all the axial groups contain an invariant subgroup of rotations and this is incompatible with the existence of any other $m$-fold axis with $m>2$, the various possibilities then determining all the axial groups.

The invariant subgroup property follows directly from (3.1.4) and (3.1.5), which together give

$$(R\,|\,t')\,(E\,|\,t)\,(R\,|\,t')^{-1} = (E\,|\,Rt) \qquad (3.5.1)$$

According to (3.4.1), the left-hand side is the " transformed operation " into which $(E\,|\,t)$ is sent on subjecting the whole lattice to the symmetry operation $(R\,|\,t')$. This result, which is true for *arbitrary* translations and rotations, here has a non-trivial significance because we require that $(E\,|\,t)$ and $(E\,|\,Rt)$ *shall both be lattice translations* (i.e. symmetry operations). Thus, in Fig. 3.14 rotation $C_4$ is compatible with a square

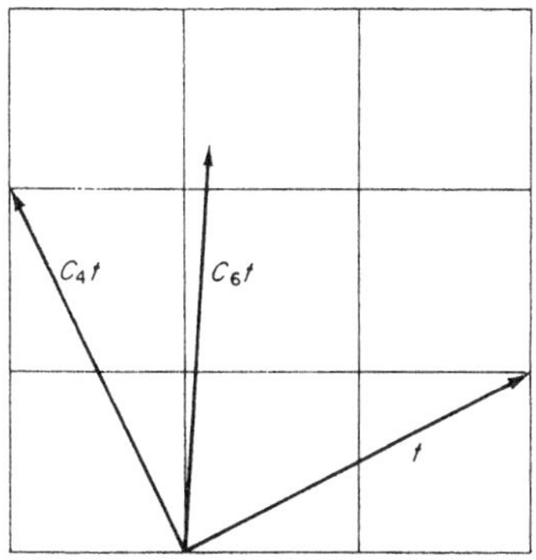


FIG. 3.14. Compatibility of rotations and translations. $C_4t$ is a lattice vector so $C_4$ is an acceptable symmetry operation : $C_6$ is not.

lattice, because any lattice translation rotated through 90° becomes another lattice translation ; but $C_6$ would *not* be a compatible operation.

Rotations on the other hand, do *not* form an invariant subgroup:

$$(S\,|\,t)\,(R\,|\,0)\,(S\,|\,t)^{-1} = (SRS^{-1}\,|\,t - SRS^{-1}t) = (R'\,|\,t') \qquad (3.5.2)$$

which is no longer a pure rotation. We note that for conjugation with a *pure translation* (3.5.2) reduces to

$$(E\,|\,t)(R\,|\,0)(E\,|\,t)^{-1} = (R\,|\,t - Rt) \tag{3.5.3}$$

This has a direct geometrical interpretation. If we choose the origin on the axis of $R$ the original operation gives

$$r \rightarrow r' = Rr$$

but the transformed operation gives $r \rightarrow r' = Rr + t - Rt$, or

$$\bar{r} \rightarrow \bar{r}' = R\bar{r} \quad (\bar{r} = r - t)$$

It therefore describes (Fig. 3.15) an identical rotation about an axis

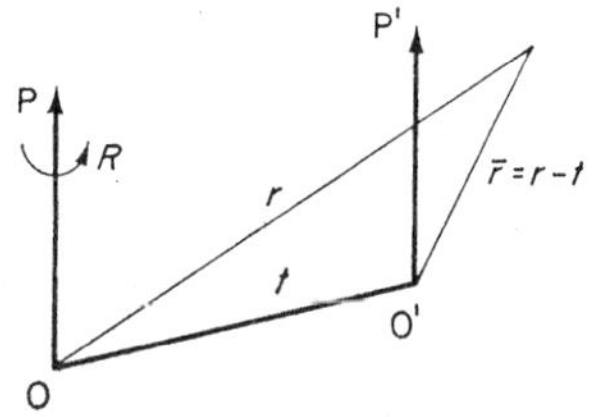


FIG. 3.15. Translation of symmetry axis. Conjugation of a rotation about OP with translation $(E \mid t)$ yields a similar rotation about the parallel axis O′P′.

translated parallel to itself by the vector $t$. Equations (3.5.1) and (3.5.3) permit the algebraic expression of two simple geometrical ideas:

(i) Rotation of a translational symmetry operation must give another translational symmetry operation, but in the rotated direction.

(ii) Translation of a rotational symmetry operation must give a similar rotational symmetry operation but about an axis which has suffered the translation.

Briefly, translation or rotation of a symmetry *element* (axis of rotation or direction of translation) must yield another symmetry element.

### *The Crystal Classes*

To make use of these ideas we first show how the number of acceptable point groups is limited by translational symmetry. Suppose we generate a lattice from primitive translations $a_1$, $a_2$, $a_3$ and assume the existence of a symmetry operation $R$ which rotates the lattice through angle $\theta$ about some axis. In the $\boldsymbol{a}$ basis we associate with

$R$ a matrix $\boldsymbol{R}$, such that the components of a lattice vector are changed according to

$$\boldsymbol{t} \to \boldsymbol{t}' = \boldsymbol{R}\boldsymbol{t}$$

Now $\boldsymbol{t}$ may be any one of the primitive translations (each component 0 or 1) and if $\boldsymbol{t}'$ is to be a lattice vector (with integral components) the matrix $\boldsymbol{R}$ must have integral elements, and, in particular, an integral *trace*. Now from (3.1.14) the trace, which is independent of choice of axes, is $\pm 1 + 2 \cos \theta$ and the requirement that this be integral gives

$$\cos \theta = 0, \quad \pm \tfrac{1}{2}, \quad \pm 1 \tag{3.5.4}$$

The only permissible rotation angles (proper or improper) for a lattice are thus $\theta = k(2\pi/n)$ where $n$, the order of the axis, is 1, 2, 3, 4 or 6. Reference to Fig. 3.2 shows that there are just 27 axial groups satisfying this requirement, and on adding the 5 cubic groups we obtain 32 acceptable point groups. This allows us to define 32 *crystal* classes. The corresponding (non-translational) symmetry types are revealed in the external forms of well developed crystals.

*The Crystal Systems and Bravais Lattices*

So far we have used the condition (3.5.1) in a weak form, considering merely the trace of $\boldsymbol{R}$. For any choice of point group, we must examine the implications of (3.5.1), and if necessary (3.5.3), in more detail; for this will show what *kind* of lattice is compatible with the full set of point group operations. Different lattice types are distinguished by the relative lengths and mutual inclinations of the vectors $\boldsymbol{a} = (a_1\ a_2\ a_3)$. Each possible type is therefore characterized by its metrical matrix $\boldsymbol{M} = \boldsymbol{a}^\dagger\boldsymbol{a}$. The fact that lattice vectors are sent into lattice vectors by every point group operation is described metrically by saying that the scalar product of *rotated* lattice vectors, $r' = Rr$ and $s' = Rs$, must be given (with a fixed basis) by the same formula as for the unrotated vectors: $r^*s = \boldsymbol{r}^\dagger\boldsymbol{M}\boldsymbol{s}$ implies

$$r'^*s' = \boldsymbol{r}'^\dagger\boldsymbol{M}\boldsymbol{s}' = \boldsymbol{r}^\dagger\boldsymbol{R}^\dagger\boldsymbol{M}\boldsymbol{R}\boldsymbol{s}$$

But a point group operation leaves any scalar product unchanged, $r'^*s' = r^*s$. On combining these results, and remembering that in real space $\boldsymbol{R}^\dagger = \tilde{\boldsymbol{R}}$, we obtain

$$\boldsymbol{M} = \tilde{\boldsymbol{R}}\boldsymbol{M}\boldsymbol{R} \quad \text{(all } \boldsymbol{R} \text{ in point group)} \tag{3.5.5}$$

The metric must be invariant under *congruent transformation* by all the matrices of a point group. If the primitive translations are chosen as basis vectors, the matrices $\boldsymbol{R}$ have integral elements and the determination of essentially distinct $\boldsymbol{M}$'s becomes a problem in number theory.

EXAMPLE 3.5. In 2-dimensions the point group $C_{2v}$ contains two generators $C_2$ and $\sigma_v$ with the properties $C_2^2 = \sigma_v^2 = E$, $C_2\sigma_v = \sigma_v C_2$. The only integral matrices which possess the same properties and also conform to (3.1.7) are

$$\boldsymbol{D}(C_2) = \begin{pmatrix} -1 & 0 \\ 0 & -1 \end{pmatrix}, \quad \boldsymbol{D}(\sigma_v) = \begin{pmatrix} 1 & 0 \\ 0 & -1 \end{pmatrix} \text{ or } \begin{pmatrix} 0 & 1 \\ 1 & 0 \end{pmatrix}$$

With the second choice for $\boldsymbol{D}(\sigma_v)$, (3.5.4) requires

$$\begin{pmatrix} M_{11}M_{12} \\ M_{21}M_{22} \end{pmatrix} = \begin{pmatrix} M_{22}M_{21} \\ M_{12}M_{11} \end{pmatrix}$$ Hence $M_{11} = M_{22}$. Basis vectors equal in length, inclination arbitrary.

With the first choice we obtain

$$\begin{pmatrix} M_{11}M_{12} \\ M_{21}M_{22} \end{pmatrix} = \begin{pmatrix} M_{11} & -M_{12} \\ -M_{21} & M_{22} \end{pmatrix}$$ Hence $M_{12} = 0$. Basis vectors may differ in length, inclination $\pi/2$.

The associated lattices, which are compatible with point group $C_{2v}$, are shown in Fig. 3.16.

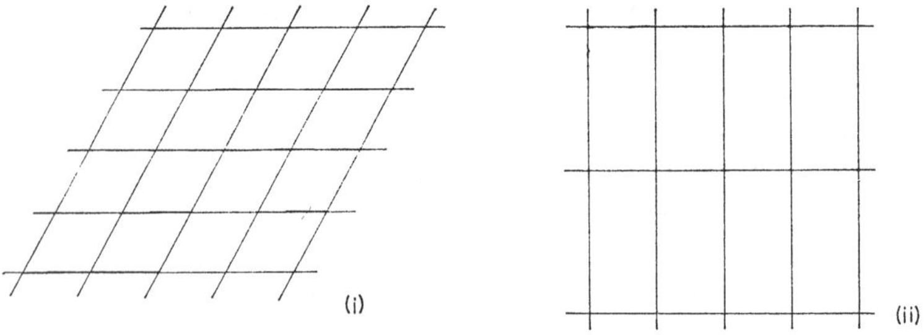


FIG. 3.16. Lattices compatible with $C_{2v}$ (two dimensions).

The set of all operations $G = \{(R \mid 0)\}$ which, by reflection or rotation about the origin bring the lattice into self-coincidence, is the *point group of the lattice* or its *holohedry*. By combining these operations with translations we can, of course, find an isomorphic point group for every lattice point; and we can find point group operations at points *inside* the cells formed by the lattice points. But our concern is with the holohedry G. In three dimensions it is found (usually by geometrical arguments) that there are just *seven* point groups which can be the holohedry of a lattice. Distinct lattices may have the same holohedry G and are then said to belong to the same *crystal system* or *syngony*. The lattices in Fig. 3.16 therefore belong to the same (2-dimensional) syngony with holohedry $C_{2v}$.

The (3-dimensional) holohedries, and the names of the crystal systems which they define, are $S_2$ (triclinic), $C_{2h}$ (monoclinic), $D_{2h}$ (orthorhombic), $D_{3d}$ (trigonal), $D_{4h}$ (tetragonal), $D_{6h}$ (hexagonal) and $O_h$ (cubic). The metrical matrices of the lattices which they leave invariant are listed in Table 3.13. In view of Example 3.5, it

TABLE 3.13

*Metric types and admissible point groups for the seven crystal systems*

| | |
|---|---|
| $\begin{pmatrix} M_{11} & M_{12} & M_{13} \\ M_{12} & M_{22} & M_{23} \\ M_{13} & M_{32} & M_{33} \end{pmatrix}$ | *Triclinic :* $a_1, a_2, a_3$ unequal in length, inclined at arbitrary angles. Compatible only with $i$. Point group $C_1$ or $S_2$. |
| $\begin{pmatrix} M & m & 0 \\ m & M' & 0 \\ 0 & 0 & M'' \end{pmatrix}$ | *Monoclinic :* $a_1, a_2, a_3$ unequal in length, two ($a_1, a_2$) lying in a plane perpendicular to the third ($a_3$). Compatible with $i$, $C_2$, $\sigma_h$. New point groups $C_{1h}$, $C_2$, $C_{2h}$. |
| $\begin{pmatrix} M & 0 & 0 \\ 0 & M' & 0 \\ 0 & 0 & M'' \end{pmatrix}$ | *Orthorhombic:* $a_1, a_2, a_3$ unequal in length, mutually orthogonal. Compatible with $i$, $C_2$, $C_2'$ (about perpendicular axis), $\sigma_h$, $\sigma_v$. New point groups $C_{2v}$, $D_2$, $D_{2h}$. |
| $\begin{pmatrix} M & m & m \\ m & M & m \\ m & m & M \end{pmatrix}$ | *Trigonal (or Rhombohedral) :* $a_1, a_2, a_3$ equal in length, equally inclined. Compatible with $i$, $C_3$, $C_2'$, $\sigma_v$. New point groups $C_3$, $S_6$, $C_{3v}$, $D_3$, $D_{3d}$. |
| $\begin{pmatrix} M & -M/2 & 0 \\ -M/2 & M & 0 \\ 0 & 0 & M' \end{pmatrix}$ | *Hexagonal :* $a_1, a_2$ equal in length, inclined at 120°, and in plane perpendicular to $a_3$. Compatible with $i$, $C_6$, $C_2'$, $\sigma_v$, $\sigma_h$. New point groups $C_6$, $C_{3h}$, $C_{6h}$, $C_{6v}$, $D_6$, $D_{3h}$, $D_{6h}$. |
| $\begin{pmatrix} M & 0 & 0 \\ 0 & M & 0 \\ 0 & 0 & M' \end{pmatrix}$ | *Tetragonal :* $a_1, a_2$ equal in length, $a_3$ different, mutually orthogonal. Compatible with $i$, $C_4$, $C_2'$, $\sigma_v$, $\sigma_h$. New point groups $C_4$, $C_{4v}$, $C_{4h}$, $S_4$, $D_4$, $D_{2d}$, $D_{4h}$. |
| $\begin{pmatrix} M & 0 & 0 \\ 0 & M & 0 \\ 0 & 0 & M \end{pmatrix}$ | *Cubic :* $a_1, a_2, a_3$ equal in length, mutually perpendicular. Compatible with the generators of $O_h$. New point groups $T$, $T_h$, $T_d$, $O$, $O_h$. |

may seem surprising that only *one* metric is given for each holohedry: this is because distinct lattices in the same system can be described by the same type of metrical matrix, provided the basis is properly chosen. To see how this comes about, we observe that the holohedry of a lattice is the point group which brings into self-coincidence the set of primitive vectors $\{a_1, a_2, a_3, -a_1, -a_2, -a_3\}$, or more generally the set connecting the origin to *all* its nearest neighbours, and that this vector set exhibits *full* point group symmetry which may not be obvious from $a_1$, $a_2$, $a_3$ alone. The most appropriate basis vectors are those which behave most simply under the point group operations one set for each holohedry—and these define the form of the metric.

EXAMPLE 3.6. Let us add a primitive vector $a_3$ normal to each lattice in Fig. 3.16 to obtain Fig. 3.17. Reference to Table 3.13 shows that the second lattice ($b$) is orthorhombic; but the first ($a$) would be monoclinic were it not for the fact that $a_1$ and $a_2$ have the same length.

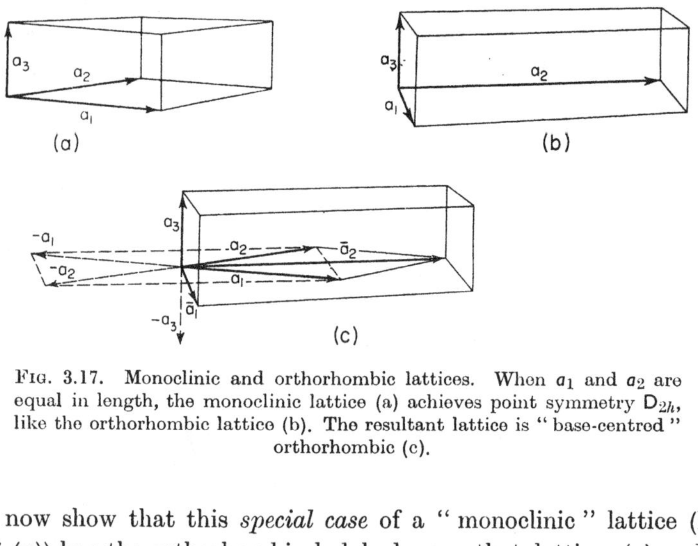


FIG. 3.17. Monoclinic and orthorhombic lattices. When $a_1$ and $a_2$ are equal in length, the monoclinic lattice (a) achieves point symmetry $D_{2h}$, like the orthorhombic lattice (b). The resultant lattice is " base-centred " orthorhombic (c).

We now show that this *special case* of a " monoclinic " lattice (Fig. 3.17 ($a$)) has the orthorhombic holohedry, so that lattices ($a$) and ($b$) *both* belong to the same syngony, and that with a new choice of basis the " monoclinic " metric also assumes the orthorhombic form. To do this we merely add the vectors $-a_1$, $-a_2$, $-a_3$ and note that, owing to the equality of length of $a_1$ and $a_2$, the set then fits into a *rectangular* prism (Fig. 3.17(c)), which exhibits the full $D_{2h}$ symmetry. Consequently, vectors $\bar{a}_1 = a_1 - a_2$, $\bar{a}_2 = a_1 + a_2$, and $\bar{a}_3 = a_3$, parallel to the edges of the prism, have the same behaviour under point group operations as the $a_1$, $a_2$, $a_3$ of the orthorhombic lattice, and give a metrical matrix of the same orthorhombic form. The only new features are the appearance of a lattice point at the *centre of the base* of the cell defined by the orthogonal vectors $\bar{a}_1$, $\bar{a}_2$, $\bar{a}_3$, and the fact that the rectangular cell has twice the volume of the cell with edges $a_1$, $a_2$, $a_3$. The lattice based on $\bar{a}_1$, $\bar{a}_2$, $\bar{a}_3$ is said to be *base-centred orthorhombic.*

Such considerations show that there are *fourteen* distinct lattices belonging to the seven syngonies. Each syngony admits a *primitive lattice*, based on the primitive vectors $a_1$, $a_2$, $a_3$ with the metrical matrix indicated in Table 3.13, but in addition it may contain one or more *centred lattices*, which arise from the primitive lattices in *other* syngonies when special length or angle relationships are imposed and result in the point symmetry characterizing the higher syngony. The 14 *Bravais lattices* which are recognized in this way are distributed among the crystal systems according to Table 3.14: the various types (body-centred, face-centred, base-centred) are discussed in any book on crystallography (*e.g.* Buerger, 1956).

### 3.6. Derivation of Space Groups

So far we have been concerned with an " empty " lattice of points. We have found 14 such lattices which are brought into self-coincidence by the operations (of type $(R \mid 0)$) of one or other of seven point groups —the seven holohedries. But in *space group* theory we are concerned with the symmetry types which arise when symmetrical *objects*, such as molecules, are put into the lattice. These may or may not have the full holohedral symmetry, and the crystal which results may consequently be invariant only under a *sub*group of the operations in its holohedry. The *space group* of a crystal is the full group of operations of type $(R \mid t)$ which bring the crystal (i.e. lattice and contents) into self-coincidence: and the derivation of all possible space groups amounts to a systematic study of the reduction of the holohedry of each type of lattice (Table 3.14) to its subgroups, by the insertion of symmetrical

TABLE 3.14

*Holohedry of the Bravais lattices*

| Lattice | System | Triclinic | Monoclinic | Orthorhombic | Trigonal | Tetragonal | Hexagonal | Cubic |
|---|---|---|---|---|---|---|---|---|
| Simple | | $S_2$ | $C_{2h}$ | $D_{2h}$ | $D_{3d}$ | $D_{4h}$ | $D_{6h}$ | $O_h$ |
| Centred | face | | | $D_{2h}$ | | | | $O_h$ |
| | base | | $C_{2h}$ | $D_{2h}$ | | | | |
| | body | | | $D_{2h}$ | | $D_{4h}$ | | $O_h$ |

objects in the lattice cells. By the progressive removal of symmetry operations in a given holohedry a considerable number of space groups may be distinguished within each crystal *system*. The crystals which they describe belong to different crystal *classes*, having different point groups, and are described as *hemihedral* because their point groups are *sub*groups of the given holohedry. When the point group is reduced to one which has appeared in an earlier system of Table 3.13 the space group falls into that earlier system. Consequently, the 32 crystal classes are divided out among the 7 crystal systems in the manner shown in Table 3.13: the point groups admitted in any given system are

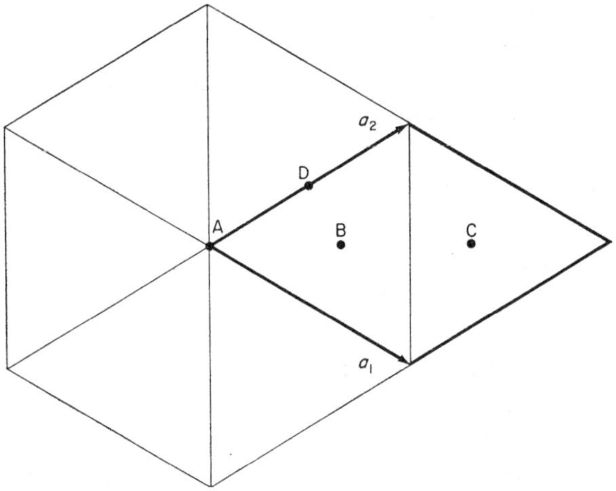


FIG. 3.18. Symmetry points in a unit cell (two-dimensional hexagonal).
A has holohedral symmetry $C_{6v}$
B, C have hemihedral symmetry $C_{3v}$
D has hemihedral symmetry $C_{2v}$

subgroups of the holohedry which have not appeared previously and are therefore peculiar to that system.

In considering the insertion into the lattice of objects with point group symmetry it is useful to define the *site symmetry*. If a point is left unmoved under a group of operations H, which is a (point) subgroup of the full space group, the point is a *symmetry point* and defines a *site* with symmetry group H. The lattice points themselves, for instance, define sites of holohedral symmetry: sites with hemihedral symmetry are illustrated in Fig. 3.18. Clearly objects of given point symmetry, placed at all sites of the same symmetry preserve the full symmetry of the empty lattice: but if the objects are of lower symmetry than their sites, the crystal space group will be a *sub*group of that of the empty lattice—for some of the operations which bring the *lattice* into self-coincidence will not do likewise with its *contents*, and must therefore be rejected. Since the whole crystal may be built up by translation of a *unit cell*, it is sufficient to consider the insertion of " units of pattern " centred on the lattice points themselves, other symmetry points arising as a necessary consequence of the translational symmetry.

If we take each lattice in Table 3.14 and put objects of holohedral or hemihedral symmetry at the lattice points, a cursory investigation‡ would suggest 61 possible space groups. This number is raised to 73 when it is recognized that the symmetry of a base-centred lattice may be reduced in more than one way by alternative orientation of a given hemihedral object. But many more types of crystal symmetry are known. These have been missed because we have considered only the combination of operations $(R \mid 0)$ and $(E \mid t)$, overlooking the possibility that there may be generating elements of the form $(R \mid v_R)$, where the rotation $R$ appears *only in conjunction with a translation* $v_R$. Generally, the rotation may be proper or improper (reflection), and the corresponding symmetry elements are called " screw axes " and " glide planes " respectively. The enumeration of all space groups when such operations are admitted is a considerable task: one of the earliest detailed accounts is due to Hilton (1906) and an excellent recent compilation, showing unit cells and symmetry operations for all the 230 space groups which result, appears in the book by Lyubarskii (1960). Reference may also be made to the International Tables for X-Ray Crystallography (1959). Here we shall merely discuss the origin of the two new types of space group element.

### *Glide Planes and Screw Axes*

First we note that an operation of the form $(R \mid v_R)$ *may* describe simply a point group operation about *some centre* other than the origin.

We have admitted all such operations which are consistent with the existence of a lattice: in seeking *new* operations they must be excluded. We therefore require that $(R \mid v_R)$ shall *not* be equivalent to a pure

‡ For example, there are 7 permitted point groups in the tetragonal system and 2 lattices—suggesting 14 tetragonal space groups.

rotation, translated from the origin to some other point $r$. The trivial operations, which will be discarded, must clearly fulfil ‡

$$(E \mid -r)(R \mid v_R)(E \mid -r)^{-1} = (R \mid v_R - r + Rr) = (R \mid 0)$$

or

$$(E - R)r = v_R$$

If we introduce a basis this equation takes the matrix form

$$(\mathbf{1} - \boldsymbol{R})\boldsymbol{r} = \boldsymbol{v}_R \qquad (3.6.1)$$

Thus, $(R \mid v_R)$ will be equivalent to a pure rotation if, and only if, we can find a vector $\boldsymbol{r}$ satisfying (3.8.1). Formally the solution may be written $\boldsymbol{r} = (\mathbf{1} - \boldsymbol{R})^{-1}\boldsymbol{v}_R$, but if $(\mathbf{1} - \boldsymbol{R})$ has any zero eigenvalues (i.e. if $\boldsymbol{R}$ has eigenvalues $+1$) the matrix will be singular, and its inverse will not exist. Now $\boldsymbol{r}$ and $\boldsymbol{v}_R$ can be expressed in terms of the eigenvectors $\boldsymbol{v}_1$, $\boldsymbol{v}_2$, $\boldsymbol{v}_3$ of $\boldsymbol{R}$ and we know (p. 57) that eigenvalues $+1$ occur in three cases:

(i) $R =$ identity ($+1$ three times)

(ii) $R -$ reflection across the plane of the two eigenvectors ($v_1$, $v_2$ say) with eigenvalue $+1$

(iii) $R =$ rotation about the eigenvector ($v_3$ say) with eigenvalue $+1$

Case (i) is trivial, and we can therefore always solve (3.6.1) except possibly in the cases (ii) and (iii), where the matrix is singular. These cases require

(ii) $\boldsymbol{v}_R = (\mathbf{1} - \boldsymbol{R})(r_1\boldsymbol{v}_1 + r_2\boldsymbol{v}_2 + r_3\boldsymbol{v}_3) = (\mathbf{1} - \boldsymbol{R})r_3\boldsymbol{v}_3 = c\boldsymbol{v}_3$, say.
Hence $v_R$ is normal to the reflection plane.

(iii) $\boldsymbol{v}_R = (\mathbf{1} - \boldsymbol{R})(r_1\boldsymbol{v}_1 + r_2\boldsymbol{v}_2 + r_3\boldsymbol{v}_3) = (\mathbf{1} - \boldsymbol{R})(r_1\boldsymbol{v}_1 + r_2\boldsymbol{v}_2) =$
$= a\boldsymbol{v}_1 + b\boldsymbol{v}_2$, say.
Hence $v_R$ is normal to the rotation axis.

Solutions of these types will be discarded, since $(R \mid v_R)$ then describes a simple point group operation. Solutions *fail* to exist (and this is the case of interest) only when

(a) $R$ is a reflection and $v_R$ contains components *in* the reflection plane. In this case we can reduce $v_R - r + Rr$ only to the form $av_1 + bv_2$

(b) $R$ is a rotation and $v_R$ contains a component *along* the axis. In this case we can reduce $v_R - r + Rr$ only to the form $cv_3$.

The two new types of symmetry operation which we must admit are thus

(a) The *glide plane*: $(R \mid v_R)$, $R$ a reflection, and $v_R$ a displacement which may be chosen parallel to the plane. Since repetition

‡ As may be seen by transferring the rotation back to the origin, using (3.5.3) *et seq.* with $t = -r$.

gives $(E\,|\,2v_R)$, $v_R$ must be *half* a lattice vector. Such operations can be generated from a basic operation in which $v_R$ is half a *primitive* translation.

(b) The *screw axis*: $(R\,|\,v_R)$, $R$ a rotation, $v_R$ a displacement along the axis. Since, with an $n$-fold axis, $n$-fold repetition gives $(E\,|\,nv_R)$, $v_R$ must be one $n$th of a lattice vector. Such operations can be generated from a basic operation in which $v_R$ is one $n$th of a *primitive* translation.

These operations can be reduced to their simplest forms by appropriate choice of origin: but they can *never* be expressed as a product of a point group operation and a *lattice* translation. We conclude this section with two examples illustrating the nature of the new operations.

EXAMPLE 3.7. The orthorhombic lattice is compatible with symmetry operations $C_2$, $\sigma$ and $\sigma'$ (Fig. 3.19(a)). If we introduce objects of

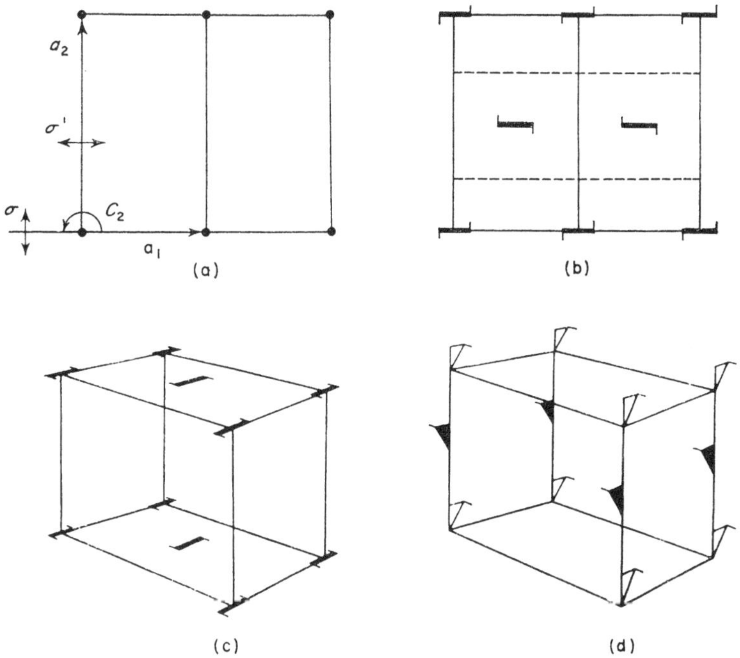


FIG. 3.19. Illustrating glide plane and screw axis.

(a) Orthorhombic lattice, looking down $a_3$ axis
(b) Insertion of objects of hemihedral symmetry at symmetry points, to give glide plane
(c) Unit cell in perspective
(d) Insertion of objects to give screw axis

hemihedral symmetry, $C_2$, as for instance in (b), the reflections $\sigma$ and $\sigma'$ are ruled out. The resultant pattern has only $C_2$ point symmetry. But $\sigma$ now appears *in conjunction with a translation*, giving a glide plane operation. Thus in pattern (b), a corner object is sent into a base-centre object by $(\sigma \mid \frac{1}{2}a_1 + \frac{1}{2}a_2)$. The glide plane operation is reduced to " standard form " if the translational part is replaced by $\frac{1}{2}a_1$ (parallel to the reflection plane) and the origin for reflections is moved through $\frac{1}{4}a_1$, as indicated by the broken lines.

EXAMPLE 3.8. Another new space group may be associated with the orthorhombic lattice of the last example by discarding the reflections and introducing the screw operation $(C_2 \mid \frac{1}{2}a_3)$ in place of $(C_2 \mid 0)$. The new symmetry is indicated in Fig. 3.19(c).

In conclusion, it should be noted that the classification of crystals in terms of the 32 acceptable point groups is not dependent on whether the point group operations occur alone or only in screw or glide operations. This is because the translations involved are so minute that they have no recognizable effect on the macroscopic properties (e.g. the development of crystal faces) associated with point symmetry. Screw axes and glide planes can of course be distinguished by X-ray and electron diffraction, which depend on the precise arrangement of the molecules within each unit cell.

## REFERENCES

1. SEITZ, F., *Z. Kristallogr.* 1934, **88,** 433; 1935, **90,** 289; 1936, **94,** 100.
2. HERZBERG, G., *Infra-red and Raman Spectra of Polyatomic Molecules*, Introduction, Van Nostrand, 1945.

## BIBLIOGRAPHY

BUERGER, M. J., *Elementary Crystallography*, Wiley, 1956. (This book provides an elementary account, without group theory, of the symmetry of crystals. It also discusses the relationship between Schoenflies notation and the Hermann–Mauguin (International) notation used by crystallographers.)

HILTON, H., *Mathematical Crystallography*, Oxford University Press, 1906. (One of the earliest detailed derivations of the 230 space groups.)

LYUBARSKII, G. YA., *The Application of Group Theory in Physics*, Pergamon Press, 1960. (Contains a fairly advanced group theoretical discussion of space groups, together with basic information about all 230 groups.)

*International Table for X-Ray Crystallography*, Kynoch Press, 1959. (Volume 1 is the crystallographer's standard reference work and contains a vast amount of information connected with all the 230 space groups.)

CHAPTER 4

# REPRESENTATIONS OF POINT AND TRANSLATION GROUPS

## 4.1. Matrices for Point Group Operations

In this chapter we shall first construct matrix representations for the 32 point groups, following the procedure indicated in previous examples (e.g. example 2.7). Sophisticated methods of constructing representations are available but are quite unnecessary in dealing with the point groups. It is in fact possible to construct all the required representations explicitly from elementary principles—considering rotations of a vector basis in ordinary 3-space and using one or two simple devices to extend the list. We shall do this partly to become thoroughly familiar with basic ideas, before passing to the underlying theory, and partly because these particular representations are required again and again in physics and chemistry.

Since the actual matrices will depend on choice of basis, though representations which differ merely through this choice are not to be considered distinct (section 2.7), it is necessary to adopt a standard convention. We continue to use the right-handed Cartesian system whose $z$ axis coincides with the principal axis (in the case of the axial groups, Fig. 3.5), or passes through a cube face in the case of the cubic groups (Fig. 3.11). We note that the 3-dimensional representation carried by $\mathbf{e} = (\mathbf{e}_1\ \mathbf{e}_2\ \mathbf{e}_3)$ will then often break down into a 1-dimensional representation (the identity) carried by $\mathbf{e}_3$ and a 2-dimensional representation carried by $\mathbf{e}_1$ and $\mathbf{e}_2$. In fact all the distinct representations of the point groups are 1-, 2-, or 3-dimensional. We also observe that a representation is determined entirely by the matrices representing the *generators* of the group, since the matrices must satisfy the same multiplication table as the group elements. First we consider the uni-dimensional representations, where each " matrix " has only one element.

### *Uni-dimensional Representations*

All Abelian groups possess nontrivial uni-dimensional representations. Those of the cyclic group, which may be a subgroup of the full point group, may be obtained by the following argument (cf. p. 58). If we associate with $C_n$ the number $\omega$ we must associate with $C_n^k$ the number $\omega^k$. Any 1-dimensional representation must therefore be of the form

$$\begin{array}{cccccc} C_n & C_n^2 & \dots & C_n^{n-1} & C_n^n(=E) & \\ \omega & \omega^2 & \dots & \omega^{n-1} & \omega^n(=1) & \qquad (4.1.1) \end{array}$$

where $\omega^n = 1$ and $\omega$ is consequently an $n$th root of unity. Thus

$$\omega = e^{i\theta}, \quad \theta = 2\pi p/n \ \ (p \text{ integral}) \qquad (4.1.2)$$

If we increase $p$ by $n$, $\theta$ will be increased by $2\pi$ and the numbers associated with the group elements will be unchanged. We can therefore find only $n$ different representations. These correspond to, say

$$p = 0, \quad 1, \quad 2, \quad \dots, \quad (n-1) \qquad (4.1.3)$$

and it will appear later that no other distinct representations of a cyclic group can be found. The representations of the cyclic groups of order $n = 1, 2, 3, 4, 6$ appear in the tables given in the next section. All representations of a given group may be written in terms of a single $\omega$, $\omega = \exp(2\pi i/n)$, by using the fact that $\exp(2\pi ik/n) = \omega^k$ and $\omega^{n+k} = \omega^k = (\omega^*)^{-k}$.

It will be noticed that the representations occur in conjugate pairs: but we already know from section 3.1 that two such representations, in which corresponding elements are complex conjugate, are equivalent to a *real two*-dimensional representation if we change the basis. In the tables, conjugate representations are usually taken together and described as a "degenerate" 2-dimensional representation. This equivalence is often used; we therefore consider an example.

EXAMPLE 4.1. The conjugate representations of $C_3$ (Table 4.5) may be put together in the diagonal matrices

$$\begin{pmatrix} 1 & 0 \\ 0 & 1 \end{pmatrix} \quad \begin{pmatrix} \omega & 0 \\ 0 & \omega^* \end{pmatrix} \quad \begin{pmatrix} \omega^* & 0 \\ 0 & \omega \end{pmatrix} \qquad (\omega = \exp(2\pi i/3))$$

Each set of diagonal elements multiplies according to the group multiplication table; so, then, do the matrices which therefore provide a 2-dimensional representation. By similarity transformation (section 2.7) with $\boldsymbol{U} = \begin{pmatrix} \frac{1}{2}\sqrt{2} & \frac{1}{2}i\sqrt{2} \\ \frac{1}{2}\sqrt{2} & -\frac{1}{2}i\sqrt{2} \end{pmatrix}$ we obtain the real form

$$\begin{pmatrix} 1 & 0 \\ 0 & 1 \end{pmatrix} \quad \begin{pmatrix} c & -s \\ s & c \end{pmatrix} \quad \begin{pmatrix} c & s \\ -s & c \end{pmatrix} \quad \begin{pmatrix} c = \cos 2\pi/3 \\ s = \sin 2\pi/3 \end{pmatrix}$$

This is the representation carried by the vectors $e_1$ and $e_2$, normal to the rotation axis. It will be recognized that this process is the reverse of that used in example 2.10, where spherical basis vectors were introduced in order to simplify the transformation formulae by sacrificing their real form.

*Matrices of the Generators*

We next consider the 3-dimensional representations carried by the basis $\mathbf{e} = (e_1\ e_2\ e_3)$. The matrix $\boldsymbol{D}(R)$ associated with a rotation $R$ is defined by (2.6.4). For an arbitrary vector $r$, with components $\boldsymbol{r}$,

$$r \to r' = Rr, \qquad \boldsymbol{r} \to \boldsymbol{r}' = \boldsymbol{D}(R)\boldsymbol{r} \tag{4.1.4}$$

With our present choice of basis, the matrices of the generators (Figs. 3.4, 3.5) are easily found to be

$$\left.\begin{aligned}
\boldsymbol{D}(C_n^z) &= \begin{pmatrix} c_n & -s_n & 0 \\ s_n & c_n & 0 \\ 0 & 0 & 1 \end{pmatrix} \qquad & \boldsymbol{D}(\sigma^{(1)}) &= \begin{pmatrix} 1 & 0 & 0 \\ 0 & -1 & 0 \\ 0 & 0 & 1 \end{pmatrix} \\
\boldsymbol{D}(\sigma_h) &= \begin{pmatrix} 1 & 0 & 0 \\ 0 & 1 & 0 \\ 0 & 0 & -1 \end{pmatrix} \qquad & \boldsymbol{D}(C_2^x) &= \begin{pmatrix} 1 & 0 & 0 \\ 0 & -1 & 0 \\ 0 & 0 & -1 \end{pmatrix} \\
\boldsymbol{D}(C_3^{xyz}) &= \begin{pmatrix} 0 & 0 & 1 \\ 1 & 0 & 0 \\ 0 & 1 & 0 \end{pmatrix} \qquad & \boldsymbol{D}(i) &= \begin{pmatrix} -1 & 0 & 0 \\ 0 & -1 & 0 \\ 0 & 0 & -1 \end{pmatrix}
\end{aligned}\right\} \tag{4.1.5}$$

Here, as usual, $c_n = \cos 2\pi/n$ and $s_n = \sin 2\pi/n$. When $C_3^{xyz}$ does not occur (i.e. in the axial groups) all matrix products have block form and the representations break down into one 2-dimensional and one 1-dimensional (or sometimes into three 1-dimensional). The 3-dimensional representations occur only for the cubic groups. Most of the representations given in the tables were obtained in this way (i.e. by considering rotations of a basis in ordinary 3-space—or the corresponding change of vector components).

*Other Representations*

Two simple devices may be used to get new representations from existing ones:

(i) If D is any representation and $\mathrm{D}_1$ is a 1-dimensional representation, it is clear that the collection of matrices $\boldsymbol{D}_1(R)\boldsymbol{D}(R)$ (in which $\boldsymbol{D}_1(R)$ is simply a numerical multiplier) is also a representation, since it still satisfies the multiplication table. This is a special case of a

" direct product " representation (section 5.9), written $D_1 \times D$, and may or may not differ trivially (i.e. by basis change) from D.

EXAMPLE 4.2. From the 1-dimensional representations of $C_{4v}$ denoted in Table 4.15 by $A_1$, $A_2$, $B_1$, we obtain direct products merely by multiplying together corresponding elements. Thus $A_1 \times B_1 = B_1$ (nothing new), but $A_2 \times B_1 = B_2$ (new representation).

(ii) In rotating a vector about the $z$ axis, its components change according to

$$x \to x' = cx - sy, \quad y \to y' = sx + cy \quad (c = \cos\theta,\ s = \sin\theta)$$

Hence

$$x^2 \to c^2x^2 - 2csxy + s^2y^2, \quad y^2 \to s^2x^2 + 2scxy + c^2y^2$$
$$xy \to cs(x^2 - y^2) + (c^2 - s^2)\,xy$$

If now we put $x^2 = X_1$, $y^2 = X_2$, $xy = X_3$, and regard these three quantities as vector components in a *new representation space*, the " induced " rotation is described by

$$\begin{pmatrix} X_1 \\ X_2 \\ X_3 \end{pmatrix} \to \begin{pmatrix} X_1' \\ X_2' \\ X_3' \end{pmatrix} = \begin{pmatrix} c^2 & s^2 & -2cs \\ s^2 & c^2 & 2cs \\ cs & -cs & c^2 - s^2 \end{pmatrix} \begin{pmatrix} X_1 \\ X_2 \\ X_3 \end{pmatrix}$$

In this way we obtain the matrices of a new representation.

EXAMPLE 4.3. If in the equations above we put $\theta = 60°$, we obtain the effect of a generator $C_6$ of the group $C_{6v}$, say. The coordinates $x$, $y$ then carry the $E_1$ representation of Table 4.10. In this case it is profitable to introduce

$$X_1 = (x^2 + y^2), \quad X_2 = (x^2 - y^2), \quad X_3 = -2xy$$

for then the induced rotation gives

$$X_1 \to X_1' = X_1 \text{ (invariant)}, \quad \begin{pmatrix} X_2 \\ X_3 \end{pmatrix} \to \begin{pmatrix} X_2' \\ X_3' \end{pmatrix} = \begin{pmatrix} -c & s \\ -s & -c \end{pmatrix} \begin{pmatrix} X_2 \\ X_3 \end{pmatrix}$$

When the second generator (reflection) is considered similarly, we find that the coordinate $X_1$, carries the identity representation, while the coordinates $X_2$, $X_3$ carry a new representation. This is not equivalent to $E_1$: it is the representation denoted by $E_2$. The coordinates have been so chosen that the matrices differ from those of $E_1$ only in sign: in fact $E_2 = B_1 \times E_1$.

Fortunately, our power to invent essentially new representations is severely limited: the same ones (except perhaps for a basis change) appear again and again. The representations given in the tables which follow do, in fact, exhaust all possibilities. The theoretical basis of this remarkable fact will be discussed in chapter 5.

### 4.2. Nomenclature. Representations

Different types of representation are given standard names (Herzberg, Ref. 1). Before presenting the tables it is therefore necessary to explain the nomenclature. Each representation is named by a letter, in accordance with the following scheme:

A  A one-dimensional representation in which 1 is associated with each rotation $C_n{}^k$ (or $S_n{}^k$) about the principal axis (or the cube diagonals in the cubic groups).

B  A one-dimensional representation in which $(-1)^k$ is associated with each rotation $C_n{}^k$ (or $S_n{}^k$) about the principal axis (not used in the cubic groups).

E  A two-dimensional representation.

T  A three-dimensional representation.‡

Often there is more than one representation of given type. These may be distinguished as follows:

*A and B Representations*

Subscript 1 or 2 indicates that 1 or $-1$ is associated with the generator $\sigma_v$ (reflection through the plane containing the $x$ and $z$ axes) for the $C_{nv}$ groups, or with $C_2^x$ (rotation about the $x$-axis) for the $D_n$ groups. For the groups $D_2$ and $D_{2h}$, where there are three axes of the same order, $B_1$, $B_2$ and $B_3$ are used when 1 is associated with $C_2^z$, with $C_2^y$ or with $C_2^x$, respectively ($-1$ being associated with each of the other two rotations). When the inversion operation appears it is usual to add a subscript $g$ or $u$ (gerade—even, ungerade—odd) according as 1 or $-1$ is associated with $i$. Any further classification is indicated by primes. Thus, if the operation $\sigma_h$ occurs, an even or odd character under this operation is indicated by a single or a double prime.

‡ Some authors (including Herzberg) use F. We continue to use T since this is more usual in ligand field theory (Griffith, Ref. 2), where the cubic groups are of special importance.

*E and T Representations*

For the axial groups only the E type occurs. With $C_n$ is associated $e^{\pm i\theta}$ ($\theta = 2\pi p/n$) or a corresponding *real* $2 \times 2$ matrix (as in example 4.1): $E_1$ and $E_2$ then correspond to $p = 1, 2$. In the cubic groups, different E representations are distinguished by $g$ or $u$ character. Different T representations occur only in O, $O_h$ and $T_d$. Since $O_h = S_2 \times O$ and O is isomorphic with $T_d$ (so that its representations are obtained simply by renaming the operations of $T_d$) it is sufficient to consider $T_d$ alone: $T_1$ and $T_2$ then correspond, respectively, to even and odd behaviour under the $\sigma_d$-type reflection $\sigma^{x\bar{y}}$.

## The Axial Groups

With the above conventions we can readily construct and classify all the representations. First we list those of the 27 axial groups belonging to the seven crystal systems. In many cases isomorphisms exist and the same table may serve for several groups: in other cases the group is a direct product of two simpler groups and it is unnecessary to give the full table since this may be written down by inspection. The elements following a subgroup usually form its left coset with the new generator (for details see Tables 3.1–3.12). The row headed by $ij$ (left-hand column) gives the $ij$-elements of the representation matrices.

*Tables* 4.1–4.4. *Representations of Point Groups* (*Triclinic, Monoclinic, Orthorhombic*)

Table 4.1

$C_1$

| $C_1$ | $E$ |
|---|---|
| A | 1 |

Table 4.2

$C_{1h}$, $C_2$, $S_2$

| $S_2$ | | | $E$ | $i$ |
|---|---|---|---|---|
| | $C_{1h}$ | | $E$ | $\sigma_h$ |
| | | $C_2$ | $E$ | $C_2$ |
| $A_g$ | A′ | A | 1 | 1 |
| $A_u$ | A″ | B | 1 | −1 |

TABLE 4.3

$C_{2h}$, $C_{2v}$, $D_2$

| $D_2$ | | | $E$ | $C_2$ | $C_2^{(1)}$ | $C_2^{(12)}$ |
|---|---|---|---|---|---|---|
| | $C_{2v}$ | | $E$ | $C_2$ | $\sigma^{(1)}$ | $\sigma^{(12)}$ |
| | | $C_{2h}$ | $E$ | $C_2$ | $i$ | $\sigma_h$ |
| A | $A_1$ | $A_g$ | 1 | 1 | 1 | 1 |
| $B_1$ | $A_2$ | $A_u$ | 1 | 1 | −1 | −1 |
| $B_3$ | $B_1$ | $B_g$ | 1 | −1 | 1 | −1 |
| $B_2$ | $B_2$ | $B_u$ | 1 | −1 | −1 | 1 |

TABLE 4.4

$D_{2h}$

Table follows from that for $D_2$ since $D_{2h} = S_2 \times D_2$

| | | | |
|---|---|---|---|
| $D_2 = E$ | $C_2$ | $C_2^{(1)}$ | $C_2^{(12)}$ |
| $iD_2 = i$ | $\sigma_h$ | $\sigma^{(12)}$ | $\sigma^{(1)}$ |

Each representation of $D_2$ splits into a $g$-type and a $u$-type. With the new elements we associate $\pm 1$ times the entries for corresponding $D_2$ elements to get $g$ and $u$ types respectively.

## *Tables 4.5–4.13. Representations of Point Groups (Trigonal and Hexagonal)*

TABLE 4.5

$C_3$

| $C_3$ | $E$ | $C_3$ | $\bar{C}_3$ |
|---|---|---|---|
| A | 1 | 1 | 1 |
| E {11 | 1 | $\omega$ | $\omega^*$ |
| E {22 | 1 | $\omega^*$ | $\omega$ |

$(\omega = \exp(2\pi i/3))$

TABLE 4.6

$C_{3v}$, $D_3$

| $D_3$ | $E$ | $C_3$ | $\bar{C}_3$ | $C_2^{(1)}$ | $C_2^{(2)}$ | $C_2^{(3)}$ |
|---|---|---|---|---|---|---|
| $C_{3v}$ | $E$ | $C_3$ | $\bar{C}_3$ | $\sigma^{(1)}$ | $\sigma^{(2)}$ | $\sigma^{(3)}$ |
| $A_1$ | 1 | 1 | 1 | 1 | 1 | 1 |
| $A_2$ | 1 | 1 | 1 | $-1$ | $-1$ | $-1$ |
| E 11 | 1 | $-c$ | $-c$ | 1 | $-c$ | $-c$ |
| E 12 | 0 | $-s$ | $s$ | 0 | $-s$ | $s$ |
| E 21 | 0 | $s$ | $-s$ | 0 | $-s$ | $s$ |
| E 22 | 1 | $-c$ | $-c$ | $-1$ | $c$ | $c$ |

($s = \sin 2\pi/6 = \frac{1}{2}\sqrt{3}$, $c = \cos 2\pi/6 = \frac{1}{2}$)

TABLE 4.7

$C_6$, $C_{3h}$, $S_6$

| $S_6$ | | | $E$ | $C_3$ | $\bar{C}_3$ | $i$ | $\bar{S}_6$ | $S_6$ |
|---|---|---|---|---|---|---|---|---|
| | $C_{3h}$ | | $E$ | $C_3$ | $\bar{C}_3$ | $\sigma_h$ | $S_3$ | $\bar{S}_3$ |
| | | $C_6$ | $E$ | $C_3$ | $\bar{C}_3$ | $C_2$ | $\bar{C}_6$ | $C_6$ |
| $A_g$ | $A'$ | A | 1 | 1 | 1 | 1 | 1 | 1 |
| $A_u$ | $A''$ | B | 1 | 1 | 1 | $-1$ | $-1$ | $-1$ |
| $E_u$ | $E''$ | $E_1$ 11 | 1 | $-\omega^*$ | $-\omega$ | $-1$ | $\omega^*$ | $\omega$ |
| | | $E_1$ 22 | 1 | $-\omega$ | $-\omega^*$ | $-1$ | $\omega$ | $\omega^*$ |
| $E_g$ | $E'$ | $E_2$ 11 | 1 | $-\omega^*$ | $-\omega$ | 1 | $-\omega^*$ | $-\omega$ |
| | | $E_2$ 22 | 1 | $-\omega$ | $-\omega^*$ | 1 | $-\omega$ | $-\omega^*$ |

($\omega = \exp(2\pi i/6)$)

TABLE 4.8

$D_{3h}$

Table follows from that for $D_3$ since $D_{3h} = C_h \times D_3$

| | | | | | | |
|---|---|---|---|---|---|---|
| $D_3 =$ | $E$ | $C_3$ | $\bar{C}_3$ | $C_2^{(1)}$ | $C_2^{(2)}$ | $C_2^{(3)}$ |
| $\sigma_h D_3 =$ | $\sigma_h$ | $S_3$ | $\bar{S}_3$ | $\sigma^{(1)}$ | $\sigma^{(2)}$ | $\sigma^{(3)}$ |

Each representation of $D_3$ splits into a primed and a double-primed type. With the new elements we associate $\pm 1$ times the entries for corresponding $D_3$ elements to get primed and double-primed types, respectively.

TABLE 4.9

$D_{3d}$

Table follows from that for $D_3$ since $D_{3d} = S_2 \times D_3$.

| | | | | | | |
|---|---|---|---|---|---|---|
| $D_3 =$ | $E$ | $C_3$ | $\bar{C}_3$ | $C_2^{(1)}$ | $C_2^{(2)}$ | $C_2^{(3)}$ |
| $i\,D_3 =$ | $i$ | $\bar{S}_6$ | $S_6$ | $\sigma^{(23)}$ | $\sigma^{(31)}$ | $\sigma^{(12)}$ |

Each representation of $D_3$ splits into a $g$-type and a $u$-type. With the new elements we associate $\pm 1$ times the entries for corresponding $D_3$ elements to get $g$ and $u$ types respectively.

## TABLE 4.10

### $C_{6v}$

| $C_{6v}$ | $\sigma^{(1)}$ $E$ | $\sigma^{(3)}$ $C_3$ | $\sigma^{(2)}$ $\bar{C}_3$ | $\sigma^{(23)}$ $C_2$ | $\sigma^{(34)}$ $C_6$ | $\sigma^{(12)}$ $\bar{C}_6$ | For top row multiply entries by |
|---|---|---|---|---|---|---|---|
| $A_1$ | 1 | 1 | 1 | 1 | 1 | 1 | 1 |
| $A_2$ | 1 | 1 | 1 | 1 | 1 | 1 | $-1$ |
| $B_1$ | 1 | 1 | 1 | $-1$ | $-1$ | $-1$ | 1 |
| $B_2$ | 1 | 1 | 1 | $-1$ | $-1$ | $-1$ | $-1$ |
| $E_1$ 11 | 1 | $-c$ | $-c$ | $-1$ | $c$ | $c$ | 1 |
| $E_1$ 12 | 0 | $-s$ | $s$ | 0 | $-s$ | $s$ | 1 |
| $E_1$ 21 | 0 | $s$ | $-s$ | 0 | $s$ | $-s$ | $-1$ |
| $E_1$ 22 | 1 | $-c$ | $-c$ | $-1$ | $c$ | $c$ | $-1$ |
| $E_2$ 11 | 1 | $-c$ | $-c$ | 1 | $-c$ | $-c$ | 1 |
| $E_2$ 12 | 0 | $-s$ | $s$ | 0 | $s$ | $-s$ | 1 |
| $E_2$ 21 | 0 | $s$ | $-s$ | 0 | $-s$ | $s$ | $-1$ |
| $E_2$ 22 | 1 | $-c$ | $-c$ | 1 | $-c$ | $-c$ | $-1$ |

($s = \sin 2\pi/6 = \frac{1}{2}\sqrt{3}$, $c = \cos 2\pi/6 = \frac{1}{2}$)

## TABLE 4.11

### $D_6$

Table follows from that for $C_{6v}$, the groups being isomorphic, on replacing

| | | | | | | |
|---|---|---|---|---|---|---|
| | $\sigma^{(1)}$ | $\sigma^{(2)}$ | $\sigma^{(3)}$ | $\sigma^{(23)}$ | $\sigma^{(34)}$ | $\sigma^{(12)}$ |
| by | $C_2^{(1)}$ | $C_2^{(2)}$ | $C_2^{(3)}$ | $C_2^{(23)}$ | $C_2^{(34)}$ | $C_2^{(12)}$ |

## TABLE 4.12

### $C_{6h}$

Table follows from that for $C_6$, since $C_{6h} = S_2 \times C_6$

| | | | | | | |
|---|---|---|---|---|---|---|
| $C_6 =$ | $E$ | $C_3$ | $\bar{C}_3$ | $C_2$ | $C_6$ | $\bar{C}_6$ |
| $iC_6 =$ | $i$ | $\bar{S}_3$ | $S_3$ | $\sigma_h$ | $\bar{S}_6$ | $S_6$ |

Each representation of $C_6$ splits into a $g$-type and a $u$-type. With the new elements we associate $\pm 1$ times the entries for corresponding $C_6$ elements to get $g$ and $u$ types, respectively.

## TABLE 4.13

### $D_{6h}$

Table follows from that for $D_6$ since $D_{6h} = S_2 \times D_6$

| | | | | | | | | | | | | |
|---|---|---|---|---|---|---|---|---|---|---|---|---|
| $D_6 =$ | $E$ | $C_3$ | $\bar{C}_3$ | $C_2$ | $C_6$ | $\bar{C}_6$ | $C_2^{(1)}$ | $C_2^{(3)}$ | $C_2^{(2)}$ | $C_2^{(23)}$ | $C_2^{(34)}$ | $C_2^{(12)}$ |
| $iD_6 =$ | $i$ | $\bar{S}_3$ | $S_3$ | $\sigma_h$ | $\bar{S}_6$ | $S_6$ | $\sigma^{(23)}$ | $\sigma^{(12)}$ | $\sigma^{(34)}$ | $\sigma^{(1)}$ | $\sigma^{(2)}$ | $\sigma^{(3)}$ |

Each representation of $D_6$ splits into a $g$-type and a $u$-type. With the new elements we associate $\pm 1$ times the entries for the corresponding $D_6$ elements to get $g$ and $u$ types, respectively.

*Tables 4.14–4.17. Representations of Point Groups (Tetragonal)*

## Table 4.14

### $C_4$, $S_4$

| $S_4$ | $E$ | $C_2$ | $S_4$ | $\bar{S}_4$ |
|---|---|---|---|---|
| $C_4$ | $E$ | $C_2$ | $C_4$ | $\bar{C}_4$ |
| A | 1 | 1 | 1 | 1 |
| B | 1 | 1 | −1 | −1 |
| E 11 | 1 | −1 | $i$ | $-i$ |
| E 22 | 1 | −1 | $-i$ | $i$ |

## Table 4.15

### $C_{4v}$, $D_4$, $D_{2d}$

| $D_{2d}$ | $E$ | $C_2$ | $S_4$ | $\bar{S}_4$ | $C_2^{(1)}$ | $C_2^{(2)}$ | $\sigma^{(23)}$ | $\sigma^{(12)}$ |
|---|---|---|---|---|---|---|---|---|
| $D_4$ | $E$ | $C_2$ | $C_4$ | $\bar{C}_4$ | $C_2^{(1)}$ | $C_2^{(2)}$ | $C_2^{(23)}$ | $C_2^{(12)}$ |
| $C_{4v}$ | $E$ | $C_2$ | $C_4$ | $\bar{C}_4$ | $\sigma^{(1)}$ | $\sigma^{(2)}$ | $\sigma^{(23)}$ | $\sigma^{(12)}$ |
| $A_1$ | 1 | 1 | 1 | 1 | 1 | 1 | 1 | 1 |
| $A_2$ | 1 | 1 | 1 | 1 | −1 | −1 | −1 | −1 |
| $B_1$ | 1 | 1 | −1 | −1 | 1 | 1 | −1 | −1 |
| $B_2$ | 1 | 1 | −1 | −1 | −1 | −1 | 1 | 1 |
| E 11 | 1 | −1 | 0 | 0 | 1 | −1 | 0 | 0 |
| E 12 | 0 | 0 | −1 | 1 | 0 | 0 | −1 | 1 |
| E 21 | 0 | 0 | 1 | −1 | 0 | 0 | −1 | 1 |
| E 22 | 1 | −1 | 0 | 0 | −1 | 1 | 0 | 0 |

## Table 4.16

### $C_{4h}$

Table follows from that for $C_4$ since $C_{4h} = S_2 \times C_4$:

| | | | | |
|---|---|---|---|---|
| $C_4 =$ | $E$ | $C_2$ | $C_4$ | $\bar{C}_4$ |
| $iC_4 =$ | $i$ | $\sigma_h$ | $\bar{S}_4$ | $S_4$ |

Each representation of $C_4$ splits into a $g$-type and a $u$-type. With the new elements we associate $\pm 1$ times the entries for corresponding $C_4$ elements to get $g$ and $u$ types, respectively.

## Table 4.17

### $D_{4h}$

Table follows from that for $D_4$ since $D_{4h} = S_2 \times D_4$.

| | | | | | | | | |
|---|---|---|---|---|---|---|---|---|
| $D_4 =$ | $E$ | $C_2$ | $C_4$ | $C_4$ | $C_2^{(1)}$ | $C_2^{(2)}$ | $C_2^{(23)}$ | $C_2^{(12)}$ |
| $iD_4 =$ | $i$ | $\sigma_h$ | $S_4$ | $S_4$ | $\sigma^{(2)}$ | $\sigma^{(1)}$ | $\sigma^{(12)}$ | $\sigma^{(23)}$ |

Each representation of $D_4$ splits into a $g$-type and a $u$-type. With the new elements we associate $\pm 1$ times the entries for corresponding $D_4$ elements to get $g$ and $u$ types, respectively.

CUBIC GROUPS

*The Group* T. This is the basic group which appears as a subgroup in all the others. The E representation is taken so that the subgroup of each 3-fold axis is represented by the corresponding matrices of $C_3$: in the simplest possible (complex) form this consists, as usual, of a conjugate pair of 1-dimensional representations. The T representation describes the rotation of the basis vectors $\mathbf{e}_1$, $\mathbf{e}_2$, $\mathbf{e}_3$ (or, equivalently, the change of vector components $x$, $y$, $z$).

The matrices of the generators ($C_2^z$ and $C_3^{xyz}$), are given in (4.15). Those for other elements follow by multiplication, using Table 3.8. The representations are collected in Table 4.18.

*The Group* O. From T we obtain the basic octohedral group O on replacing the generator $C_2^z$ by $C_4^z$. This gives the twelve new operations of the coset $C_4^z$O (Table 3.11). According to the conventions, we obtain representations $A_1$ and $A_2$ according as 1 or $-1$ is associated with $C_4^z$.

The E representation no longer consists of a degenerate pair and is more conveniently chosen in real form. If we examine the products of the coordinates (as in example 4.3) we find that the generating operations $C_4^z$ and $C_3^{xyz}$, send $(x^2-y^2)$ and $(2z^2-x^2-y^2)$ into new linear combinations of each other. These quantities may therefore be regarded as components of a vector in a 2-dimensional carrier space. If we put

$$X_1 = (2z^2 - x^2 - y^2), \quad X_2 = \sqrt{3}\,(x^2 - y^2) \tag{4.2.1}$$

we obtain for the generators of the E representation

$$\boldsymbol{D}_{\mathrm{E}}(C_4^z) = \begin{pmatrix} 1 & 0 \\ 0 & -1 \end{pmatrix} \quad \boldsymbol{D}_{\mathrm{E}}(C_3^{xyz}) = \begin{pmatrix} -\frac{1}{2} & -\frac{1}{2}\sqrt{3} \\ \frac{1}{2}\sqrt{3} & -\frac{1}{2} \end{pmatrix} \tag{4.2.2}$$

This new representation has been so chosen that the matrices for the three-fold axis appear in the standard (real) form for the $C_3$ subgroup with the $xyz$ axial direction. From the generators we easily determine the other matrices.

The $T_1$ representation describes the basis vector transformation, the matrices being taken from (4.1.5). The $T_2$ representation is obtained as the direct product $T_2 = A_2 \times T_1$.

These results are embodied in Table 4.19, the same table entries serve for both the subgroup T and its coset $C_4^z$T (appearing in the top row) when the headings for the latter are read from the right of the table.

TABLE 4.18

*Representations of the Group* T

| T | $E$ | $C_2^x$ | $C_2^y$ | $C_2^z$ | $C_3^{xyz}$ | $C_3^{x\bar{y}\bar{z}}$ | $C_3^{\bar{x}y\bar{z}}$ | $C_3^{\bar{x}\bar{y}z}$ | $\bar{C}_3^{xyz}$ | $\bar{C}_3^{x\bar{y}\bar{z}}$ | $\bar{C}_3^{\bar{x}y\bar{z}}$ | $\bar{C}_3^{\bar{x}\bar{y}z}$ |
|---|---|---|---|---|---|---|---|---|---|---|---|---|
| A | 1 | 1 | 1 | 1 | 1 | 1 | 1 | 1 | 1 | 1 | 1 | 1 |
| E 11 | 1 | 1 | 1 | 1 | $\omega$ | $\omega$ | $\omega$ | $\omega$ | $\omega^*$ | $\omega^*$ | $\omega^*$ | $\omega^*$ |
| E 22 | 1 | 1 | 1 | 1 | $\omega^*$ | $\omega^*$ | $\omega^*$ | $\omega^*$ | $\omega$ | $\omega$ | $\omega$ | $\omega$ |
| T 11 | 1 | 1 | −1 | −1 | 0 | 0 | 0 | 0 | 0 | 0 | 0 | 0 |
| T 12 | 0 | 0 | 0 | 0 | 0 | 0 | 0 | 0 | 1 | −1 | −1 | 1 |
| T 13 | 0 | 0 | 0 | 0 | 1 | −1 | 1 | −1 | 0 | 0 | 0 | 0 |
| T 21 | 0 | 0 | 0 | 0 | 1 | −1 | −1 | 1 | 0 | 0 | 0 | 0 |
| T 22 | 1 | −1 | 1 | −1 | 0 | 0 | 0 | 0 | 0 | 0 | 0 | 0 |
| T 23 | 0 | 0 | 0 | 0 | 0 | 0 | 0 | 0 | 1 | 1 | −1 | −1 |
| T 31 | 0 | 0 | 0 | 0 | 0 | 0 | 0 | 0 | 1 | −1 | 1 | −1 |
| T 32 | 0 | 0 | 0 | 0 | 1 | 1 | −1 | −1 | 0 | 0 | 0 | 0 |
| T 33 | 1 | −1 | −1 | 1 | 0 | 0 | 0 | 0 | 0 | 0 | 0 | 0 |

($\omega = \exp(2\pi i/3)$. Vertical lines separate the classes)

TABLE 4.19

*Representations of the Group* O

| O | | $C_4^z$ <br> $E$ | $C_2^{xy}$ <br> $C_2^x$ | $C_2^{x\bar{y}}$ <br> $C_2^y$ | $\bar{C}_4^z$ <br> $C_2^z$ | $C_2^{yz}$ <br> $C_3^{xyz}$ | $C_4^x$ <br> $C_3^{x\bar{y}\bar{z}}$ | $\bar{C}_4^x$ <br> $C_3^{\bar{x}y\bar{z}}$ | $C_2^{y\bar{z}}$ <br> $C_3^{\bar{x}\bar{y}z}$ | $\bar{C}_4^y$ <br> $\bar{C}_3^{xyz}$ | $C_2^{z\bar{x}}$ <br> $\bar{C}_3^{x\bar{y}\bar{z}}$ | $C_2^{zx}$ <br> $\bar{C}_3^{\bar{x}y\bar{z}}$ | $C_4^y$ <br> $\bar{C}_3^{\bar{x}\bar{y}z}$ | For top row read from this side and multiply table entries by | | |
|---|---|---|---|---|---|---|---|---|---|---|---|---|---|---|---|---|
| $A_1$, $A_2$ | | 1 | 1 | 1 | 1 | 1 | 1 | 1 | 1 | 1 | 1 | 1 | 1 | | $A_1$ <br> 1 | $A_2$ <br> −1 |
| E | 11 | 1 | 1 | 1 | 1 | $-c$ | $-c$ | $-c$ | $-c$ | $-c$ | $-c$ | $-c$ | $-c$ | | 1 | |
| | 12 | 0 | 0 | 0 | 0 | $-s$ | $-s$ | $-s$ | $-s$ | $s$ | $s$ | $s$ | $s$ | | 1 | |
| | 21 | 0 | 0 | 0 | 0 | $s$ | $s$ | $s$ | $s$ | $-s$ | $-s$ | $-s$ | $-s$ | | −1 | |
| | 22 | 1 | 1 | 1 | 1 | $-c$ | $-c$ | $-c$ | $-c$ | $-c$ | $-c$ | $-c$ | $-c$ | | −1 | |
| | | | | | | | | | | | | | | | $T_1$ | $T_2$ |
| $T_1$, $T_2$ | 11 | 1 | 1 | −1 | −1 | 0 | 0 | 0 | 0 | 0 | 0 | 0 | 0 | 21 | 1 | −1 |
| | 12 | 0 | 0 | 0 | 0 | 0 | 0 | 0 | 0 | 1 | −1 | −1 | 1 | 22 | 1 | −1 |
| | 13 | 0 | 0 | 0 | 0 | 1 | −1 | 1 | −1 | 0 | 0 | 0 | 0 | 23 | 1 | −1 |
| | 21 | 0 | 0 | 0 | 0 | 1 | −1 | −1 | 1 | 0 | 0 | 0 | 0 | 11 | −1 | 1 |
| | 22 | 1 | −1 | 1 | −1 | 0 | 0 | 0 | 0 | 0 | 0 | 0 | 0 | 12 | −1 | 1 |
| | 23 | 0 | 0 | 0 | 0 | 0 | 0 | 0 | 0 | 0 | 0 | −1 | −1 | 13 | −1 | 1 |
| | 31 | 0 | 0 | 0 | 0 | 0 | 0 | 0 | 0 | 1 | −1 | 1 | −1 | 31 | 1 | −1 |
| | 32 | 0 | 0 | 0 | 0 | 1 | 1 | −1 | −1 | 0 | 0 | 0 | 0 | 32 | 1 | −1 |
| | 33 | 1 | −1 | −1 | 1 | 0 | 0 | 0 | 0 | 0 | 0 | 0 | 0 | 33 | 1 | −1 |

($s = \sin 2\pi/6 = \frac{1}{2}\sqrt{3}$, $c = \cos 2\pi/6 = \frac{1}{2}$. Note that vertical lines are omitted as the classes are not separated in the top-row coset)

The representations for the groups $\mathsf{T}_d$, $\mathsf{T}_h$ and $\mathsf{O}_h$ may be derived easily from those of $\mathsf{T}$ and $\mathsf{O}$.

*The Group* $\mathsf{T}_d$. The group is isomorphic with $\mathsf{O}$ and the same table may therefore be used if we rename the operations and representations. $\mathsf{T}_d$ contains an *improper* 4-fold axis in place of each 4-fold axis in $\mathsf{O}$ and isomorphism is ensured by associating $S_4^z$ in the new group with $C_4^z$ in the old. As in section 3.3, we therefore take $C_3^{xyz}$ and $S_4^z$ as the generators. The matrices associated with $S_4^z$ in the representations carried by $(x, y, z)$ and by $(X_1, X_2)$ of (4.2.1) are, respectively,

$$\boldsymbol{D}(S_4^z) = \begin{pmatrix} 0 & 1 & 0 \\ -1 & 0 & 0 \\ 0 & 0 & -1 \end{pmatrix} \qquad \boldsymbol{D}'(S_4^z) = \begin{pmatrix} 1 & 0 \\ 0 & -1 \end{pmatrix}$$

The second matrix is identical with that associated with $C_4^z$ in the E representation of $\mathsf{O}$: so the same table applies if we merely change the coset $C_4^z\mathsf{T}$ to $S_4^z\mathsf{T}$. The first matrix differs from that for $C_4^z$ in the $\mathrm{T}_1$ representation of $\mathsf{O}$ by a factor $-1$; in other words it is the matrix associated with $C_4^z$ in the $\mathrm{T}_2$ representation of $\mathsf{O}$. Thus, on renaming the operations, we find the $\mathrm{T}_2$ representation of $\mathsf{O}$ gives the $(xyz)$-representation of $\mathsf{T}_d$. If, therefore, we leave the naming of the representations unchanged we must remember that in the $\mathsf{T}_d$ group the coordinates $(x, y, z)$ carry the $\mathrm{T}_2$ representation—not, as is more usual, the $\mathrm{T}_1$. With this convention (which agrees with Herzberg (Ref. 1), Griffith (Ref. 2) and others, and has been recommended for general adoption by Mulliken (Ref. 3)) we obtain Table 4.20:

TABLE 4.20

*Representations of the Group* $\mathsf{T}_d$

Table follows from that for $\mathsf{O}$ on replacing the coset $C_4^z\mathsf{T}$ (top row in Table 4.19) by $S_4^z\mathsf{T}$, given in Table 3.9.

*The Group* $\mathsf{T}_h$. The group is a direct product, $\mathsf{S}_2 \times \mathsf{T}$ and the representations follow from those of $\mathsf{T}$:

TABLE 4.21

*Representations of the Group* $\mathsf{T}_h$

Table follows from that for $\mathsf{T}$, since $\mathsf{T}_h = \mathsf{S}_2 \times \mathsf{T}$. Elements of coset $i\mathsf{T}$ appear in Table 3.10. Each representation of $\mathsf{T}$ splits into a $g$-type and a $u$-type. With the elements of $i\mathsf{T}$ we associate $\pm 1$ times the entries for corresponding $\mathsf{T}$ elements to get $g$ and $u$ types, respectively.

*The Group* $O_h$. The group is a direct product, $O_h = S_2 \times O$, the 24 new operations produced by $i$ being listed in Table 3.12. The representations of O then give rise to $g$ and $u$ types in the usual way:

TABLE 4.22

*Representations of the Group* $O_h$

Table follows from that for O, since $O_h = S_2 \times O$. Elements of coset $iO$ appear in Table 3.12. Each representation of O splits into a $g$-type and a $u$-type. With the elements of $iO$ we associate $\pm 1$ times the entries for corresponding O elements to get $g$ and $u$ types, respectively.

### 4.3. Translation Groups. Representations and Reciprocal Space

The group of translations which bring a 3-dimensional lattice into self-coincidence was introduced in section 2.2. With the notation of space group theory (section 3.1) the translational subgroup comprises operations $(E\,|\,t)$ where

$$t = t_1 a_1 + t_2 a_2 + t_3 a_3 \qquad (t_1,\ t_2,\ t_3 \text{ integers}) \tag{4.3.1}$$

is a lattice translation. The generators of the group are $(E\,|\,a_1)$, $(E\,|\,a_2)$, and $(E\,|\,a_3)$, which all commute, and the general element is

$$(E\,|\,t) = (E\,|\,a_1)^{t_1}\,(E\,|\,a_2)^{t_2}\,(E\,|\,a_3)^{t_3} \tag{4.3.2}$$

The full group is thus the direct product of three groups, each of the form

$$\{(E\,|\,a)^t\} \quad t = 0, \pm 1, \pm 2, \ldots \tag{4.3.3}$$

Now such a group is cyclic and accordingly has only 1-dimensional representations. If the number $\omega_{\kappa_i}{}^{t_i}$ is associated with $(E\,|\,a_i)^{t_i}$ in a representation $D_{\kappa_i}$ (cf. (4.1.1)) then the association

$$(E\,|\,t) \to \omega_{\kappa_1}{}^{t_1}\ \omega_{\kappa_2}{}^{t_2}\ \omega_{\kappa_3}{}^{t_3}$$

will give a representation $D_{\kappa_1\kappa_2\kappa_3}$ of the full group. First, therefore, we seek representations of the group (4.3.3) which describes a 1-dimensional lattice.

To avoid a discussion of infinite groups we use the simple device of counting $(E\,|\,a)^G$, where $G$ is some large integer, as identical with the unit operator (zero displacement).‡ The group then becomes finite, containing only the first $G$ distinct operations $\{(E\,|\,a),\ (E\,|\,a)^2,\ \ldots,\ (E\,|\,a)^G = (E\,|\,0)\}$; but $G$ may be made indefinitely large without

‡ This device corresponds to the well-known "cyclic lattice" condition in solid state theory. If we take a chain of $G$ atoms and join its ends to make a ring it becomes equivalent to an infinite chain in which the condition of the $(G + k)$th atom is identical with that of the $k$th. If the atoms are numbered 1, 2, . . . , $G$ the primitive displacement is equivalent to $C_G$ (a rotation through $2\pi/G$ about the ring centre) and $C_G{}^{k+G} = C_G{}^k$.

affecting our conclusions. Now for a cyclic group of order $G$ we know the representations are

$$\mathrm{D}_\kappa: \qquad (E\,|\,a) \to \omega_\kappa = e^{i\theta_\kappa}, \qquad \theta_\kappa = (2\pi\kappa/G) \tag{4.3.4}$$

Distinct representations occur for $\kappa = 0, 1, \ldots, (G-1)$. Other choices of $\kappa$ would of course be equally acceptable—for example $\kappa = 0, \pm 1, \ldots, G/2$ or $\pm\{(G-1)/2\}$‡—but would give nothing new since $\mathrm{D}_{\kappa+G} = \mathrm{D}_\kappa$. Each representation is then completely determined

$$\mathrm{D}_\kappa: \qquad (E\,|\,a)^t \to \exp\,(2\pi i\kappa t/G) \tag{4.3.5}$$

We now turn to the 3-dimensional case. The general group element is labelled by a triplet of integers $(t_1, t_2, t_3)$ and the available representations will also be labelled by triplets $(\kappa_1, \kappa_2, \kappa_3)$. Thus (4.3.4) becomes

$$\mathrm{D}_{\kappa_1\kappa_2\kappa_3}: \quad (E\,|\,t) = (E\,|\,a_1)^{t_1}\,(E\,|\,a_2)^{t_2}\,(E\,|\,a_3)^{t_3} \to \exp\left(\frac{2\pi i}{G}\,(\kappa_1 t_1 + \kappa_2 t_2 + \kappa_3 t_3)\right) \tag{4.3.6}$$

We now set up a geometrical description of the representations by introducing the so-called "reciprocal lattice". Basis vectors $b_1$, $b_2$, $b_3$ are defined in such a way that $(\kappa_1 t_1 + \kappa_2 t_2 + \kappa_3 t_3)$ in (4.3.6) becomes the scalar product of the displacement $t$ with a vector

$$\kappa = \kappa_1 b_1 + \kappa_2 b_2 + \kappa_3 b_3$$

Since‡‡ $\kappa \,.\, t = \sum_{i,j} \kappa_i t_j (b_i \,.\, a_j)$ this requires that

$$b_i \,.\, a_j = \delta_{ij} \quad (i, j = 1, 2, 3) \tag{4.3.7}$$

In other words each $b$-vector must have unit scalar product with the corresponding $a$-vector, whilst being perpendicular to the other two $a$-vectors. This is easily ensured by choosing each $b$-vector as a vector product (in the sense of elementary vector algebra):

$$b_i = \frac{a_j \times a_k}{(a_i \,.\, a_j \times a_k)} \quad (ijk = 123,\ 231,\ 312) \tag{4.3.8}$$

Strictly, the components of a $b$-vector are *reciprocal* lengths, with different physical dimensions from those of $a$-vectors, and the introduction of reciprocal space is simply a geometrical device. But it is a familiar device: the numerical *magnitudes* of physical quantities of various dimensions are all represented formally by lengths whenever we draw a graph.

‡ According as $G$ is even or odd.

‡‡ We use the dot notation of elementary vector algebra.

The connection between vectors and "reciprocal vectors" of the form (4.3.8) is clearer with the more general notation of section 2.8. Any vector of reciprocal space can be written in terms of the basis $\mathbf{a}$ (with metric $\boldsymbol{M}$) thus:

$$\mathsf{r} = \mathbf{b}\boldsymbol{r} = \mathbf{a}\bar{\boldsymbol{r}}$$

where the reciprocal basis is yet to be defined. The scalar product of two vectors $\mathsf{r}$ and $\mathsf{s}$ is then

$$\mathsf{r}^{*}\mathsf{s} = \bar{\boldsymbol{r}}^{\dagger}\boldsymbol{M}\boldsymbol{s}$$

If now we define $\boldsymbol{r}^{\dagger} = \bar{\boldsymbol{r}}^{\dagger}\boldsymbol{M}$ we may write

$$\mathsf{r}^{*}\mathsf{s} = \boldsymbol{r}^{\dagger}\boldsymbol{s} = \sum_i r_i^{*} s_i$$

and, taking the Hermitian transpose (with $\boldsymbol{M}^{\dagger} = \boldsymbol{M}$), $\boldsymbol{r} = \boldsymbol{M}\bar{\boldsymbol{r}}$. Now the basis $\mathbf{b}$ in which $\mathsf{r}$ takes components $\boldsymbol{r}$ is obtained from $\mathbf{a}$ by the reciprocal transformation (see (2.5.6)): $\mathbf{b} = \mathbf{a}\boldsymbol{M}^{-1}$. The reciprocal basis is thus related to the lattice basis by

$$(\mathsf{b}_1\ \mathsf{b}_2\ \mathsf{b}_3) = (\mathsf{a}_1\ \mathsf{a}_2\ \mathsf{a}_3)\boldsymbol{M}^{-1} \tag{4.3.9}$$

where $\boldsymbol{M}$ is the matrix of scalar products of the lattice basis vectors.

For many purposes it is convenient to absorb the $2\pi$ in (4.3.6) into the reciprocal vectors. The basis $(2\pi\mathsf{b}_1,\ 2\pi\mathsf{b}_2,\ 2\pi\mathsf{b}_3)$ defines the so-called " $\mathsf{k}$-space " in which a typical vector is

$$\mathsf{k} = k_1(2\pi\mathsf{b}_1) + k_2(2\pi\mathsf{b}_2) + k_3(2\pi\mathsf{b}_3) \tag{4.3.10}$$

and with this notation the representations finally appear in the form

$$\begin{aligned} \mathrm{D}_k:\quad & (\mathsf{E}\,|\,\mathsf{t}) \rightarrow \exp(i\mathsf{k}\,.\,\mathsf{t}) \\ & k_i = \kappa_i/G \quad (\kappa_i = 0,\ 1,\ 2,\ \ldots,\ (G-1);\ i = 1,\ 2,\ 3) \end{aligned} \tag{4.3.11}$$

The geometrical interpretation is then as follows. The three translations $2\pi\mathsf{b}_1$, $2\pi\mathsf{b}_2$, $2\pi\mathsf{b}_3$, which may be repeated indefinitely to give a lattice, define a parallelepiped in $\mathsf{k}$-space. This basic cell may be filled with points whose coordinates satisfy the conditions $0 \leqslant k_1,\ k_2,\ k_3 < 1$ and each of the $G^3$ points with $\mathsf{k}$ values allowed in (4.3.11) will be associated with a representation. The points in other cells in $\mathsf{k}$-space which have components differing by integers from those of point $\mathsf{k}$ in the basic cell therefore correspond to representations indistinguishable from $\mathrm{D}_k$. When used merely as a means of classifying representations, points of $\mathsf{k}$-space which differ by a lattice translation are thus precisely equivalent. There are only $G^3$ *distinct* representations, labelled by the triplets $(k_1,\ k_2,\ k_3)$ corresponding to allowed points in the basic cell: but we note that the choice is arbitrary and that any cell containing non-equivalent points could be used. The value of this classification of representations will become apparent in example 7.17.

Finally, we note that a given translation group is only a subgroup of the space groups in the corresponding crystal system. From a knowledge of the *k*-space representations, and of the properties of the point groups, it is possible to construct explicit representations of the space groups themselves. These are sometimes required in solid state theory, but for a detailed account of such developments the reader is referred elsewhere (Ref. 4).

## REFERENCES

1. Herzberg, G., *Infra-red and Raman Spectra of Polyatomic Molecules*, chapter 2, section 3, Van Nostrand, 1945.
2. Griffith, J. S., *The Theory of Transition Metal Ions*, p. 163, Cambridge University Press, 1961.
3. Mulliken, R. S., *J. Chem. Phys.*, 1955, **23,** 1997.
4. Koster, G. F., *Solid State Physics*, 1957, **5,** 174.

CHAPTER 5

# IRREDUCIBLE REPRESENTATIONS

## 5.1. Reducibility. Nature of the Problem

In this chapter we shall study in greater generality the association of matrix representations with a finite group $\mathsf{G}$. We have already remarked that two such representations may differ in a trivial way or may differ fundamentally. If the differences correspond merely to change of basis in a carrier space (as in example 2.9) the representations are *equivalent* and are not regarded as distinct. On the other hand the 1- and 2-dimensional representations of $\mathsf{C}_{3v}$, given in (2.7.3), cannot be so related: they are *inequivalent*.

We now look for inequivalent representations of the "simplest form". To see what this means, we note that from any two representations we may easily form a third, which is of higher dimension, as in example 4.1. For if the matrices of two representations $\mathrm{D}_1 = \{\boldsymbol{A}_1, \boldsymbol{B}_1, \ldots, \boldsymbol{R}_1, \ldots\}$ and $\mathrm{D}_2 = \{\boldsymbol{A}_2, \boldsymbol{B}_2, \ldots, \boldsymbol{R}_2, \ldots\}$ satisfy the same multiplication table as the group elements $\{\mathsf{A}, \mathsf{B}, \ldots, \mathsf{R}, \ldots\}$, then so will the "block-form" matrices

$$\boldsymbol{A} = \left(\begin{array}{c|c} \boldsymbol{A}_1 & \mathbf{0} \\ \hline \mathbf{0} & \boldsymbol{A}_2 \end{array}\right), \quad \boldsymbol{B} = \left(\begin{array}{c|c} \boldsymbol{B}_1 & \mathbf{0} \\ \hline \mathbf{0} & \boldsymbol{B}_2 \end{array}\right), \quad \ldots, \quad \boldsymbol{R} = \left(\begin{array}{c|c} \boldsymbol{R}_1 & \mathbf{0} \\ \hline \mathbf{0} & \boldsymbol{R}_2 \end{array}\right), \quad \ldots$$

in which the original matrices appear as diagonal blocks and the $\mathbf{0}$'s indicate that all the remaining elements are zeros. Such a representation is called the "*direct sum*" of $\mathrm{D}_1$ and $\mathrm{D}_2$ and the process of stringing together the blocks along a diagonal is described by writing $\mathrm{D} = \mathrm{D}_1 + \mathrm{D}_2$. The representation $\mathrm{D} = \{\boldsymbol{A}, \boldsymbol{B}, \ldots, \boldsymbol{R}, \ldots\}$ may then be replaced by an equivalent representation $\bar{\mathrm{D}}$, of less simple appearance, if we mix the different blocks by means of a similarity transformation, $\boldsymbol{T}$. For

$$\bar{\boldsymbol{R}} = \bar{\boldsymbol{D}}(\mathsf{R}) = \boldsymbol{T}^{-1} \left(\begin{array}{c|c} \boldsymbol{R}_1 & \mathbf{0} \\ \hline \mathbf{0} & \boldsymbol{R}_2 \end{array}\right) \boldsymbol{T} = \left(\begin{array}{c|c} \bar{\boldsymbol{R}}_{11} & \bar{\boldsymbol{R}}_{12} \\ \hline \bar{\boldsymbol{R}}_{21} & \bar{\boldsymbol{R}}_{22} \end{array}\right)$$

will not in general have the same block form as $\boldsymbol{D}(\mathsf{R})$, though $\mathrm{D}$ and $\bar{\mathrm{D}}$ are *equivalent* and we shall write $\bar{\mathrm{D}} = \mathrm{D} = \mathrm{D}_1 + \mathrm{D}_2$ where the $=$ sign *between representation symbols* is used merely to denote equivalence.

In searching for *simple* representations we shall evidently be interested in reversing the above procedure. If we can find a similarity transformation $\boldsymbol{T}$ which brings *all* the matrices of a representation ($\bar{\mathrm{D}}$) to a similar block form (D) we imply that there are two representations ($\mathrm{D}_1$ and $\mathrm{D}_2$) of lower dimension. This process of finding representations of lower dimension, to whose direct sum a given representation is equivalent, is termed *reduction.* It may clearly be repeated until the various blocks can be reduced no further. The resultant sets of matrices,

$$\mathrm{D}_1 = \{\boldsymbol{A}_1, \boldsymbol{B}_1, \ldots, \boldsymbol{R}_1, \ldots\},\ \ \mathrm{D}_2 = \{\boldsymbol{A}_2, \boldsymbol{B}_2, \ldots, \boldsymbol{R}_2, \ldots\}, \text{ etc.}$$

are then called *irreducible representations* of the group G. The search for " simplest forms " evidently amounts to looking for irreducible representations of dimension 1, 2, 3, . . . . The remarkable fact is, however, that we need not continue indefinitely.

The number of inequivalent irreducible representations of any finite group is limited and determinable : it is in fact equal to the number of classes.

If we possess the full number of irreducible representations we therefore know that any other representation is equivalent either to one of them or to a direct sum of two or more.‡ This result, along with a considerable body of related theorems which we must establish, is of far reaching importance in physics and chemistry. The methods to be used are essentially those of Wigner (Ref. 1) but alternative treatments are available in the books listed in the Bibliography.

### 5.2. Reduction and Complete Reduction. Basic Theorems

In this section we establish the basic properties of irreducible representations. The main features of the argument are as follows. We show that

(1) Any representation of a finite group G is equivalent to a representation by *unitary* matrices (see section 2.9).

(2) If two representations are equivalent, with $\bar{\boldsymbol{D}}(R) = \boldsymbol{T}^{-1}\boldsymbol{D}(R)\boldsymbol{T}$ it is possible to choose as $\boldsymbol{T}$ a *unitary* matrix.

(3) If a matrix is reducible to the form $\left(\begin{array}{c|c} & \\ \hline \mathbf{0} & \end{array}\right)$ then it is *completely* reducible to the form $\left(\begin{array}{c|c} & \mathbf{0} \\ \hline \mathbf{0} & \end{array}\right)$.

‡ Repetitions are allowed—so different representations along the diagonal may or may not be equivalent.

These theorems‡ establish the possibility of complete reduction to *unitary* form, by *unitary* transformations. The next step is to establish certain remarkable relationships involving the matrices of any two irreducible representations of a finite group : these are usually referred to as the *orthogonality relations*, and follow directly from a preliminary result—Schur's lemma.

| | |
|---|---|
| THEOREM. Any representation of a finite group $\mathsf{G} = \{\mathsf{A}, \mathsf{B}, \ldots, \mathsf{R}, \ldots\}$ by non-singular matrices $\{\boldsymbol{A}, \boldsymbol{B}, \ldots, \boldsymbol{R}, \ldots\}$ is equivalent to a *unitary* representation. | (5.2.1) |

We must prove that a new basis can be found, $\bar{\mathbf{e}} = \mathbf{e}\boldsymbol{T}$, such that $\bar{\boldsymbol{R}} = \boldsymbol{T}^{-1}\boldsymbol{R}\boldsymbol{T}$ is unitary. The proof depends on the properties of a matrix

$$\boldsymbol{H} = \sum_{\boldsymbol{R}} \boldsymbol{R}^{\dagger}\boldsymbol{R}$$

which is Hermitian, since

$$\boldsymbol{H}^{\dagger} = \sum_{\boldsymbol{R}} \boldsymbol{R}^{\dagger}(\boldsymbol{R}^{\dagger})^{\dagger} = \sum_{\boldsymbol{R}} \boldsymbol{R}^{\dagger}\boldsymbol{R} = \boldsymbol{H}$$

and positive definite (p. 47), since for an arbitrary vector $r$, with components $\boldsymbol{r}$,

$$\boldsymbol{r}^{\dagger}\boldsymbol{H}\boldsymbol{r} = \sum_{\boldsymbol{R}} \boldsymbol{r}^{\dagger}\boldsymbol{R}^{\dagger}\boldsymbol{R}\boldsymbol{r} = \sum_{\boldsymbol{R}} (\boldsymbol{R}\boldsymbol{r})^{\dagger}(\boldsymbol{R}\boldsymbol{r}) > 0 \qquad (\boldsymbol{r} \neq 0)$$

$\boldsymbol{R}\boldsymbol{r}$ containing the components of the image of $r$ under the mapping with matrix $\boldsymbol{R}$ and the final sum containing only squared moduli. $\boldsymbol{H}$ can therefore (cf. p. 49) be written

$$\boldsymbol{H} = \boldsymbol{V}^{\dagger}\boldsymbol{V}$$

$\boldsymbol{H}$ also has the property

$$\boldsymbol{R}^{\dagger}\boldsymbol{H}\boldsymbol{R} = \sum_{\boldsymbol{S}} \boldsymbol{R}^{\dagger}\boldsymbol{S}^{\dagger}\boldsymbol{S}\boldsymbol{R} = \sum_{\boldsymbol{S}} (\boldsymbol{S}\boldsymbol{R})^{\dagger}(\boldsymbol{S}\boldsymbol{R}) = \boldsymbol{H}$$

since $\boldsymbol{S}\boldsymbol{R}$ is just the matrix associated with $\mathsf{SR}$ and as $\mathsf{S}$ runs through the group so does $\mathsf{SR}$ (by (1.3.1) *et seq.*). If then we define

$$\bar{\boldsymbol{R}} = \boldsymbol{V}\boldsymbol{R}\boldsymbol{V}^{-1}$$

we shall have

$$\bar{\boldsymbol{R}}^{\dagger}\bar{\boldsymbol{R}} = \boldsymbol{V}^{-1\dagger}\boldsymbol{R}^{\dagger}\boldsymbol{V}^{\dagger}\boldsymbol{V}\boldsymbol{R}\boldsymbol{V}^{-1} = \boldsymbol{V}^{-1\dagger}\boldsymbol{H}\boldsymbol{V}^{-1} = \mathbf{1}$$

The new matrices $\bar{\boldsymbol{A}}, \bar{\boldsymbol{B}}, \ldots, \bar{\boldsymbol{R}}, \ldots$ therefore provide a unitary representation, establishing the theorem.

| | |
|---|---|
| THEOREM. If two unitary representations $\mathrm{D} = \{\boldsymbol{A}, \boldsymbol{B}, \ldots, \boldsymbol{R}, \ldots\}$ and $\bar{\mathrm{D}} = \{\bar{\boldsymbol{A}}, \bar{\boldsymbol{B}}, \ldots, \bar{\boldsymbol{R}}, \ldots\}$ are equivalent, $\bar{\boldsymbol{R}} = \boldsymbol{T}^{-1}\boldsymbol{R}\boldsymbol{T}$, it is possible to find a *unitary* matrix $\boldsymbol{U}$ which relates them according to $\bar{\boldsymbol{R}} = \boldsymbol{U}^{\dagger}\boldsymbol{R}\boldsymbol{U}$. | (5.2.2) |

The proof is suggested by that of the previous theorem, which depended upon finding a matrix $\boldsymbol{H}$ with the property $\boldsymbol{R}\boldsymbol{H}\boldsymbol{R}^{\dagger} = \boldsymbol{H}$ (all $\boldsymbol{R}$). When the representation is unitary this means $\boldsymbol{H}$ commutes with every $\boldsymbol{R}$, $\boldsymbol{H}\boldsymbol{R} = \boldsymbol{R}\boldsymbol{H}$. If we can find *any* such matrix $\boldsymbol{M}$ then

$$\boldsymbol{M}^{-1}\boldsymbol{R}\boldsymbol{M} = \boldsymbol{R}$$

‡ The proofs of these theorems provide a useful exercise in the manipulation of matrix groups. They may, however, be omitted on first reading without serious handicap.

and we can write

$$\bar{R} = T^{-1}RT = T^{-1}M^{-1}RMT$$

The question is then whether $M$ can also be chosen to make $MT$ unitary.

In the present case a matrix with the required commutation property can be formed from $T$. For $T\bar{R} = RT$ implies, on taking the Hermitian transpose and remembering $R$ and $\bar{R}$ are unitary, $\bar{R}^{-1}T^\dagger = T^\dagger R^{-1}$. On replacing $\bar{R}^{-1}$, $R^{-1}$ by $\bar{R}$, $R$ (the equation holding for *any* matrix in the collection), we easily obtain

$$RTT^\dagger = TT^\dagger R$$

and see that $TT^\dagger$ has the required property. Now not only $TT^\dagger$ but every *function*‡ of $TT^\dagger$ will commute with the matrices of D. We must choose as $M$ that function which makes $MT$ unitary. The requirement is thus $MT\ T^\dagger M^\dagger = 1$ or, since $M$ will commute with $TT^\dagger$, $MM^\dagger = (TT^\dagger)^{-1}$. Moreover, since $(TT^\dagger)^\dagger = (TT^\dagger)$ and hence $M^\dagger = M$, we require

$$M^2 = (TT^\dagger)^{-1}$$

This serves to define $M$ as the matrix $(TT^\dagger)^{-\frac{1}{2}}$. The existence of the inverse square root is ensured by the Hermitian form of the matrix and the definition of a matrix function. The theorem is thus established with

$$U = (TT^\dagger)^{-\frac{1}{2}}T$$

Before proving the last of the three theorems we discuss the geometrical meaning of reduction and give a concrete example. We suppose that a $g$-dimensional representation is carried by the basis vectors $\mathbf{e}_1, \mathbf{e}_2, \ldots, \mathbf{e}_g$ so that $R\mathbf{e} = \mathbf{e}\boldsymbol{R}$. In looking for a reduced representation we are trying to choose the basis so that the matrices $\boldsymbol{A}, \boldsymbol{B}, \ldots, \boldsymbol{R}, \ldots$ have a certain special form. The first $g_1$ basis vectors will *by themselves* carry a representation in $g_1$-dimensions if, under the mappings $A, B, \ldots, R, \ldots$, they turn into new linear combinations of themselves *alone*, i.e. if, for every $R$, $R\mathbf{e}_i = \sum_{j=1}^{g_1} \mathbf{e}_j c_j$ $(i = 1, 2, \ldots, g_1)$. In this case the *sub*space spanned by the $g_1$ vectors $\mathbf{e}_1, \mathbf{e}_2, \ldots, \mathbf{e}_{g_1}$ is said to be *invariant* under the group of mappings. Any vector in this subspace is sent into an image lying in the same subspace. The matrices $\boldsymbol{A}, \boldsymbol{B}, \ldots, \boldsymbol{R}, \ldots$ must then all have the form

$$\boldsymbol{R} = \left(\begin{array}{c|c} \boldsymbol{R}_1 & \boldsymbol{R}_{12} \\ \hline \mathbf{0} & \boldsymbol{R}_2 \end{array}\right)$$

where $\boldsymbol{R}_1$, is $g_1 \times g_1$ and $\boldsymbol{R}_2$ is $g_2 \times g_2$ $(g_1 + g_2 = g)$. The fact that the invariant subspace carries a representation is of course reflected in the multiplication properties of the matrices: if $\boldsymbol{RS} = \boldsymbol{T}$ the partitioned product is

$$\left(\begin{array}{c|c} \boldsymbol{R}_1 & \boldsymbol{R}_{12} \\ \hline \mathbf{0} & \boldsymbol{R}_2 \end{array}\right)\left(\begin{array}{c|c} \boldsymbol{S}_1 & \boldsymbol{S}_{12} \\ \hline \mathbf{0} & \boldsymbol{S}_2 \end{array}\right) = \left(\begin{array}{c|c} \boldsymbol{R}_1\boldsymbol{S}_1 & \boldsymbol{R}_1\boldsymbol{S}_{12} + \boldsymbol{R}_{12}\boldsymbol{S}_2 \\ \hline \mathbf{0} & \boldsymbol{R}_2\boldsymbol{S}_2 \end{array}\right) = \left(\begin{array}{c|c} \boldsymbol{T}_1 & \boldsymbol{T}_{12} \\ \hline \mathbf{0} & \boldsymbol{T}_2 \end{array}\right)$$

‡ For the definition and properties of a function of a Hermitian matrix $M$, see Ref. 2. Briefly, if $U$ reduces $M$ to diagonal form, $m = U^\dagger MU$, we can define $f(M) = Uf(m)U^\dagger$.

and consequently

$$\boldsymbol{R}_1\boldsymbol{S}_1 = \boldsymbol{T}_1, \qquad \boldsymbol{R}_2\boldsymbol{S}_2 = \boldsymbol{T}_2$$

The invariant subspace *and* its "complementary" subspace *each* carry a representation and the basis is said to be "adapted to a reduction (or decomposition) of the representation". Reduction in this wider sense evidently does not require that the matrices have zeros in *both* off-diagonal blocks. It is instructive to consider a geometrical example.

EXAMPLE 5.1. Let us regard the symmetry operations of $C_{3v}$ as mappings of a *three*-dimensional space upon itself, by putting the vertices of the equilateral triangle at the extremities of three orthogonal

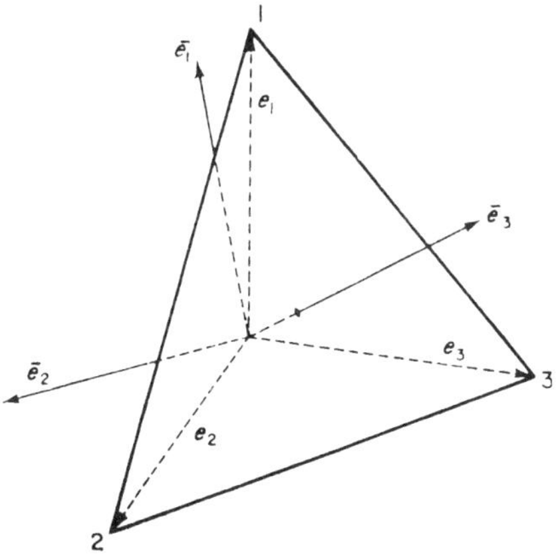


FIG. 5.1. Reduction of a representation. Basis $e_1$, $e_2$, $e_3$ carries a representation of the group of the equilateral triangle (1 2 3). The new basis $\bar{e}_1$, $\bar{e}_2$, $\bar{e}_3$ is adapted to a reduction into 1- and 2-dimensional representations.

unit vectors (Fig. 5.1). The symmetry operations then simply permute the vectors $e_1$, $e_2$, $e_3$. For example, $e_1' = C_3 e_1 = e_2$ etc., or

$$C_3(e_1\ e_2\ e_3) = (e_1\ e_2\ e_3)\begin{pmatrix}0 & 0 & 1\\ 1 & 0 & 0\\ 0 & 1 & 0\end{pmatrix}$$

This gives a 3-dimensional representation which is *not* in reduced form. To reduce the representation we must find a new set of basis vectors, some of which span an invariant subspace. It is obvious geometrically that there is a vector perpendicular to the lamina, through its centroid, which will be invariant under all symmetry operations. This single vector, $\bar{e}_3$ say, spans an invariant 1-dimensional subspace. Consequently, whatever the choice of the remaining two vectors, the matrices of the new representation will take the form

$$\left(\begin{array}{cc|c} & & 0 \\ & & 0 \\ \hline & & 1 \end{array}\right)$$

If now we choose $\bar{e}_1$ and $\bar{e}_2$ as in Fig. 5.1 the mappings which carry the lamina into self-coincidence will send $\bar{e}_1$ and $\bar{e}_2$ into new linear combinations of themselves: these vectors will therefore also span an invariant subspace and $\boldsymbol{R}$ will take the form

$$\left(\begin{array}{cc|c} & & 0 \\ & & 0 \\ \hline 0 & 0 & 1 \end{array}\right)$$

This special simplification, not *essential* for reduction, arises from the orthogonality between the vectors of the two subspaces. If we had chosen $\bar{e}_2$ and $\bar{e}_3$ as the unit vectors indicated but had taken $\bar{e}_1 = e_1$, then $\bar{e}_1$ and $\bar{e}_2$ would *not* have spanned an invariant subspace: with this choice, for example, $C_3\bar{e}_1 = e_2$, which cannot be expressed in terms of $\bar{e}_1$ and $\bar{e}_2$ alone, not lying in the plane which they define.

We summarize our conclusions as follows:

When, by means of a similarity transformation, all the matrices $\boldsymbol{R} = \boldsymbol{D}(R)$ of a group representation can be brought to the same block form

$$\left(\begin{array}{c|c} & \mathbf{0} \\ \hline \mathbf{0} & \end{array}\right) \tag{5.2.3}$$

the representation is *completely reducible*. Geometrically, we can find a basis in the representation space consisting of two sets of vectors which are not mixed by any of the mappings of the group: such a basis is " adapted " to the reduction of the representation.

Sometimes, as in example 5.1, it is possible to pick invariant subspaces by inspection. But of course, this is not generally possible.

We must now show that if a representation of a finite group can be reduced at all it can be completely reduced:

> **Theorem.** If a matrix representation D of a finite group G is reducible, then it is *completely* reducible. (5.2.4)

The proof requires two steps: we first show that a matrix of the form $\left(\begin{array}{c|c} & \\ \hline \mathbf{0} & \end{array}\right)$ can be made *unitary* by a similarity transformation and then note the unitary property to infer that the second off-diagonal block must also be a zero matrix.

From (5.2.1) the representation $\mathrm{D} = \{\boldsymbol{A}, \boldsymbol{B}, \ldots, \boldsymbol{R}, \ldots\}$ can be brought to unitary form by introducing a matrix $\boldsymbol{H} = \sum_{\boldsymbol{R}} \boldsymbol{R}^\dagger \boldsymbol{R} = \boldsymbol{V}^\dagger \boldsymbol{V}$ and taking

$$\bar{\boldsymbol{R}} = \boldsymbol{V}\boldsymbol{R}\boldsymbol{V}^{-1}$$

We now arrange that $\boldsymbol{V}$ shall be *triangular,* having only zeros below the diagonal, by the following construction. We regard $\boldsymbol{H}$ as the metric for a basis $\mathbf{e}$ and introduce a unitary basis $\bar{\mathbf{e}} = \mathbf{e}\boldsymbol{T}$: then the metric $\boldsymbol{H} = \mathbf{e}^\dagger\mathbf{e} = \boldsymbol{V}^\dagger\boldsymbol{V}$ where $\boldsymbol{V} = \boldsymbol{T}^{-1}$. A triangular form is ensured by choosing the basis in a special way: we simply take $\bar{\mathrm{e}}_1 = \mathrm{e}_1$, $\bar{\mathrm{e}}_2 = a\mathrm{e}_1 + b\mathrm{e}_2$ (choosing $a$, $b$ so that $\bar{\mathrm{e}}_2$ is normalized and orthogonal to $\bar{\mathrm{e}}_1$), $\bar{\mathrm{e}}_3 = a'\mathrm{e}_1 + b'\mathrm{e}_2 + c'\mathrm{e}_3$ (choosing the coefficients so that $\bar{\mathrm{e}}_3$ is normalized and orthogonal to $\bar{\mathrm{e}}_1$ and $\bar{\mathrm{e}}_2$), etc. Generally, the coefficients in $\bar{\mathrm{e}}_k$ are determined by one normalization and $(k-1)$ orthogonality conditions and the resultant matrix $\boldsymbol{T}$ is triangular. $\boldsymbol{V}(= \boldsymbol{T}^{-1})$ is the matrix which expresses the e's in terms of the ē's and has the same triangular form.

$\boldsymbol{V}$ and $\boldsymbol{V}^{-1}$ may therefore both be written in the form $\left(\begin{array}{c|c} & \\ \hline \mathbf{0} & \end{array}\right)$ where the partitioning corresponds to that of $\boldsymbol{R}$. And in this case

$$\bar{\boldsymbol{R}} = \boldsymbol{V}\boldsymbol{R}\boldsymbol{V}^{-1} = \left(\begin{array}{c|c} \bar{\boldsymbol{R}}_1 & \bar{\boldsymbol{R}}_{12} \\ \hline \mathbf{0} & \bar{\boldsymbol{R}}_2 \end{array}\right)$$

This establishes the first part of the theorem : any matrix $\boldsymbol{D}(R)$ can be brought to unitary form *without* changing its block structure. But now we need only observe that if $R \to \bar{\boldsymbol{R}}$ then

$$R^{-1} \to \bar{\boldsymbol{R}}^{-1} = \bar{\boldsymbol{R}}^\dagger = \left(\begin{array}{c|c} \bar{\boldsymbol{R}}_1^\dagger & \mathbf{0} \\ \hline \bar{\boldsymbol{R}}_{12}^\dagger & \bar{\boldsymbol{R}}_2^\dagger \end{array}\right)$$

which is another matrix of $\bar{\mathrm{D}}$ but is *not* of the established form unless $\bar{\boldsymbol{R}}_{12}^\dagger = \mathbf{0}$. Consequently, the rectangular matrix $\bar{\boldsymbol{R}}_{12}$ cannot be anything but a block of zeros, and the theorem is established. Moreover, by (5.2.2) the matrix which effects this complete reduction can always be replaced by a *unitary* matrix $\boldsymbol{U}$. We have therefore shown that:

> A reducible matrix representation of a finite group is *completely* reducible by a *unitary* transformation, in such a way that each component representation also appears in unitary form. (5.2.5)

By continuing the reduction process, starting afresh with the individual blocks, a representation can be completely reduced until the typical matrix takes the form (dropping the bars)

$$R = \begin{pmatrix} \boxed{R_1} & & & \\ & \boxed{R_2} & & \\ & & \boxed{R_3} & \\ & & & \ddots \end{pmatrix} \tag{5.2.6}$$

and no further reduction can be achieved. Each block is then a matrix of a corresponding *irreducible* representation. Generally, some of these irreducible representations will differ only by a similarity transformation and will not be counted as distinct. If the $\alpha$th *distinct* irreducible representation is denoted by $\mathrm{D}_\alpha$ and there are altogether $n_\alpha$ such representations in the reduced form (5.2.6), differing at most by a similarity transformation, we write

$$\mathrm{D} = n_1\mathrm{D}_1 + n_2\mathrm{D}_2 + \ldots = \sum_\alpha n_\alpha \mathrm{D}_\alpha \tag{5.2.7}$$

We repeat that the equality here means " equivalent under a similarity transformation " and the + (implied in the summation symbol) indicates the process of " stringing together " along a diagonal the matrices of $\mathrm{D}_1$ ($n_1$ times), $\mathrm{D}_2$ ($n_2$ times), etc. We must now examine the properties of the irreducible representations.

### 5.3. The Orthogonality Relations

We begin by examining the representations of the group $\mathsf{C}_{3v}$ in order to expose the properties which we wish to investigate generally.

Example 5.2. The representations $\mathrm{D}_1$ and $\mathrm{D}_2$ of (2.7.3) are simply sets of numbers, one number associated with each group element, and it is apparent that if these number sets were to be interpreted as components of a vector, referred to a unitary basis, then the vectors associated with the different representations would be orthogonal. With the 2-dimensional representation $\mathrm{D}_3$ we can associate four more vectors by interpreting *corresponding* matrix elements as the components of a vector. Altogether, we have six vectors with the components which appear in the rows of Table 4.6. It is a remarkable fact that all these vectors are unitary-orthogonal, one with another.

Example 5.2 suggests that for a group of order $g$ it may be possible to construct $g$ orthogonal vectors, spanning a $g$-dimensional space, a conjecture which we shall shortly verify. This is not the space which carries the representation: it is a space introduced purely for the purpose of discussing and relating the inequivalent irreducible representations of the given group. The dimension of this space is equal to the order of the group, $g$, and it is basically the fact that we cannot construct more than $g$ orthogonal vectors in such a space which limits the number of distinct irreducible representations. Since the number of orthogonal vectors associated with a $g_\alpha$-dimensional representation is $g_\alpha^2$, it might be conjectured further that $\sum_\alpha g_\alpha^2 = g$ (thus, in example 5.2, $1^2 + 1^2 + 2^2 = 6$). This result will presently be established.

Example 5.2 also suggests that the vectors associated with any one representation, and those associated with distinct representations of the *same* dimension, are of the same length: in this example we observe $(\text{length})^2 = (g/g_\alpha)$ for the vectors of each $g_\alpha$-dimensional representation. The orthogonality and length properties may be embodied in the statement

$$\sum_R D_\alpha(\mathsf{R})_{ij}^* D_\beta(\mathsf{R})_{kl} = (g/g_\alpha)\delta_{\alpha\beta}\delta_{ik}\delta_{jl} \tag{5.3.1}$$

where the star has been added in anticipation of the extension to complex representations.

SCHUR'S LEMMA. To establish (5.3.1) we require a lemma, due to Schur, concerning the existence of a relationship between the matrices of two irreducible collections—which may or may not be representations of a group. Thus, if $\{\boldsymbol{A}, \boldsymbol{B}, \ldots, \boldsymbol{R}, \ldots\}$ and $\{\boldsymbol{A}', \boldsymbol{B}', \ldots, \boldsymbol{R}', \ldots\}$ are irreducible and of dimension $g$ and $g'$ respectively, each $\boldsymbol{R}'$ may or may not be related to the corresponding $\boldsymbol{R}$. If $g = g'$ we are interested in the possibility of relationship by a similarity transformation, i.e. a relationship of the form

$$\boldsymbol{RX} = \boldsymbol{XR}' \qquad (\text{all } \boldsymbol{R}, \text{ fixed } \boldsymbol{X}) \tag{5.3.2}$$

More generally, we ask what possible relationship of this form can exist when $g \neq g'$ and $\boldsymbol{X}$ is therefore a $g \times g'$ matrix. Schur's Lemma answers this question:

> Given two irreducible collections of matrices $\{\boldsymbol{R}\}$ and $\{\boldsymbol{R}'\}$ and a postulated relationship $\boldsymbol{RX} = \boldsymbol{XR}'$ (all $\boldsymbol{R}$, fixed $\boldsymbol{X}$) then *either* (i) $\boldsymbol{X}$ contains only zeros *or* (ii) $\boldsymbol{X}$ is square and non-singular. (5.3.3)

The only possible non-trivial relationship between the two irreducible collections is therefore one of *equivalence*, corresponding matrices differing by a similarity transformation.

The proof depends on the idea of an invariant subspace, introduced in example 5.1. Suppose $\boldsymbol{R}$, $\boldsymbol{R}'$ describe mappings in spaces V and V′ of dimension $g$ and $g'$ ($g \geqslant g'$).‡ Then we write the postulated relationship (5.3.2)

$$\sum_j R_{ij}X_{jk} = \sum_l X_{il}R'_{lk} \quad (i, j = 1, 2, \ldots, g;\ k, l = 1, 2, \ldots, g')$$

and investigate possible forms of $\boldsymbol{X}$. Let us write $\boldsymbol{X} = (\boldsymbol{x}_1 \mid \boldsymbol{x}_2 \mid \ldots \mid \boldsymbol{x}_{g'})$ where each column represents a vector in V. Then $X_{ij} = (\boldsymbol{x}_j)_i$ and the relationship becomes $\boldsymbol{R}\boldsymbol{x}_k = \sum_l \boldsymbol{x}_l R'_{lk}$ and states that there is a mapping, with matrix $\boldsymbol{R}$, which sends each of the $g'$ vectors of V into linear combinations of themselves. This is true for all matrices of the collection $\{\boldsymbol{R}\}$ and therefore these vectors span an *invariant subspace* of V of dimension $g'(<g)$. But we assumed the collection irreducible, which amounts to denying the existence of such a subspace—apart from two trivial possibilities, the " zero space " and the whole space of dimension $g' = g$. The first possibility requires that every column of $\boldsymbol{X}$ is the zero vector, and leads to case (i) in (5.3.3). The second possibility requires that $g' = g$ and that $\boldsymbol{X}$ is non-singular (so that the columns do define an alternative linearly independent set of basis vectors and consequently do span the whole space): this leads to case (ii) in (5.3.3).

There is a useful corollary to Schur's Lemma:

> **The only (non-trivial) matrix which commutes with all the matrices of an irreducible collection of $n \times n$ matrices is a multiple of the unit matrix, $\mathbf{1}_n$.** (5.3.4)

The proof follows from the Lemma on identifying the two collections of matrices (discarding the primes). If $\boldsymbol{X}$ is non-zero it must then be non-singular (case (ii)). Consequently there are non-zero eigenvalues $\lambda_1, \lambda_2, \ldots, \lambda_n$ such that, for instance,

$$|\boldsymbol{X} - \lambda_i \mathbf{1}_n| = 0$$

But $\boldsymbol{X} - \lambda\mathbf{1}_n$ is another matrix which commutes with $\boldsymbol{A}, \boldsymbol{B}, \ldots, \boldsymbol{R}, \ldots$ (assuming $\boldsymbol{X}$ does) for any $\lambda$, in particular for $\lambda = \lambda_i$, and (5.3.3) then gives

$$\text{(i) } \boldsymbol{X} - \lambda_i\mathbf{1}_n = 0 \text{ or (ii) } |\boldsymbol{X} - \lambda_i\mathbf{1}_n| \neq 0 \text{ (matrix non-singular)}$$

The second alternative is ruled out since $\lambda_i$ is an eigenvalue; and hence $\boldsymbol{X}$ must satisfy (i). But if $\boldsymbol{X}$ is a multiple of the unit matrix *any* number is an eigenvalue and it can be said only that $\boldsymbol{X}$ is some numerical multiple of the unit matrix.

We now turn to the orthogonality relations themselves. Let us consider two irreducible representations $\mathrm{D}_\alpha$ and $\mathrm{D}_\beta$ and try to form a matrix which will play the part of $\boldsymbol{X}$ in (5.3.2). This must be a $g_\alpha \times g_\beta$ matrix and it is easy to show (cf. the proof of (5.2.1)) that

$$\boldsymbol{H} = \sum_{\mathsf{R}} \boldsymbol{D}_\alpha(\mathsf{R}^{-1})\boldsymbol{M}\boldsymbol{D}_\beta(\mathsf{R}) \tag{5.3.5}$$

($\boldsymbol{M}$ being an arbitrary $g_\alpha \times g_\beta$ matrix) has the required properties. For

$$\boldsymbol{D}_\alpha(\mathsf{R})\boldsymbol{H} = \boldsymbol{D}_\alpha(\mathsf{R}) \sum_{\mathsf{S}} \boldsymbol{D}_\alpha(\mathsf{S}^{-1})\boldsymbol{M}\boldsymbol{D}_\beta(\mathsf{S})\boldsymbol{D}_\beta(\mathsf{R}^{-1})\boldsymbol{D}_\beta(\mathsf{R}) =$$
$$= \sum_{\mathsf{T}} \boldsymbol{D}_\alpha(\mathsf{T}^{-1})\boldsymbol{M}\boldsymbol{D}_\beta(\mathsf{T})\boldsymbol{D}_\beta(\mathsf{R}) = \boldsymbol{H}\boldsymbol{D}_\beta(\mathsf{R}) \tag{5.3.6}$$

‡ This merely implies that we call the space of higher dimension V.

where $T = SR^{-1}$ and runs through the group as $S$ runs through the group. It is not necessary to the proof to assume that $D_\alpha$ and $D_\beta$ are unitary. To get the orthogonality relations we merely apply Schur's Lemma :

(i) $D_\alpha$, $D_\beta$ *inequivalent*. In this case we know $\boldsymbol{H} = \mathbf{0}$ or, in full

$$\sum_R \sum_{r,s} D_\alpha(R^{-1})_{ir} M_{rs} D_\beta(R)_{sl} = 0$$

$$(i = 1, 2, \ldots, g_\alpha ; l = 1, 2, \ldots, g_\beta) \qquad (5.3.7)$$

Now $\boldsymbol{M}$ is arbitrary and we may therefore set all elements but one equal to zero, putting $M_{rs} = 1$ $(r = j,\ s = k)$, $= 0$ otherwise. Then (5.3.7) yields, for two inequivalent irreducible representations.

$$\sum_R D_\alpha(R^{-1})_{ij} D_\beta(R)_{kl} = 0 \qquad (5.3.8)$$

(ii) $D_\alpha$, $D_\beta$ *equivalent*. In this case we may, by a similarity transformation, make the representations *identical*. We shall assume this done, for *otherwise there is no theorem*. In this case $\boldsymbol{D}_\alpha(R) = \boldsymbol{D}_\beta(R)$, and (5.3.6) reduces to

$$\boldsymbol{D}_\alpha(R)\boldsymbol{H} = \boldsymbol{H}\boldsymbol{D}_\alpha(R) \qquad (\text{all } R)$$

But by the corollary (5.3.4) this means $\boldsymbol{H}$ is a scalar multiple of the unit matrix :

$$\sum_R \sum_{r,s} D_\alpha(R^{-1})_{ir} M_{rs} D_\alpha(R)_{sl} = \lambda\delta_{il} \qquad (i, l = 1, 2, \ldots, g_\alpha) \qquad (5.3.9)$$

We now again assume $M_{jk} = 1$, all other elements vanishing, and determine $\lambda$ by putting $l = i$ in (5.3.9) and summing over all $i$ values. This gives

$$\sum_R \sum_i D_\alpha(R^{-1})_{ij} D_\alpha(R)_{ki} = \lambda g_\alpha$$

But since $\boldsymbol{D}_\alpha(R)\boldsymbol{D}_\alpha(R^{-1}) = \boldsymbol{D}_\alpha(E) = \mathbf{1}_{g_\alpha}$ the summation over $i$ gives a factor $\delta_{kj}$, while summation over the $g$ matrices of the representation yields $g\delta_{kj}$. Hence $\lambda = (g/g_\alpha)\delta_{kj}$. On putting this result in (5.3.9), again with $M_{jk} = 1$ and all other elements zero,

$$\sum_R D_\alpha(R^{-1})_{ij} D_\alpha(R)_{kl} = (g/g_\alpha)\ \delta_{kj}\delta_{il} \qquad (5.3.10)$$

The results (5.3.8) and (5.3.10) can then be combined in the form

$$\sum_R D_\alpha(R^{-1})_{ij} D_\beta(R)_{kl} = (g/g_\alpha)\ \delta_{\alpha\beta}\ \delta_{kj}\ \delta_{il} \qquad (5.3.11)$$

If the representations $D_\alpha$ and $D_\beta$ are both unitary this reduces to (5.3.1) for $\boldsymbol{D}_\alpha(R^{-1}) = \{\boldsymbol{D}_\alpha(R)\}^{-1} = \boldsymbol{D}_\alpha(R)^\dagger$ and hence we find

$$\boxed{\sum_R D_\alpha(R)^*_{ij} D_\beta(R)_{kl} = (g/g_\alpha)\ \delta_{\alpha\beta}\ \delta_{ik}\ \delta_{jl}} \qquad (5.3.12)$$

The $\delta_{ik}$ and $\delta_{jl}$ refer, it should be noted, to corresponding subscripts (rows and columns, respectively). If there is no indication to the contrary, it may be assumed in what follows that a representation is unitary, and that this form of the orthogonality relations is appropriate.

Finally, however, we express the more general equation (5.3.11) in a slightly different form by introducing a "contragredient" representation, denoted by $\check{\mathrm{D}}_\alpha$. To do this we note that $\boldsymbol{RS} = \boldsymbol{T}$ implies $\boldsymbol{S}^{-1}\boldsymbol{R}^{-1} = \boldsymbol{T}^{-1}$ and that transposition then gives‡ $\tilde{\boldsymbol{R}}^{-1}\tilde{\boldsymbol{S}}^{-1} = \tilde{\boldsymbol{T}}^{-1}$. If then the matrices $\{\boldsymbol{A}, \boldsymbol{B}, \ldots, \boldsymbol{R}, \ldots\}$ provide a representation D of the group G, the matrices $\{\tilde{\boldsymbol{A}}^{-1}, \tilde{\boldsymbol{B}}^{-1}, \ldots, \tilde{\boldsymbol{R}}^{-1}, \ldots\}$ provide another representation.

> The representation whose matrices are the transposed inverses of those of D is the contragredient representation $\check{\mathrm{D}}$. Thus, with $\boldsymbol{D}(R) = \boldsymbol{R}$ and $\tilde{\boldsymbol{D}}(R) = \tilde{\boldsymbol{R}}$,
> $$\check{\boldsymbol{D}}(R) = \tilde{\boldsymbol{D}}(R^{-1}). \qquad (5.3.13)$$

The general result (5.3.11) may therefore be written

$$\sum_R \check{D}_\alpha(R)_{ij} D_\beta(R)_{kl} = (g/g_\alpha)\, \delta_{\alpha\beta}\, \delta_{ik}\, \delta_{jl} \qquad (5.3.14)$$

EXAMPLE 5.3. From example 2.6 we obtain the following non-unitary representation of $C_{3v}$:

| $R$ | $E$ | $C_3$ | $\bar{C}_3$ | $\sigma^{(1)}$ | $\sigma^{(2)}$ | $\sigma^{(3)}$ |
|---|---|---|---|---|---|---|
| $\boldsymbol{D}(R)$ | $\begin{pmatrix}1 & 0\\ 0 & 1\end{pmatrix}$ | $\begin{pmatrix}0 & -1\\ 1 & -1\end{pmatrix}$ | $\begin{pmatrix}-1 & 1\\ -1 & 0\end{pmatrix}$ | $\begin{pmatrix}1 & -1\\ 0 & -1\end{pmatrix}$ | $\begin{pmatrix}-1 & 0\\ -1 & 1\end{pmatrix}$ | $\begin{pmatrix}0 & 1\\ 1 & 0\end{pmatrix}$ |

Since $\bar{C}_3$ is the inverse of $C_3$ and each reflection is its own inverse, we obtain from (5.3.13) the contragredient representation

$$\check{\boldsymbol{D}}(R): \begin{pmatrix}1 & 0\\ 0 & 1\end{pmatrix}\begin{pmatrix}-1 & -1\\ 1 & 0\end{pmatrix}\begin{pmatrix}0 & 1\\ -1 & -1\end{pmatrix}\begin{pmatrix}1 & 0\\ -1 & -1\end{pmatrix}\begin{pmatrix}-1 & -1\\ 0 & 1\end{pmatrix}\begin{pmatrix}0 & 1\\ 1 & 0\end{pmatrix}$$

Then, for example, $\sum_R \check{D}(R)_{11} D(R)_{11} = 3$ agreeing with (5.3.14) for $i = j = k = l$ since $\alpha = \beta$. On the other hand, $\sum_R \check{D}(R)_{12} D(R)_{11} = -1+1 = 0$ again in agreement with (5.3.14).

‡ It should be noted that the inverse of a matrix transpose is the same as the transpose of the inverse.

## 5.4. Group Characters

It is now possible to derive a number of results concerning the nature and number of inequivalent irreducible representations of any given finite group. First we note that, for a group G of order $g$ and with distinct irreducible representations $D_1$, $D_2$, ..., we can form $g_1^2 + g_2^2 + \ldots$ orthogonal vectors in a $g$-dimensional space. Since there cannot be more than $g$ such vectors, it follows that

$$\sum_{\alpha} g_{\alpha}^2 = g_1^2 + g_2^2 + \ldots \quad \leqslant g \tag{5.4.1}$$

We shall not prove immediately that only the *equality* is permitted: but (5.4.1) already shows that the number of distinct irreducible representations of any finite group is *finite*—and in fact severely limited. Thus, in the case of $C_{3v}$, representations have been found with $g_1 = g_2 = 1$ and $g_3 = 2$. There is clearly no point in looking for other representations since $1^2 + 1^2 + 2^2 = 6$—*provided* it is *known* that $D_1$, $D_2$, and $D_3$ are indeed irreducible. The simplest criterion for reducibility depends upon the idea of the *character*, which plays an important part in all that follows.

> The character of an element $R$ of a finite group G in the representation D is the *trace* of its representative matrix: $\chi(R) = \mathrm{tr}\, \boldsymbol{D}(R)$. The whole set of characters
>
> $$\chi(A), \quad \chi(B), \quad \ldots, \quad \chi(R), \quad \ldots$$
>
> is the *character system* of the group G in representation D. (5.4.2)

Since $\mathrm{tr}\, \boldsymbol{T}^{-1}\boldsymbol{R}\boldsymbol{T} = \mathrm{tr}\, \boldsymbol{R}$ the character system is a property of the representation which is invariant under similarity transformations, i.e. it does not depend on a particular choice of basis. The character system thus provides a means of *identifying* an irreducible representation: the representation can change its general appearance (matrices) but not its "fingerprints" (character system) and the representations which we regard as essentially distinct (inequivalent) will have distinct character systems. It is therefore evident by inspection that the representations of $C_{3v}$ in (2.7.3) are inequivalent.

The invariance of the trace also allows us to conclude that the characters of elements in the same class must be identical, for two elements in the same class are related by $A = RBR^{-1}$ and hence their representative matrices are related by a similarity transformation. In giving the character systems of a group it is therefore only necessary to consider a typical element of each class. The *character table* is the set of character systems associated with all the possible irreducible representations.

EXAMPLE 5.4. In the group $C_{3v}$ we consider the classes $\{E\}$, $\{C_3\}$ and $\{\sigma^{(1)}\}$—the identity, the class of $C_3$ and the class of $\sigma^{(1)}$. The character table is thus, from (2.7.3),

| | | $E$ | $2C_3$ | $3\sigma_v$ |
|---|---|---|---|---|
| (A) | $\chi_1(R)$ | 1 | 1 | 1 |
| (B) | $\chi_2(R)$ | 1 | 1 | $-1$ |
| (E) | $\chi_3(R)$ | 2 | $-1$ | 0 |

Here the columns are headed, as is customary, by the type of element prefixed by the number of elements in the class.

The character tables for all the point groups follow immediately by inspection of the tables in chapter 4. The above example suggests that the character systems also form a set of orthogonal vectors.‡ That this is true generally is a simple consequence of (5.3.12). This gives

$$\sum_R D_\alpha(R)_{ii}^* D_\beta(R)_{kk} = (g/g_\alpha)\, \delta_{\alpha\beta}\, \delta_{ik}$$

On summing over $i$ and $k$, the summand becomes a product of characters while $\sum_{i,k} \delta_{ik} = \sum_i \delta_{ii} = g_\alpha$. Hence

$$\sum_R \chi_\alpha(R)^* \chi_\beta(R) = g\, \delta_{\alpha\beta} \tag{5.4.3}$$

This result shows immediately that identity of character systems is the necessary *and sufficient* condition for two representations to be equivalent. For if $\chi_\beta(R) = \chi_\alpha(R)$ (all $R$ in G) but $D_\beta$ were *not* equivalent to $D_\alpha$ (i.e. $\alpha \neq \beta$) we should have from (5.4.3)

$$\sum_R \chi_\alpha(R)^* \chi_\alpha(R) = 0$$

which is impossible. Hence $D_\beta$ must be equivalent to $D_\alpha$ if

$$\chi_\beta(R) = \chi_\alpha(R) \quad \text{(all } R\text{)}$$

‡ It should be remembered that there are several elements (and hence several equal components) in each class, e.g. the "character vector" in E (example 5.4) has components $(2\ -1\ -1\ \ 0\ \ 0\ \ 0)$.

EXAMPLE 5.5. We verify by inspection that the representations of $C_{3v}$ in example 5.3, namely D and $\check{D}$, and the representation $D_3$ in (2.7.3) are all equivalent : their common character system is that of the E-representation in Example 5.4.

From (5.4.3) it is also possible to find whether any given representation is reducible and, if so, how many times each irreducible representation occurs. Generally, $D = n_1 D_1 + n_2 D_2 + \ldots$ means that the matrices of D are equivalent to those of the block form in which $D_\alpha$ appears $n_\alpha$ times. The trace of a matrix of D (which is invariant under a similarity transformation) is thus the sum of traces of the diagonal blocks or, in terms of characters,

$$\chi(R) = n_1\chi_1(R) + n_2\chi_2(R) + \ldots \tag{5.4.4}$$

The orthogonality property of the irreducible character systems may then be used if we multiply (5.4.4) by $\chi_\alpha(R)^*$ and sum over $R$. The result is clearly, from (5.4.3), $\sum_R \chi_\alpha(R)^*\chi(R) = n_\alpha g$. Thus

$$n_\alpha = \frac{1}{g}\sum_R \chi_\alpha(R)^*\chi(R) \tag{5.4.5}$$

This result shows at once how many times each irreducible representation must occur in the reduced form, once the irreducible character systems are known. An important special case occurs when $n_\alpha = 1$, $n_\beta = 0$ $(\beta \neq \alpha)$. In this case $\chi(R) = \chi_\alpha(R)$ and the given representation is itself irreducible :

> **If the character system in a given representation D satisfies**
>
> $$\sum_R \chi(R)^*\chi(R) = g \tag{5.4.6}$$
>
> **then D is an irreducible representation.**

This is a convenient criterion for irreducibility. It is sufficient as well as necessary because if $D = n_1 D_1 + n_2 D_2 + \ldots$ (5.4.4) and (5.4.3) give

$$\sum_R \chi(R)^*\chi(R) = (n_1^2 + n_2^2 + \ldots)g \tag{5.4.7}$$

and if (5.4.6) is satisfied it follows that there can be only *one* non-zero $n_\alpha$ and this must be unity.

EXAMPLE 5.6. The three representations of $C_{3v}$ listed in (2.7.3) are each irreducible. The sum of the squares of the characters, given in example 5.4, is 6 for each representation in accordance with (5.4.6).

EXAMPLE 5.7. We take the representation considered in example 5.1, and prove that it is reducible, being the direct sum of $D_1$ and $D_3$ in (2.7.3). Each representation matrix contains only 1's and 0's, and the number of diagonal 1's is clearly the number of vertices (1, 2, 3 in Fig. 5.1) left unmoved by the symmetry operation. The given representation therefore has the character system $\chi(R)$, compared below with those in irreducible representations $D_1$ and $D_3$:

| | $E$ | $2C_3$ | $3\sigma_v$ |
|---|---|---|---|
| $\chi(R)$ | 3 | 0 | 1 |
| $\chi_1(R)$ | 1 | 1 | 1 |
| $\chi_3(R)$ | 2 | $-1$ | 0 |

It is clear by inspection that $\chi(R) = \chi_1(R) + \chi_3(R)$ in terms of the characters, and hence that $D = D_1 + D_3$. The matrices are therefore reducible to corresponding block form by a suitable similarity transformation—the conclusion reached geometrically in example 5.1.

### 5.5. The Regular Representation

We now use the foregoing results to establish that in (5.4.1) only the equality is allowed. To do this it is necessary to introduce the *regular representation* of a group G. This is a representation of dimension $g$ formed by regarding the set of group elements themselves as a set of basis vectors $(e_1\ e_2 \ldots e_g) = (A\ B \ldots S \ldots)$. Multiplication from the left by any group element may then be regarded as a mapping: for the left coset of the whole group with respect to an element $R$ is

$$R(e_1\ e_2 \ldots e_g) = (RA\ RB \ldots RS \ldots) = (e_i\ e_j \ldots e_k) \qquad (5.5.1)$$

and consists of the same elements (still regarded as basis vectors) in a *different* order. The matrices of the regular representation therefore define permutations. Each matrix will have one unit and $(g-1)$ zero's in each column, and since only $E$ leaves an element unchanged only $\boldsymbol{D}(E)$ will have units along the diagonal—being, in fact, the $g$-dimensional unit matrix. Thus $\chi(E) = g$ and $\chi(R) = 0$ $(R \neq E)$. The number of times an irreducible representation $D_\alpha$ occurs in D is therefore, by (5.4.5)

$$n_\alpha = \frac{1}{g}\chi_\alpha(E)^*\chi(E) = \chi_\alpha(E)^* = g_\alpha$$

| | |
|---|---|
| Every $g_\alpha$-dimensional irreducible representation, $D_\alpha$, occurs just $g_\alpha$ times in the regular representation. | (5.5.2) |

Since $\chi_\alpha(E) = g_\alpha$ and $D_\alpha$ occurs $n_\alpha = g_\alpha$ times, we also have $\chi(E) = \sum_\alpha n_\alpha g_\alpha = \sum_\alpha g_\alpha^2$. But $\chi(E) = g$ and hence $\sum_\alpha g_\alpha^2 = g$.

| | |
|---|---|
| The sum of the squares of the dimensions of the distinct irreducible representations of a finite group is equal to the order of the group. | (5.5.3) |

This is the result which we were led to expect on the basis of example 5.2. The regular representation can be constructed for any finite group and the result is therefore quite general.

## 5.6. The Number of Distinct Irreducible Representations

Our ability to construct orthogonal vectors from the matrices of irreducible representations, and the fact that their total number could not exceed the number of *elements* in the group, led to (5.4.1) and ultimately to (5.5.3). A somewhat similar development allows us to show that the number of distinct irreducible representations cannot exceed the number of *classes* in the group. Thus, if we suppose there are $h$ classes with $h_1$ elements in the first class, $h_2$ in the second class, etc., and denote the common character (in $D_\alpha$) of the elements in the $i$th class by $\chi_\alpha(C_i)$, (5.4.3) becomes

$$\sum_R \chi_\alpha(R)^*\chi_\beta(R) = \sum_{i=1}^{h} h_i\chi_\alpha(C_i)^*\chi_\beta(C_i) = g\delta_{\alpha\beta}$$

and hence

$$\sum_i \left[\left(\frac{h_i}{g}\right)^{\frac{1}{2}}\chi_\alpha(C_i)\right]^*\left[\left(\frac{h_i}{g}\right)^{\frac{1}{2}}\chi_\beta(C_i)\right] = \delta_{\alpha\beta}$$

The *normalized character vectors*, with components $(h_i/g)^{\frac{1}{2}}\chi_\alpha(C_i)$, thus form an orthonormal set in a space of $h$ dimensions. There is one such vector for each distinct irreducible representation and therefore the number of distinct irreducible representations *cannot exceed* the number of classes. Once again, it can be shown that only the equality is permissible (see, for example, Speiser, 1937) and hence that:

| The number of distinct irreducible representations of a finite group is equal to the number of classes. | (5.6.1) |
|---|---|

EXAMPLE 5.8. The classes of $C_{3v}$ and the irreducible character systems are given in example 5.4. The normalized character vectors are thus

| | $E$ | $2C_3$ | $3\sigma_v$ |
|---|---|---|---|
| $v_1$ | $1/\sqrt{6}$ | $1/\sqrt{3}$ | $1/\sqrt{2}$ |
| $v_2$ | $1/\sqrt{6}$ | $1/\sqrt{3}$ | $-1/\sqrt{2}$ |
| $v_3$ | $\sqrt{(2/3)}$ | $-1/\sqrt{3}$ | 0 |

$v_1$, $v_2$ and $v_3$ are therefore orthonormal vectors in a unitary 3-space, 3 being the number of classes. We cannot find more than 3 such vectors and so there cannot be more than 3 irreducible representations.

With this result we conclude our study of the basic properties of irreducible representations of finite groups. The main results, which are used again and again in applications of group theory, are (5.3.12), (5.4.3)–(5.4.6), (5.5.3) and (5.6.1). It is instructive to use these results to verify that the point group representations obtained in chapter 4 are irreducible and inequivalent and that they exhaust all possibilities. One important example is provided by the cyclic groups:

EXAMPLE 5.9. In section 4.1 we obtained $g$ distinct irreducible representations of the cyclic group of order $g$. On summing over the $g$ representations we obtain $\sum_\alpha g_\alpha^2 = g$. But by (5.5.3) the sum over *all* distinct irreducible representations must also be $g$ (the order of the group): it follows that all possibilities are exhausted by the representations already obtained.

### 5.7. Reduction of Representations

The following problem—with which this chapter opened—will arise again and again. Given a basis $\mathbf{e}$, carrying a representation D of a group G, how can we find a basis $\bar{\mathbf{e}} = \mathbf{e}U$, adapted to a decomposition into *irreducible* representations? Before considering methods of solution let us agree that $\boldsymbol{U}$ shall be so chosen that irreducible representations which occur more than once shall appear every time in *identical* form. This means that if

$$R \to \bar{\boldsymbol{D}}(R) = \begin{bmatrix} \boldsymbol{D}_{\alpha,1}(R) & & & & \\ & \boldsymbol{D}_{\alpha,2}(R) & & & \\ & & \ddots & & \\ & & & \boldsymbol{D}_{\beta,1}(R) & \\ & & & & \ddots \end{bmatrix} = \boldsymbol{U}^{\dagger}\boldsymbol{D}(R)\boldsymbol{U} \quad (5.7.1)$$

then $\boldsymbol{D}_{\alpha,1}(R) = \boldsymbol{D}_{\alpha,2}(R) = \ldots = \boldsymbol{D}_{\alpha}(R)$, say, where the second subscript distinguishes different sets of basis vectors (and hence different blocks) carrying the same representation: the actual matrices for representations differing only in the second label are thus identical—not merely equivalent. In this case we shall say the basis $\bar{\mathbf{e}}$ is " symmetry adapted ". When a comprehensive notation is required we shall write

$$\bar{\mathbf{e}} = (\underbrace{\bar{\mathsf{e}}_1^{(\alpha,1)}\ \mathsf{e}_2^{(\alpha,1)}\ \ldots}_{\mathsf{D}_\alpha \text{ (first occurrence)}}\ \underbrace{\mathsf{e}_1^{(\alpha,2)}\ \mathsf{e}_2^{(\alpha,2)}\ \ldots}_{\mathsf{D}_\alpha \text{ (second occurrence)}}\ \ldots\ \underbrace{\mathsf{e}_1^{(\beta,1)}\ \mathsf{e}_2^{(\beta,1)}\ \ldots}_{\mathsf{D}_\beta \text{ (first occurrence)}}\ \underset{\text{etc.}}{\ldots}) \quad (5.7.2)$$

where the basis vectors fall into sets spanning the representations indicated. The vectors of the symmetry adapted basis therefore require three labels: $\bar{\mathsf{e}}_i^{(\alpha,\mu)}$ is the $i$th basis vector of irreducible representation $\mathsf{D}_\alpha$ in its $\mu$th occurrence. Since the representations of different $\mu$ are identical we may say $\bar{\mathsf{e}}_i^{(\alpha,\mu)}$ transforms like the $i$th basis vector, $\mathsf{e}_i^{(\alpha)}$ say, of the " standard " irreducible representation $\mathsf{D}_\alpha$. The pair $(\alpha, i)$ indicates the *symmetry species*‡ of the vector while the " occurrence index " $\mu$ distinguishes different vectors of the same species.

We now show how, given an irreducible representation $\mathsf{D}_\alpha$, it is possible to construct from an *arbitrary vector* $\mathsf{v}$, in any space in which the symmetry operations induce mappings, a vector of symmetry species $(\alpha, i)$. The components of each resultant vector referred to the original basis, will then furnish one column of $\boldsymbol{U}$, the matrix which brings about the reduction, and by repetition the whole matrix can be constructed.

‡ The term is often applied in a broader sense, to *any* member of a set of quantities transforming like the basis vectors of a given irreducible representation $\mathsf{D}_\alpha$ (i.e. without reference to $i$): we prefer to reserve the term for the more complete specification.

Let $\mathsf{v}$ be an arbitrary vector and let us express $\mathsf{v}$ in terms of the symmetry adapted vectors,

$$\mathsf{v} = \sum_{\beta,\mu,k} v_k^{(\beta,\mu)} \bar{\mathsf{e}}_k^{(\beta,\mu)} \tag{5.7.3}$$

One way of obtaining a vector transforming like any given basis vector $\mathsf{e}_i^{(\alpha)}$ is to annihilate all components for which $\beta \neq \alpha$, $k \neq i$. Geometrically, this corresponds to *projection*: for if we write

$$\mathsf{v} = \sum_{\beta,k} \mathsf{v}_k^{(\beta)}, \qquad \mathsf{v}_k^{(\beta)} = \sum_{\mu} v_k^{(\beta,\mu)} \bar{\mathsf{e}}_k^{(\beta,\mu)} \tag{5.7.4}$$

$\mathsf{v}$ is expressed as a sum of components lying in different subspaces (classified according to symmetry species) and we wish to extract the part which lies in a particular subspace (Fig. 5.2). This may be done

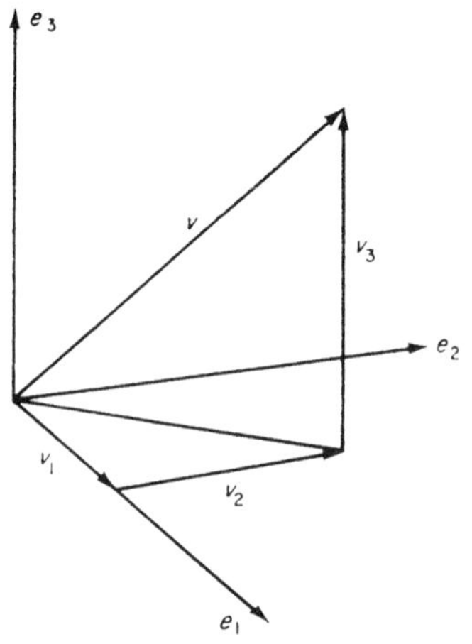


FIG. 5.2. Projection of a component of given symmetry. Here we suppose $\mathsf{e}_1$ and $\mathsf{e}_2$ are of symmetry species $(\alpha, i)$ and define a 2-dimensional subspace, while $\mathsf{e}_3$ is of species $(\beta, k)$ and defines a 1-dimensional subspace. Annihilation of the $\mathsf{v}_3$ component of an arbitrary vector $\mathsf{v} = \mathsf{v}_1 + \mathsf{v}_2 + \mathsf{v}_3$ leaves a vector $\mathsf{v}_1 + \mathsf{v}_2$, in the plane and of species $(\alpha, i)$.

by subjecting $\mathsf{v}$ to all the mappings of the group and combining the resultant vectors with suitably chosen coefficients. The choice of coefficients is suggested by the orthogonality relations. Let us consider

$$\mathsf{v}' = \sum_{R} \check{D}_\alpha(\mathsf{R})_{ij} \mathsf{R}\mathsf{v} \tag{5.7.5}$$

where, for generality, $\check{\mathrm{D}}_\alpha$ is the contragredient representation introduced in (5.3.13). Then by definition

$$\mathsf{R}\bar{\mathsf{e}}_k^{(\beta,\mu)} = \sum_{m} \bar{\mathsf{e}}_m^{(\beta,\mu)} D_\beta(\mathsf{R})_{mk} \tag{5.7.6}$$

and on inserting (5.7.3) in (5.7.5) we obtain, using (5.7.6)

$$\begin{aligned} \mathsf{v}' &= \sum_{\mathsf{R}} \sum_{\beta,\mu,k,m} \check{D}_\alpha(\mathsf{R})_{ij}\, v_k^{(\beta,\mu)}\, \bar{\mathsf{e}}_m^{(\beta,\mu)}\, D_\beta(\mathsf{R})_{mk} \\ &= \sum_{\beta,\mu,k,m} v_k^{(\beta,\mu)}\, \bar{\mathsf{e}}_m^{(\beta,\mu)}\, (g/g_\alpha)\, \delta_{\alpha\beta}\, \delta_{im}\, \delta_{jk} \end{aligned}$$

or

$$\mathsf{v}' = \sum_{\mathsf{R}} \check{D}_\alpha(\mathsf{R})_{ij}\mathsf{R}\mathsf{v} = (g/g_\alpha) \sum_{\mu} v_j^{(\alpha,\mu)}\, \bar{\mathsf{e}}_i^{(\alpha,\mu)} \tag{5.7.7}$$

The result is therefore a linear combination of vectors of species $(\alpha, i)$. We note that this is not simply the unmodified component $\mathsf{v}_i^{(\alpha)}$ in (5.7.4) or a multiple of it, except in the case $j = i$. For fixed $j$, but with $i$ running down the $j$th column of $\check{\boldsymbol{D}}_\alpha(\mathsf{R})$ we get a whole set of partners (omitting a common factor $g/g_\alpha$ and putting $v_j^{(\alpha,\mu)} = C_\mu$)

$$(\sum_\mu C_\mu\, \bar{\mathsf{e}}_1^{(\alpha,\mu)}),\quad (\sum_\mu C_\mu\, \bar{\mathsf{e}}_2^{(\alpha,\mu)}),\quad \ldots,\quad (\sum_\mu C_\mu\, \bar{\mathsf{e}}_{g_\alpha}^{(\alpha,\mu)})$$

which provide a basis for $\mathrm{D}_\alpha$—each being an identical mixture of vectors of given species. To confirm that the sums transform just like their individual members $\bar{\mathsf{e}}_1^{(\alpha,\mu)}$, $\bar{\mathsf{e}}_2^{(\alpha,\mu)}$, ..., $\bar{\mathsf{e}}_{g_\alpha}^{(\alpha,\mu)}$ (which by definition transform like the standard basis vectors $\mathsf{e}_1^{(\alpha)}$, $\mathsf{e}_2^{(\alpha)}$, ..., $\mathsf{e}_{g_\alpha}^{(\alpha)}$) we need only observe that

$$\mathsf{R}(\sum_\mu C_\mu \bar{\mathsf{e}}_i^{(\alpha,\mu)}) = \sum_\mu C_\mu \sum_k \bar{\mathsf{e}}_k^{(\alpha,\mu)} D_\alpha(\mathsf{R})_{ki} = \sum_k (\sum_\mu C_\mu \bar{\mathsf{e}}_{g_\alpha}^{(\alpha,\mu)}) D_\alpha(\mathsf{R})_{ki}$$

To summarize:

> The quantities
>
> $$\sum_{\mathsf{R}} \check{D}_\alpha(\mathsf{R})_{1j}\,\mathsf{R}\mathsf{v},\quad \sum_{\mathsf{R}} \check{D}_\alpha(\mathsf{R})_{2j}\mathsf{R}\mathsf{v},\quad \ldots,\quad \sum_{\mathsf{R}} \check{D}_\alpha(\mathsf{R})_{g_\alpha j}\mathsf{R}\mathsf{v}$$
>
> provide the same basis as
>
> $$\mathsf{e}_1^{(\alpha)},\quad \mathsf{e}_2^{(\alpha)},\quad \ldots,\quad \mathsf{e}_{g_\alpha}^{(\alpha)}$$

(5.7.8)

We adopt them as one of the $g_\alpha$-dimensional bases contained in $\bar{\mathsf{e}}$ of (5.7.2). Usually, some choices of $j$ will give repetitions of a set of vectors while others will give no vectors at all (i.e. zero vectors). The further index $\mu$ is therefore added to label the *distinct* bases which arise from various choices of $j$ and the arbitrary vector $\mathsf{v}$ in (5.7.8). The content of (5.7.8) may be expressed

$$\sum_{\mathsf{R}} \check{D}_\alpha(\mathsf{R})_{ij}\,\mathsf{R}\mathsf{v} \sim \mathsf{e}_i^{(\alpha)} \quad (\text{given } j;\quad i = 1,\ 2,\ \ldots,\ g_\alpha) \tag{5.7.9}$$

where the sign $\sim$ means "transforms like". If the representation is unitary, this reduces to

$$\sum_{\mathsf{R}} D_\alpha(\mathsf{R})_{ij}^{*}\,\mathsf{R}\mathsf{v} \sim \mathsf{e}_i^{(\alpha)} \quad (\text{given } j;\quad i = 1,\ 2,\ \ldots,\ g_\alpha) \tag{5.7.10}$$

EXAMPLE 5.10. Let us try to adapt the 3-dimensional basis provided by $e_1$, $e_2$, $e_3$ of example 5.1 to a decomposition into irreducible representations. It is known from examples 5.1 and 5.7 that $D = D_1 + D_3$ in terms of the irreducible representations of (2.7.3). Let us choose $v = e_1$: then the vector transforming like $e_1^{(1)}$ of the identity representation $D_1$ follows from (5.7.7) with $D_1(R)^*_{11} = 1$ (all $R$)

$$\sum_R Rv = e_1 + e_2 + e_3 + e_1 + e_3 + e_2 = 2(e_1 + e_2 + e_3)$$

This is a vector normal to the plane of the lamina and provides the invariant subspace (example 5.1) which carries the representation $D_1$. Similarly, using the matrices of $D_3$ with $j = 1$ in (5.7.10),

$$\sum_R D_3(R)^*_{11} Rv = e_1 - \tfrac{1}{2}e_2 - \tfrac{1}{2}e_3 + e_1 - \tfrac{1}{2}e_3 - \tfrac{1}{2}e_2 = (2e_1 - e_2 - e_3)$$

$$\sum_R D_3(R)^*_{21} Rv = 0 + \tfrac{1}{2}\sqrt{3}\, e_2 - \tfrac{1}{2}\sqrt{3}\, e_3 + 0 - \tfrac{1}{2}\sqrt{3}\, e_3 + \tfrac{1}{2}\sqrt{3}\, e_2 = \\ = \sqrt{3}\,(e_2 - e_3)$$

These vectors may be normalized and we then define the symmetry adapted basis (providing the standard irreducible representations of (2.7.3)) by

$$D_1: \quad e_1^{(1)} = (e_1 + e_2 + e_3)/\sqrt{3}$$

$$D_3: \begin{cases} e_1^{(3)} = (2e_1 - e_2 - e_3)/\sqrt{6} \\ e_2^{(3)} = (e_2 - e_3)/\sqrt{2} \end{cases}$$

Since each representation occurs only once, no occurrence index ($\mu$ in $\bar{e}_i^{(\alpha,\mu)}$) is necessary. It is readily verified that the projections defined by the *second* column ($j = 2$) of the matrices of $D_3$ simply annihilate the vector $v = e_1$. The matrix $\boldsymbol{U}$, which brings about the reduction according to (5.7.1), is thus

$$\boldsymbol{U} = \begin{pmatrix} 1/\sqrt{3} & 2/\sqrt{6} & 0 \\ 1/\sqrt{3} & -1/\sqrt{6} & 1/\sqrt{2} \\ 1/\sqrt{3} & -1/\sqrt{6} & -1/\sqrt{2} \end{pmatrix}$$

Sometimes it is more convenient to work with the *components* of $v$ in the given basis rather than with the vector itself. But we know that $v' = Rv$ is then equivalent to $\boldsymbol{v}' = \boldsymbol{R}\boldsymbol{v}$ where $\boldsymbol{R}$ is the matrix associated with mapping $R$ in the basis $\mathbf{e}$. In this case then (5.7.9) takes the form

$$\sum_R \breve{D}_\alpha(R)_{ij} \boldsymbol{R}\boldsymbol{v} = \begin{pmatrix} \text{column of components of a vector} \\ \text{which transforms like } e_i^{(\alpha)} \end{pmatrix} \qquad (5.7.11)$$

EXAMPLE 5.11. We return to example 5.9 but project from the less special vector $v$ with components (1, 2, 1). The matrices $\boldsymbol{R}$ are then the permutation matrices of example 5.1, and the rotated vectors are described by the following columns, $\boldsymbol{v}' = \boldsymbol{R}\boldsymbol{v}$ :

| $R$ | $E$ | $C_3$ | $C_3$ | $\sigma^{(1)}$ | $\sigma^{(2)}$ | $\sigma^{(3)}$ |
|---|---|---|---|---|---|---|
| $\boldsymbol{R}\boldsymbol{v}$ | $\begin{pmatrix}1\\2\\1\end{pmatrix}$ | $\begin{pmatrix}1\\1\\2\end{pmatrix}$ | $\begin{pmatrix}2\\1\\1\end{pmatrix}$ | $\begin{pmatrix}1\\1\\2\end{pmatrix}$ | $\begin{pmatrix}1\\2\\1\end{pmatrix}$ | $\begin{pmatrix}2\\1\\1\end{pmatrix}$ |

Hence, using (5.7.8) with $\alpha = 1$ and $i = j = 1$, the column with elements (8, 8, 8) represents a $D_1$ vector. With $\alpha = 3$ and $j = 1$, $i = 1, 2$ we get $(-1, \frac{1}{2}, \frac{1}{2})$ and $(0, -\frac{1}{2}\sqrt{3}, \frac{1}{2}\sqrt{3})$. On normalizing, and multiplying by a common factor $-1$ (the sign being arbitrary), the symmetry adapted basis vectors are seen to have exactly the components found in example 5.9.

### 5.8. Idempotents and Projection Operators

The construction of a vector of species $(\alpha, i)$, given an arbitrary vector $v$, has been compared with the geometrical operation of projection (Fig. 5.2). This interpretation must now be considered in more detail.

It follows from (5.7.7) and (5.7.4) that

$$(g_\alpha/g) \sum_R \check{D}_\alpha(R)_{ii} R v = v_i^{(\alpha)}$$

where $v_i^{(\alpha)}$ is the component of an arbitrary vector $v$ which belongs to symmetry species $(\alpha, i)$. The quantity which preceeds the $v$ has a meaning of its own as an element of the group algebra (section 1.10) : regarded as a single operator ‡ it is a true projection operator, since it destroys every component of an arbitrary $v$ except that of species $(\alpha, i)$, which is left unchanged. Thus, defining

$$\rho_{ii}^{(\alpha)} = (g_\alpha/g) \sum_R \check{D}_\alpha(R)_{ii} R \tag{5.8.1}$$

it follows that $\qquad \rho_{ii}^{(\alpha)} v = v_i^{(\alpha)}$

and by repetition $\qquad \rho_{ii}^{(\alpha)2} v = \rho_{ii}^{(\alpha)} v$

Hence, since $v$ is arbitrary, $\qquad \rho_{ii}^{(\alpha)2} = \rho_{ii}^{(\alpha)} \qquad$ (5.8.2)

This *idempotency* may also be established directly from the group algebra and the orthogonality relations. The other operators used in

‡ For example, $A + B$ means operate with $A$ and with $B$ and add the results together.

section 5.7 are included by defining generally

$$\rho_{ij}^{(\alpha)} = (g_\alpha/g) \sum_R \check{D}_\alpha(R)_{ij} R \tag{5.8.3}$$

It is then easy to show that

$$\rho_{ij}^{(\alpha)} \rho_{kl}^{(\beta)} = \delta_{\alpha\beta} \delta_{jk} \rho_{il}^{(\alpha)} \tag{5.8.4}$$

The quantities $\rho_{ij}^{(\alpha)}$ therefore possess the characteristic projection operator property of idempotency only for $j = i$. Different projection operators are seen to be *mutually exclusive* in the sense

$$\rho_{ii}^{(\alpha)} \rho_{jj}^{(\beta)} = 0 \quad \text{unless } \alpha = \beta, i = j \tag{5.8.5}$$

and together they form a *complete* set

$$\sum_\alpha \sum_i \rho_{ii}^{(\alpha)} = E \tag{5.8.6}$$

Equation (5.8.5) means simply that projection of one species $(\beta, j)$ leaves no component of a different species $(\alpha, i)$, the second projection then necessarily yielding the zero vector. Equation (5.8.6) means that if we project out all components and add the results together we shall obtain the original vector, the summed projections then being equivalent to the identity operator. Sets of operators with the properties (5.8.2), (5.8.5) and (5.8.6) occur frequently in quantum mechanics, where they are known as " spectral sets ". In the mathematical literature the operators $\rho_{ii}^{(\alpha)}$ are called primitive idempotents of the group algebra, and (5.8.6) is often referred to as the " resolution of the identity ".

The operators with $j \neq i$ also have interesting properties. Thus, with $\mathsf{v}_j^{(\alpha)} = \rho_{jj}^{(\alpha)}\mathsf{v}$, $\rho_{ij}^{(\alpha)}\mathsf{v}_j^{(\alpha)} \neq 0$, in general, and from (5.8.4) we obtain

$$\rho_{ii}^{(\alpha)} (\rho_{ij}^{(\alpha)}\mathsf{v}_j^{(\alpha)}) = (\rho_{ij}^{(\alpha)}\mathsf{v}_j^{(\alpha)})$$

In other words, $\rho_{ij}^{(\alpha)}\mathsf{v}_j^{(\alpha)}$ is a vector of species $(\alpha, i)$, being an eigenvector of the projection operator $\rho_{ii}^{(\alpha)}$. It is not in general *equal to* $\mathsf{v}_i^{(\alpha)}$, the component which would remain after applying $\rho_{ii}^{(\alpha)}$ directly to the original vector $\mathsf{v}$, for there may be many linearly independent vectors of the same species—and *all* linear combinations would have the same eigenvalue 1. But we know from section 5.7 that it is exactly the right partner to serve along with $\mathsf{v}_j^{(\alpha)}$ in a basis carrying a representation. Thus, $\rho_{ij}^{(\alpha)}$ takes us from the $j$th basis vector of a given representation to the $i$th : if we put $\rho_{jj}^{(\alpha)}\mathsf{v} = \mathsf{v}_j^{(\alpha)} = \mathbf{e}_j^{(\alpha)}$, then

$$\rho_{ij}^{(\alpha)} \mathbf{e}_j^{(\alpha)} = \mathbf{e}_i^{(\alpha)} \tag{5.8.7}$$

and the full set of e's, which may be generated by applying the $\rho_{ij}^{(\alpha)}$ with $i = 1, 2, \ldots, g_\alpha$, provide the representation. The quantities $\rho_{ij}^{(\alpha)}$

$(i \neq j)$ which carry us from one basis vector of a representation to another are frequently encountered in quantum mechanics, where they are called "shift operators".

Finally, we note that knowledge of the shift operators (i.e. of *all* matrix elements in an irreducible representation) is not essential: a basis carrying a given irreducible representation can be found using the primitive idempotents alone (i.e. the diagonal elements of the matrices). Thus, if we take an arbitrary vector and decide to adopt its $(\alpha, i)$ projection as the $i$th basis vector $\mathbf{e}_i^{(\alpha)}$ of a representation, a further operation $R$ gives

$$R\mathbf{e}_i^{(\alpha)} = c_1\,\mathbf{e}_1^{(\alpha)} + c_2\,\mathbf{e}_2^{(\alpha)} + \ldots + c_{g_\alpha}\,\mathbf{e}_{g_\alpha}^{(\alpha)}$$

where $\mathbf{e}_1^{(\alpha)}, \mathbf{e}_2^{(\alpha)}, \ldots$ are by definition partners of $\mathbf{e}_i^{(\alpha)}$ in a representation space which is closed under the group operations. Any basis vector $\mathbf{e}_j^{(\alpha)}$ (with non-zero coefficient) can then be projected out by using the $j$th idempotent. With one or more choices of element $R$ all basis vectors may be obtained and subsequently normalized.

EXAMPLE 5.12. Suppose $\mathbf{e}_1^{(3)}$ has been generated as in example 5.10, using the first idempotent of $D_3$. We now construct its partner $\mathbf{e}_2^{(3)}$ *without* using $\rho_{21}^{(3)}$. To do this we apply, say, $C_3$ to $\mathbf{e}_1^{(3)}$ (unnormalized)

$$C_3(2\mathbf{e}_1 - \mathbf{e}_2 - \mathbf{e}_3) = (2\mathbf{e}_2 - \mathbf{e}_3 - \mathbf{e}_1) = v'$$

and then use the second idempotent to project out $\mathbf{e}_2^{(3)}$. Thus

$$\rho_{22}^{(3)} = \tfrac{1}{3}(E - \tfrac{1}{2}C_3 - \tfrac{1}{2}\bar{C}_3 - \sigma_1 + \tfrac{1}{2}\sigma_2 + \tfrac{1}{2}\sigma_3)$$

and

$$\rho_{22}^{(3)}\mathbf{e}_1 = \tfrac{1}{3}(\mathbf{e}_1 - \tfrac{1}{2}\mathbf{e}_2 - \tfrac{1}{2}\mathbf{e}_3 - \mathbf{e}_1 + \tfrac{1}{2}\mathbf{e}_3 + \tfrac{1}{2}\mathbf{e}_2) = 0$$

$$\rho_{22}^{(3)}\mathbf{e}_2 = \tfrac{1}{3}(\mathbf{e}_2 - \tfrac{1}{2}\mathbf{e}_1 - \tfrac{1}{2}\mathbf{e}_3 - \mathbf{e}_2 + \tfrac{1}{2}\mathbf{e}_3 + \tfrac{1}{2}\mathbf{e}_1) = 0$$

$$\rho_{22}^{(3)}\mathbf{e}_3 = \tfrac{1}{3}(\mathbf{e}_3 - \tfrac{1}{2}\mathbf{e}_1 - \tfrac{1}{2}\mathbf{e}_2 - \mathbf{e}_2 + \tfrac{1}{2}\mathbf{e}_1 + \tfrac{1}{2}\mathbf{e}_3) = \tfrac{1}{2}(\mathbf{e}_3 - \mathbf{e}_2)$$

Consequently

$$\rho_{22}^{(3)}v' = \tfrac{1}{2}(\mathbf{e}_2 - \mathbf{e}_3)$$

and on normalizing we obtain $\mathbf{e}_2$ as in example 5.10.

### 5.9. The Direct Product

It was noted in section 5.2 that reduction is the converse of a process in which two representations are put together to give a third —the *direct sum*. There is another, less trivial, way of combining two representations, in which we form the *direct product*. The origin of this concept is perhaps best indicated by an example.

Example 5.13. We recall from example 2.8 that a set of *functions* may provide a basis for a representation of a symmetry group: such functions occur when we solve the Schrödinger equation for an atomic or molecular system and are discussed in more detail in chapter 7 (also Ref. 3). For one electron in the presence of two nuclei (lying, say, on the $z$ axis, at points $z = \pm a$) solutions occur in pairs, each pair being of the form

$$\psi_1(x, y, z) = xf(x^2 + y^2, z), \qquad \psi_2(x, y, z) = yf(x^2 + y^2, z)$$

($x, y, z$ being electron coordinates), both corresponding to the same electronic energy, $E$ say. Under rotations about the $z$ axis, $f(x^2 + y^2, z)$ is invariant and the functions therefore behave exactly like $\phi_1$ and $\phi_2$ in example 2.8. They span a space which provides a representation of the group of rotations about the axis—and with $\mathsf{R}_\phi$ we associate a matrix:

$$\mathsf{R}_\phi(\psi_1\ \psi_2) = (\psi_1\ \psi_2)\begin{pmatrix} \cos\phi & -\sin\phi \\ \sin\phi & \cos\phi \end{pmatrix}$$

Now it is shown in textbooks on quantum theory (Ref. 3) that the wave functions describing *two* electrons (with coordinates $x_1, y_1, z_1$ and $x_2, y_2, z_2$, say) may be built up from *products* of one-electron functions. If, in fact, electrons were non-interacting, the four products

$$\psi_1(x_1, y_1, z_1)\psi_1(x_2, y_2, z_2), \qquad \psi_1(x_1, y_1, z_1)\psi_2(x_2, y_2, z_2),$$
$$\psi_2(x_1, y_1, z_1)\psi_1(x_2, y_2, z_2), \qquad \psi_2(x_1, y_1, z_1)\psi_2(x_2, y_2, z_2)$$

would each describe a two-electron state in which each electron was in one of the states $\psi_1$ or $\psi_2$, the total electronic energy being $2E$. Evidently, we shall be interested in three distinct function spaces:

(i) that spanned by functions $\psi_1(x_1, y_1, z_1)$, $\psi_2(x_1, y_1, z_1)$—describing electron 1 with energy $E$;

(ii) that spanned by functions $\psi_1(x_2, y_2, z_2)$, $\psi_2(x_2, y_2, z_2)$—describing electron 2 with energy $E$;

(iii) that spanned by the four products, $\psi_i(x_1, y_1, z_1)$, $\psi_j(x_2, y_2, z_2)$, $(i, j = 1, 2)$—describing (with neglect of interaction) the two electrons with total energy $2E$.

The third space is the direct product of the other two. If we dispense with the variables, indicating functions of the first two spaces by superscript $\alpha$ or $\beta$ and those of the third space by superscript $\alpha\beta$, the three function spaces would be

$$\{\psi_i^{(\alpha)}\}, \qquad \{\psi_i^{(\beta)}\}, \qquad \{\psi_{ij}^{(\alpha\beta)}\} \qquad (i, j = 1, 2)$$

where we have used a single function symbol with a double subscript to describe each of the four products: $\psi_{ij}^{(\alpha\beta)} = \psi_i^{(\alpha)}\psi_j^{(\beta)}$.

Suppose now, quite generally, that we have two spaces $\mathsf{V}^{(\alpha)}$ and $\mathsf{V}^{(\beta)}$, spanned by basis vectors

$$\{\mathsf{e}_1^{(\alpha)}, \mathsf{e}_2^{(\alpha)}, \ldots, \mathsf{e}_{g_\alpha}^{(\alpha)}\}, \qquad \{\mathsf{e}_1^{(\beta)}, \mathsf{e}_2^{(\beta)}, \ldots, \mathsf{e}_{g_\beta}^{(\beta)}\}$$

and that these carry representations $\mathrm{D}_\alpha$ and $\mathrm{D}_\beta$, respectively, of a group $\mathsf{G}$, so that

$$R \to R^{(\alpha)}(\text{a mapping in } \mathsf{V}^{(\alpha)}), \qquad R \to R^{(\beta)}(\text{a mapping in } \mathsf{V}^{(\beta)}) \quad (5.9.1)$$

And consider the set of $g_\alpha \times g_\beta$ basis vector *pairs*

$$\{\mathsf{e}_i^{(\alpha)}\mathsf{e}_j^{(\beta)}\} \quad (i = 1, 2, \ldots g_\alpha;\ j = 1, 2, \ldots g_\beta)$$

in which the "product" is interpreted commutatively, $\mathsf{e}_j^{(\beta)}\mathsf{e}_i^{(\alpha)}$ not being regarded as distinct from $\mathsf{e}_i^{(\alpha)}\mathsf{e}_j^{(\beta)}$. These entities may be regarded as the basis vectors of a new space, of $g_\alpha g_\beta$ dimensions, and are conveniently labelled using a double-subscript notation, just as in example 5.13:

$$\left.\begin{aligned} &\{\mathsf{e}_{11}^{(\alpha\beta)}, \quad \mathsf{e}_{12}^{(\alpha\beta)}, \quad \ldots, \quad \mathsf{e}_{ij}^{(\alpha\beta)}, \quad \ldots, \quad \mathsf{e}_{g_\alpha g_\beta}^{(\alpha\beta)}\} \\ \text{where} \quad &\mathsf{e}_{ij}^{(\alpha\beta)} = \mathsf{e}_i^{(\alpha)}\mathsf{e}_j^{(\beta)} \end{aligned}\right\} \quad (5.9.2)$$

The space so defined is called the *direct product* of those which carry the representations $\mathrm{D}_\alpha$ and $\mathrm{D}_\beta$. Its value, from the point of view of group theory, is revealed by the following example.

EXAMPLE 5.14. If we rotate the system considered in example 5.13, the functions $\{\psi_i^{(\alpha)}\}$ and $\{\psi_i^{(\beta)}\}$ will suffer a common transformation

$$R_\phi^{(\alpha)}(\psi_1^{(\alpha)}\psi_2^{(\alpha)}) = (\psi_1^{(\alpha)}\psi_2^{(\alpha)})\begin{pmatrix} c & -s \\ s & c \end{pmatrix}, \qquad R_\phi^{(\beta)}(\psi_1^{(\beta)}\psi_2^{(\beta)}) = (\psi_1^{(\beta)}\psi_2^{(\beta)})\begin{pmatrix} c & -s \\ s & c \end{pmatrix}$$

where, strictly, we must add the superscript to $R_\phi$ to indicate the space in which it operates, but where $R_\phi^{(\alpha)}$ and $R_\phi^{(\beta)}$ are both associated with the same physical rotation $R_\phi$. Naturally, we are interested also in the behaviour of the products $\{\psi_i^{(\alpha)}\psi_j^{(\beta)}\}$, which span a product space $\mathsf{V}^{(\alpha\beta)}$, for these describe the whole *two*-electron system. Rotation of the system induces a corresponding rotation in $\mathsf{V}^{(\alpha\beta)}$ which may be written $R_\phi^{(\alpha\beta)} = R_\phi^{(\alpha)}R_\phi^{(\beta)}$, the first factor describing the rotation of $\psi_1^{(\alpha)}$ and $\psi_2^{(\alpha)}$ and the second that of $\psi_1^{(\beta)}$ and $\psi_2^{(\beta)}$. Thus, for example,

$$\begin{aligned} R_\phi^{(\alpha\beta)}(\psi_1^{(\alpha)}\psi_2^{(\beta)}) &= (c\psi_1^{(\alpha)} + s\psi_2^{(\alpha)})(-s\psi_1^{(\beta)} + c\psi_2^{(\beta)}) \\ &= -cs\psi_1^{(\alpha)}\psi_1^{(\beta)} + c^2\psi_1^{(\alpha)}\psi_2^{(\beta)} - s^2\psi_2^{(\alpha)}\psi_1^{(\beta)} + sc\psi_2^{(\alpha)}\psi_2^{(\beta)} \end{aligned}$$

On denoting the product-space vectors by $\psi_{ij}^{(\alpha\beta)} = \psi_i^{(\alpha)}\psi_j^{(\beta)}$, we see

$$R_\phi^{(\alpha\beta)}\psi_{12}^{(\alpha\beta)} = -cs\psi_{11}^{(\alpha\beta)} + c^2\psi_{12}^{(\alpha\beta)} - s^2\psi_{21}^{(\alpha\beta)} + sc\psi_{22}^{(\alpha\beta)}$$

and by examining the effect of rotation on all four basis vectors we obtain a 4-dimensional representation. The mapping $R_\phi^{(\alpha\beta)}$ is associated with the physical rotation $R_\phi$, and with $R_\phi^{(\alpha\beta)}$ we associate a matrix

$$R_\phi^{(\alpha\beta)}\boldsymbol{\psi}^{(\alpha\beta)} = \boldsymbol{\psi}^{(\alpha\beta)}\begin{pmatrix} c^2 & -cs & -cs & s^2 \\ cs & c^2 & -s^2 & -cs \\ cs & -s^2 & c^2 & -cs \\ s^2 & cs & cs & c^2 \end{pmatrix}$$

where, as usual, the *row* matrix of basis vectors is abbreviated to a single bold symbol. The square matrix is $\boldsymbol{D}_{\alpha\beta}(R_\phi)$, the matrix associated with rotation $R_\phi$ in the space spanned by the functions $\{\psi_{ij}^{(\alpha\beta)}\}$ : in this way then we can obtain a new representation of the group of operations $\{R_\phi\}$. It will be recognized that this is essentially the device used in section 4.1—though here we are considering not vector components but the basis vectors themselves.

Again we express the content of the example in general form. We suppose that $R^{(\alpha)}$ is the mapping associated with a group element $R$ in one space $V^{(\alpha)}$, and that $R^{(\beta)}$ is that associated with $R$ in space $V^{(\beta)}$. We formally define the mapping associated with $R$ in the product space $V^{(\alpha\beta)}$, by

$$R^{(\alpha\beta)} = R^{(\alpha)}R^{(\beta)}$$

where the symbols are to be interpreted commutatively, $R^{(\alpha)}$ working only on the $e_i^{(\alpha)}$ and $R^{(\beta)}$ only on the $e_i^{(\beta)}$. To associate a matrix $\boldsymbol{R}^{(\alpha\beta)}$ with the mapping $R^{(\alpha\beta)}$ in product space we observe that

$$R^{(\alpha\beta)}e_{ij}^{(\alpha\beta)} = R^{(\alpha)}e_i^{(\alpha)}R^{(\beta)}e_j^{(\beta)} = \sum_{k,l} e_k^{(\alpha)}R_{ki}^{(\alpha)}e_l^{(\beta)}R_{lj}^{(\beta)} = \sum_{k,l} e_{kl}^{(\alpha\beta)}R_{kl,ij}^{(\alpha\beta)}$$

where the elements of $\boldsymbol{R}^{(\alpha\beta)}$ are most naturally labelled using double subscripts and the $kl$, $ij$-element is

$$R_{kl,ij}^{(\alpha\beta)} = R_{ki}^{(\alpha)}R_{lj}^{(\beta)} \tag{5.9.3}$$

The matrix element of the product-space mapping is a product of matrix elements of the individual mappings : the matrix with elements defined by (5.9.3) is called the *direct* or *outer product* of matrices $\boldsymbol{R}^{(\alpha)}$ and $\boldsymbol{R}^{(\beta)}$ and is written

$$\boldsymbol{R}^{(\alpha\beta)} = \boldsymbol{R}^{(\alpha)} \times \boldsymbol{R}^{(\beta)} \tag{5.9.4}$$

In this way we distinguish three representations of the group whose general element is $\mathsf{R}$:

(i) $\mathrm{D}_\alpha$ in which $\mathsf{R} \rightarrow \boldsymbol{R}^{(\alpha)} = \boldsymbol{D}_\alpha(\mathsf{R})$
(ii) $\mathrm{D}_\beta$ in which $\mathsf{R} \rightarrow \boldsymbol{R}^{(\beta)} = \boldsymbol{D}_\beta(\mathsf{R})$
(iii) $\mathrm{D}_{\alpha\beta}$ in which $\mathsf{R} \rightarrow \boldsymbol{D}_{\alpha\beta}(\mathsf{R}) = \boldsymbol{D}_\alpha(\mathsf{R}) \times \boldsymbol{D}_\beta(\mathsf{R})$

The representation $\mathrm{D}_{\alpha\beta}$ is called the *direct* or *Kronecker product* of representations $\mathrm{D}_\alpha$ and $\mathrm{D}_\beta$ and is written $\mathrm{D}_{\alpha\beta} = \mathrm{D}_\alpha \times \mathrm{D}_\beta$. In example 5.14, the two representations $\mathrm{D}_\alpha$ and $\mathrm{D}_\beta$ were identical, but in section 4.1 we met instances in which $\mathrm{D}_\alpha$ and $\mathrm{D}_\beta$ even differed in dimension (one being uni-dimensional), and there are in fact no restrictions on $g_\alpha$ and $g_\beta$. To summarize:

> **From any two representations of a group $\mathsf{G}$, $\mathrm{D}_\alpha$ and $\mathrm{D}_\beta$ (of dimension $g_\alpha$ and $g_\beta$, respectively), we can construct a third, the direct product $\mathrm{D}_{\alpha\beta} = \mathrm{D}_\alpha \times \mathrm{D}_\beta$ (of dimension $g_{\alpha\beta} = g_\alpha g_\beta$), by forming the outer product of corresponding matrices:**
>
> $$\boldsymbol{D}_{\alpha\beta}(\mathsf{R}) = \boldsymbol{D}_\alpha(\mathsf{R}) \times \boldsymbol{D}_\beta(\mathsf{R})$$ (5.9.5)

The outer products are readily written down. If, for example, we have two $2 \times 2$ matrices, $\boldsymbol{A}$ and $\boldsymbol{B}$, their $4 \times 4$ outer product is

$$\boldsymbol{A} \times \boldsymbol{B} = \begin{array}{c} \\ 11 \\ 12 \\ 21 \\ 22 \end{array} \!\!\begin{array}{c} \begin{array}{cccc} 11 & 12 & 21 & 22 \end{array} \\ \left(\begin{array}{cc|cc} A_{11}B_{11} & A_{11}B_{12} & A_{12}B_{11} & A_{12}B_{12} \\ A_{11}B_{21} & A_{11}B_{22} & A_{12}B_{21} & A_{12}B_{22} \\ \hline A_{21}B_{11} & A_{21}B_{12} & A_{22}B_{11} & A_{22}B_{12} \\ A_{21}B_{21} & A_{21}B_{22} & A_{22}B_{21} & A_{22}B_{22} \end{array}\right) \end{array} = \left(\begin{array}{c|c} A_{11}\boldsymbol{B} & A_{12}\boldsymbol{B} \\ \hline A_{21}\boldsymbol{B} & A_{22}\boldsymbol{B} \end{array}\right)$$

Generally, the *first* letter of each double subscript labels a block, in the partitioned product, the *second* letter indicating a row or column *within* that block. Since $(\boldsymbol{A} \times \boldsymbol{B})_{ij,kl} = A_{ik}B_{jl}$ the elements of the $i, k$-block have a common factor $A_{ik}$ multiplying the elements of $\boldsymbol{B}$. The matrix in example 5.14 could evidently have been written down by inspection, using this simple rule. An important property of the outer product, following directly from the definition, is that

$$(\boldsymbol{AB}) \times (\boldsymbol{CD}) = (\boldsymbol{A} \times \boldsymbol{C})(\boldsymbol{B} \times \boldsymbol{D}) \tag{5.9.6}$$

where absence of the $\times$ indicates ordinary matrix multiplication. From this property it is easy to prove directly that the outer products formed from two representations do form another representation:

for if $RS = T$ implies $\boldsymbol{R}^{(\alpha)}\boldsymbol{S}^{(\alpha)} = \boldsymbol{T}^{(\alpha)}$ and $\boldsymbol{R}^{(\beta)}\boldsymbol{S}^{(\beta)} = \boldsymbol{T}^{(\beta)}$, it follows that

$$(\boldsymbol{R}^{(\alpha)} \times \boldsymbol{R}^{(\beta)})(\boldsymbol{S}^{(\alpha)} \times \boldsymbol{S}^{(\beta)}) = (\boldsymbol{R}^{(\alpha)}\boldsymbol{S}^{(\alpha)}) \times (\boldsymbol{R}^{(\beta)}\boldsymbol{S}^{(\beta)}) = (\boldsymbol{T}^{(\alpha)} \times \boldsymbol{T}^{(\beta)})$$

and consequently the outer products also have the same multiplication table as the group elements.

Finally, we note that the direct product of two representations is *not*, in general, irreducible. This simply means that we can find linear combinations of product space vectors (e.g. of the product functions in example 5.12) which bring about a reduction—and which therefore transform among themselves more simply, like the basis vectors of the various irreducible representations.

EXAMPLE 5.15. Let us consider the vector $\psi_{11}^{(\alpha\beta)} + \psi_{22}^{(\alpha\beta)}$ in the product space of example 5.14. Under the mapping induced by rotation $R_\phi$ this goes over into

$$\begin{aligned} R_\phi^{(\alpha\beta)}(\psi_{11}^{(\alpha\beta)} + \psi_{22}^{(\alpha\beta)}) &= c^2\psi_{11}^{(\alpha\beta)} + cs\psi_{12}^{(\alpha\beta)} + cs\psi_{21}^{(\alpha\beta)} + s^2\psi_{22}^{(\alpha\beta)} \\ &\quad + s^2\psi_{11}^{(\alpha\beta)} - cs\psi_{12}^{(\alpha\beta)} - cs\psi_{21}^{(\alpha\beta)} + c^2\psi_{22}^{(\alpha\beta)} \\ &= (c^2 + s^2)(\psi_{11}^{(\alpha\beta)} + \psi_{22}^{(\alpha\beta)}) = (\psi_{11}^{(\alpha\beta)} + \psi_{22}^{(\alpha\beta)}) \end{aligned}$$

Thus, the linear combination $(\psi_{11}^{(\alpha\beta)} + \psi_{22}^{(\alpha\beta)})$ carries an irreducible representation—the identity.

In later chapters we shall often be concerned with the reduction of the direct product of two irreducible representations; and it is therefore of interest to ask how many times each irreducible representation occurs in the direct product. This question is easily answered by taking the traces of the matrices, using the definition (5.9.3). Thus

$$\chi_{\alpha\beta}(R) = \operatorname{tr} \boldsymbol{R}^{(\alpha\beta)} = \sum_{i,j} R_{ij,ij}^{(\alpha\beta)} = \sum_{i,j} R_{ii}^{(\alpha)} R_{jj}^{(\beta)} = \chi_\alpha(R)\chi_\beta(R)$$

> The character of an element in the direct product representation is the product of its characters in the individual representations. (5.9.7)

The number of occurrences of each irreducible representation then follows from (5.4.5).

EXAMPLE 5.16. We cannot apply (5.4.5) directly to example 5.14 because the rotations form a continuous group (infinite order): but we can consider a finite subgroup $C_n$ and ultimately let $n \to \infty$. Thus, in the product representation of example 5.14,

$$\chi_{\alpha\beta}(R_\phi) = 4\ \cos^2(2\pi k/n)$$

To find, for example, the number of occurrences of the identity representation, we have from (5.4.5)

$$\begin{aligned} n_A &= \lim_{n\to\infty} \frac{1}{n} \sum_{k=1}^{n} 4\cos^2(2\pi k/n) \\ &= \lim_{n\to\infty} \frac{1}{n} \sum_{k=1}^{n} (2 + 2\cos(4\pi k/n)) \\ &= \lim_{n\to\infty} \frac{1}{n}(2n) \qquad (n\text{-independent}) \\ &= 2 \end{aligned}$$

One of the functions carrying the identity representation is $(\psi_{11}^{(\alpha\beta)} + \psi_{22}^{(\alpha\beta)})$ (example 5.14). The other is easily seen to be $(\psi_{12}^{(\alpha\beta)} - \psi_{21}^{(\alpha\beta)})$. In the original context of example 5.13 the two functions are

$$\psi_1(x_1, y_1, z_1)\,\psi_1(x_2, y_2, z_2) + \psi_2(x_1, y_1, z_1)\,\psi_2(x_2, y_2, z_2) = (x_1x_2 + y_1y_2) \times \text{invariant}$$

$$\psi_1(x_1, y_1, z_1)\,\psi_2(x_2, y_2, z_2) - \psi_2(x_1, y_1, z_1)\,\psi_1(x_2, y_2, z_2) = (x_1y_2 - y_1x_2) \times \text{invariant}$$

## REFERENCES

1. WIGNER, E. P., *Group Theory and its Application to the Quantum Mechanics of Atomic Spectra*, pp. 72–87, Academic Press, 1959.
2. HALL, G. G., *Matrices and Tensors*, p. 58, Pergamon Press, 1963.
3. PAULING, L., and WILSON, E. B., *Introduction to Quantum Mechanics*, chapter 8, McGraw-Hill, 1935.

## BIBLIOGRAPHY

The following books establish, using a considerable variety of methods, all the basic properties of irreducible representations of finite groups.

BAUER, E., *Introduction à la Théorie des Groupes et ses Applications à la Physique Quantique*, Presses Universitaires de France, 1933.

LYUBARSKII, G. Ya, *The Application of Group Theory in Physics*, Pergamon Press, 1960. (Translated from the Russian by S. Dedijer.)

SPEISER, A., *Die Theorie der Gruppen von Endlicher Ordnung*, 3rd Ed., Springer, 1937.

VAN DER WAERDEN, B. L., *Die Gruppentheoretische Methode in der Quantenmechanik*, Springer, 1932.

WIGNER, E. P., *Gruppentheorie und ihre Anwendung auf die Quantenmechanik der Atomspektren*, Vieweg, 1931. For a revised and enlarged edition, in English, see Ref. 1 (above).

CHAPTER 6

# APPLICATIONS INVOLVING ALGEBRAIC FORMS

## 6.1. Nature of Applications

The symmetry of certain physical systems has been discussed from a purely geometrical standpoint in earlier chapters. In many cases, however, the condition of a physical system is described algebraically or analytically and the existence of symmetry greatly restricts the mathematical form of such descriptions.

Suppose, for example, that a system is polarized by a uniform electric field $X$ with Cartesian components $X_1$, $X_2$, $X_3$. An electric moment $Y$ is set up with components $Y_1$, $Y_2$, $Y_3$ given by

$$\boldsymbol{Y} = \boldsymbol{k}\boldsymbol{X} \tag{6.1.1}$$

where $\boldsymbol{Y}$ and $\boldsymbol{X}$ are, as usual, columns of components and $\boldsymbol{k}$ is a symmetric $3 \times 3$ matrix describing the *polarizability* of the system. The polarization *energy* is then a scalar product and is given in terms of components by

$$2E = Y.X = \tilde{\boldsymbol{Y}}\boldsymbol{X} = \tilde{\boldsymbol{X}}\boldsymbol{k}\boldsymbol{X} = \sum_{i,j} X_i k_{ij} X_j \tag{6.1.2}$$

The energy is thus a *quadratic form* in a set of vector components, the nature of the system being completely described by the coefficients in this form collected into the matrix $\boldsymbol{k}$. If the system is symmetrical, restrictions will be placed upon the form of $\boldsymbol{k}$. We shall find later, for example, that for a system with tetrahedral or cubic symmetry $\boldsymbol{k} = k\mathbf{1}_3$ so that the polarization vector can only be a scalar multiple of the field vector; $Y = kX$. We shall therefore be interested in the symmetry properties of *algebraic forms*. More general situations, in which $\boldsymbol{k}$ describes an arbitrary *tensor* property, are also of great importance and are discussed in chapter 8. Here we shall consider only simple quadratic forms such as (6.1.2).

On the other hand, the temperature inside a continuous solid is a continuous function of position, and if the solid has a symmetrical form and symmetrical boundary conditions the temperature function will be expected to exhibit a similar symmetry. We shall therefore be interested also in the symmetry properties of *functions*.

In this chapter we shall illustrate the application of group theory to problems involving quadratic forms, and to the related eigenvalue problem. Some of the most fruitful of these applications occur in molecular vibration theory. In the next chapter we shall return to the symmetry properties of functions, indicating some of the applications in quantum mechanics. The reader interested mainly in applications of that kind may pass directly to chapter 7.

### 6.2. Invariant Forms. Symmetry Restrictions

First we take up the study of forms such as (6.1.2) which may be regarded as a special case of the *Hermitian form*

$$f = \boldsymbol{X}^\dagger \boldsymbol{F} \boldsymbol{X} \tag{6.2.1}$$

Here $\boldsymbol{X}$ is a column of components and $\boldsymbol{F}^\dagger = \boldsymbol{F}$. For 3-dimensional vectors with real components $\boldsymbol{X}^\dagger = \tilde{\boldsymbol{X}}$ and (6.2.1) is equivalent to (6.1.2). To be specific we consider the case where $\boldsymbol{X}$ represents a field vector, $\boldsymbol{F}$ describes a property of the system (e.g. the polarizability of a molecule or crystal) and $f$ is a field-dependent scalar quantity (e.g. energy). If we now rotate both the system and the applied field, $f$ (which is a physical quantity determined by the *relationship* between system and field) must be unchanged, but $\boldsymbol{F}$ will have to be replaced by a new matrix, $\boldsymbol{F}'$, giving us a *new form*. If, in fact, under a rotation $R$, $\boldsymbol{X} \to \boldsymbol{X}' = \boldsymbol{R}\boldsymbol{X}$, then we can say

$$f = \boldsymbol{X}'^\dagger \boldsymbol{F}' \boldsymbol{X}' = \boldsymbol{X}^\dagger \boldsymbol{R}^\dagger \boldsymbol{F}' \boldsymbol{R} \boldsymbol{X} = \boldsymbol{X}^\dagger \boldsymbol{F} \boldsymbol{X} \tag{6.2.2}$$

and this requires

$$\boldsymbol{R}^\dagger \boldsymbol{F}' \boldsymbol{R} = \boldsymbol{F} \tag{6.2.3}$$

Now if the rotation described by $\boldsymbol{R}$ brings the physical system into self-coincidence, then $\boldsymbol{F}'$, which relates the energy of the rotated system to an arbitrary field vector, cannot be distinguished from $\boldsymbol{F}$. In this case the form (6.2.1), determined by the set of coefficients collected in $\boldsymbol{F}$, is said to be *invariant under the rotation* $R$. It is important to distinguish the two types of invariance. Scalar quantities (such as the energy of the system) remain unchanged unless the relationship between the system and its environment is disturbed. This "numerical" invariance is (for present purposes) trivial; a common rotation of system and environment is formally equivalent to a change of coordinates throughout and corresponds merely to a new mathematical description of the same physical situation. But the forms which describe intrinsic properties of the system itself (e.g. its electric polarizability) are invariant only under the *special* rotations which bring the system into self-coincidence. The equation $\boldsymbol{F}' = \boldsymbol{F}$ or, from (6.2.3)

$$\boldsymbol{F} = \boldsymbol{R}^\dagger \boldsymbol{F} \boldsymbol{R} \tag{6.2.4}$$

thus describes a symmetry property. Since the symmetry operations form a group, $\mathsf{G} = \{\mathsf{A}, \mathsf{B}, \ldots, \mathsf{R}, \ldots\}$, and $\mathrm{D} = \{\boldsymbol{A}, \boldsymbol{B}, \ldots, \boldsymbol{R}, \ldots\}$ is a representation, $\boldsymbol{F}$ will in general be invariant in the sense of (6.2.4) under a whole group of transformations. We may, without loss of generality, assume an orthonormal coordinate basis in which $\boldsymbol{A}, \boldsymbol{B}, \ldots, \boldsymbol{R}, \ldots$ are unitary: and in this case (6.2.4) becomes, since $\boldsymbol{R}^\dagger = \boldsymbol{R}^{-1}$,

$$\boldsymbol{R}\boldsymbol{F} = \boldsymbol{F}\boldsymbol{R} \tag{6.2.5}$$

Before considering the general case, it is worth illustrating the implications of (6.2.5) by means of two simple examples:

EXAMPLE 6.1. Suppose $\boldsymbol{F}$ describes the polarizability of a *tetrahedral* molecule, $\boldsymbol{X}$ being a set of field components. In a rotation $\mathsf{R}$ which sends the basis vectors $\mathbf{e}$ into $\mathsf{R}\mathbf{e} = \mathbf{e}\boldsymbol{R}$ the field vector with components $\boldsymbol{X}$ is sent into a rotated vector with components $\boldsymbol{R}\boldsymbol{X}$. Now for a tetrahedral system (e.g. the methane molecule) the basis $\mathbf{e}$ carries the *irreducible representation* T given in Table 4.18. Hence $\boldsymbol{F}\boldsymbol{R} = \boldsymbol{R}\boldsymbol{F}$ for every matrix $\boldsymbol{R}$ of an *irreducible* collection. But (5.3.4) then allows us to conclude that $\boldsymbol{F}$ is a multiple of the unit matrix; $\boldsymbol{F} = k\mathbf{1}_3$. Thus, although the symmetric $3 \times 3$ matrix $\boldsymbol{F}$ apparently contains six independent elements, the existence of tetrahedral symmetry requires that the polarizability is described by a *single* parameter $k$. The polarization vector $\boldsymbol{Y} = \boldsymbol{F}\boldsymbol{X}$ is thus merely a scalar multiple of the field vector and the energy is proportional to the squared magnitude of the field vector

$$2E = k(X_1^2 + X_2^2 + X_3^2) = k\,|\mathsf{X}|^2$$

The quadratic form (6.2.1) thus necessarily reduces to a sum of squares. Inspection of the other tables in chapter 4 shows that this result applies to all systems with cubic symmetry (point groups $\mathsf{T}$, $\mathsf{T}_d$, $\mathsf{T}_h$, $\mathsf{O}$, $\mathsf{O}_h$); and since the group considered may be merely a subgroup of a higher symmetry group (e.g. a space group) the result applies to all *crystals* whose point groups contain $\mathsf{T}$ or $\mathsf{O}$ subgroups. Systems of this kind, in which the " proportionality factor " relating vectors $\boldsymbol{Y}$ and $\boldsymbol{X}$ is simply a scalar multiplier, are said to be *isotropic*.

EXAMPLE 6.2. Suppose now that $\boldsymbol{F}$ refers to a system which has $C_{3v}$ symmetry (or, in the case of a trigonal crystal, contains a $C_{3v}$ subgroup). The matrices $\{\boldsymbol{A}, \boldsymbol{B}, \ldots, \boldsymbol{R}, \ldots\}$ describing the rotation of the basis are then reducible: if the basis is chosen as in Fig. 5.1, the matrices take the form $\left(\begin{array}{c|c} \boldsymbol{R}_{11} & \mathbf{0} \\ \hline \mathbf{0} & \boldsymbol{R}_{22} \end{array}\right)$ where $\boldsymbol{R}_{11}$, $\boldsymbol{R}_{22}$ yield the representations $D_3$ and $D_1$ (respectively) of (2.7.3). With this choice $\boldsymbol{F}$ may be partitioned in a corresponding way and (6.2.5) then breaks into four matrix equations (one for each block):

$$\boldsymbol{F}_{11}\boldsymbol{R}_{11} = \boldsymbol{R}_{11}\boldsymbol{F}_{11} \qquad \boldsymbol{F}_{12}\boldsymbol{R}_{22} = \boldsymbol{R}_{11}\boldsymbol{F}_{12}$$
$$\boldsymbol{F}_{21}\boldsymbol{R}_{11} = \boldsymbol{R}_{22}\boldsymbol{F}_{21} \qquad \boldsymbol{F}_{22}\boldsymbol{R}_{22} = \boldsymbol{R}_{22}\boldsymbol{F}_{22}$$

Schur's Lemma and its corollary, (5.3.3) (5.3.4), then show that

$$\boldsymbol{F}_{11} = k_1\mathbf{1}_2, \quad \boldsymbol{F}_{12} = \mathbf{0}, \quad \boldsymbol{F}_{21} = \mathbf{0}, \quad \boldsymbol{F}_{22} = k_2\mathbf{1}_1$$

since the relationship holds for all corresponding matrices, $\boldsymbol{R}_{11}$ and $\boldsymbol{R}_{22}$, of two irreducible collections. The proportionality matrix $\boldsymbol{F}$ therefore depends on *two* independent parameters, and the form itself becomes the *weighted* sum of squares

$$2E = k_1(X_1^2 + X_2^2) + k_2X_3^2$$

Molecules such as $NH_3$ and crystals such as graphite are therefore *anisotropic*: $\boldsymbol{Y} = \boldsymbol{F}\boldsymbol{X}$ reduces to simple proportionality when $\mathsf{X}$ is along the principal symmetry axis ($X_1 = X_2 = 0$, proportionality constant $k_2$) and when $\mathsf{X}$ lies in a plane *normal* to the axis ($X_3 = 0$, proportionality constant $k_1$), but in general $\mathsf{Y}$ will *not* be merely a scalar multiple of $\mathsf{X}$.

Forms of the kind just considered arise not only in 3 but in $n$ dimensions, as the following example shows.

EXAMPLE 6.3. Suppose the atoms of a diatomic molecule are displaced from their equilibrium positions as in Fig. 6.1(i). The potential energy, relative to that in the equilibrium position, is then given to second order‡ by

$$V = (a_{xx}x_1^2 + a_{yy}y_1^2 + a_{zz}z_1^2 + 2a_{xy}x_1y_1 + \ldots) + \\ + (b_{xx}x_2^2 + \ldots) + (2c_{xx}x_1x_2 + \ldots)$$

If we adopt the more convenient notation $(x_1, y_1, z_1, x_2, y_2, z_2) \rightarrow$

‡ Constant terms are absent if we take $V = 0$ when all displacements are zero: first degree terms are absent if the undisplaced configuration corresponds to *equilibrium* $(\partial V/\partial x_1 = \partial V/\partial y_1 = \ldots = 0)$.

$(X_1, X_2, \ldots, X_6)$ and arrange the coefficients in a symmetric matrix

$$\boldsymbol{F} = \begin{pmatrix} F_{11} & F_{12} & \ldots & F_{16} \\ F_{21} & F_{22} & \ldots & F_{26} \\ \ldots & \ldots & & \ldots \\ F_{61} & F_{62} & \ldots & F_{66} \end{pmatrix} = \begin{pmatrix} a_{xx} & a_{xy} & a_{xz} & c_{xx} & c_{xy} & c_{xz} \\ a_{yx} & a_{yy} & a_{yz} & c_{yx} & c_{yy} & c_{yz} \\ \ldots & \ldots & \ldots & \ldots & \ldots & \ldots \\ c_{zx} & c_{zy} & c_{zz} & b_{zx} & b_{zy} & b_{zz} \end{pmatrix}$$

we can then write

$$V = \tilde{\boldsymbol{X}}\boldsymbol{F}\boldsymbol{X}$$

This is a 6-dimensional quadratic form. To see how the invariance problem arises we submit the deformed system to a rotation $R$ which

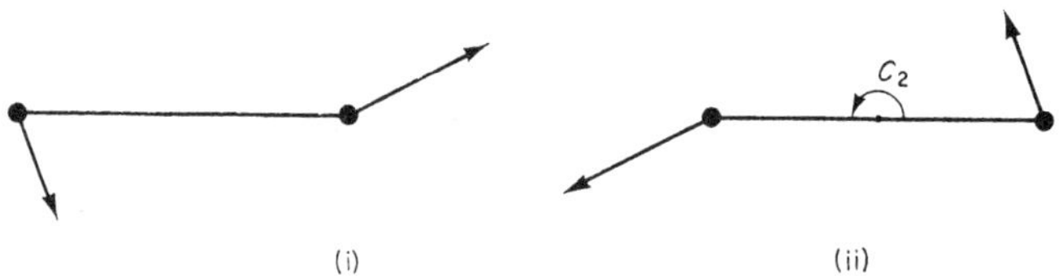


FIG. 6.1. Atomic displacements in moving molecule (i) before rotation (ii) after rotation by $C_2$.

would bring the undeformed molecule into self-coincidence: this yields the configuration shown in Fig. 6.1(ii), with the same energy as that in Fig. 6.1(i) (since the relationship of the parts of the system is undisturbed by the common rotation). But this new configuration may be achieved alternatively by applying *a new set of displacements* to the *unrotated* molecule (i). The new set of displacements is

$$\begin{pmatrix} X_1' \\ X_2' \\ \ldots \\ X_6' \end{pmatrix} = \left(\begin{array}{c|c} \mathbf{0} & \boldsymbol{R}_c \\ \hline \boldsymbol{R}_c & \mathbf{0} \end{array}\right) \begin{pmatrix} X_1 \\ X_2 \\ \ldots \\ X_6 \end{pmatrix}$$

where $\boldsymbol{R}_c$ is the matrix describing the change of vector components under a rotation $R$ and the subscript c indicates the basic Cartesian or $(xyz)$-representation which we shall now denote by $D_c$. Thus $R \rightarrow \boldsymbol{D}_c(R) = \boldsymbol{R}_c$. The off-diagonal position of each $\boldsymbol{R}_c$ indicates that the new displacement on centre 1 is obtained by rotating the displacement on centre 2 and then applying it to centre 1, and vice versa (Fig. 6.2). The new displacements $\boldsymbol{X}'$ are therefore defined by a *virtual* operation rather than by a real rotation of the system: this imagined rotation and re-identification of the *displacements alone,* is what is meant by a *symmetry operation* in vibration theory. There is clearly an isomorphism between the symmetry operations on the displacements and the corresponding rotations of the molecular point group: but since the former involve *two* vectors they are described as mappings in a *six*-dimensional space. Once this relationship is understood,

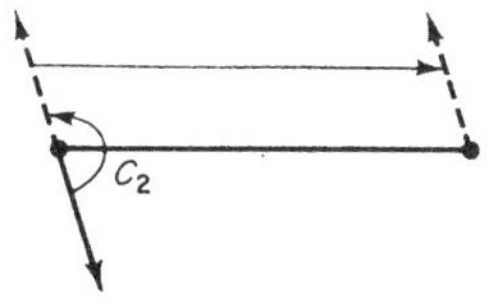


FIG. 6.2. Symmetry operation associated with $C_2$. Each displacement (cf. Fig. 6.1) is rotated by $C_2$ and transferred to another atom—to give a new set of displacements, of the *unrotated* molecule, with the same potential energy as originally.

there is no harm in using the same symbol ($R$) both for a rotation of the *molecule* in real 3-space and for the " induced " rotation in a representation space of (for an $N$-atomic molecule) $3N$ dimensions. If we denote the partitioned matrix of the last equation by $\boldsymbol{R}$, equality of the potential energy of the two configurations requires in general that‡ $V = \mathbf{X}^\dagger \boldsymbol{F} \mathbf{X} = \mathbf{X}^\dagger \boldsymbol{R}^\dagger \boldsymbol{F} \boldsymbol{R} \mathbf{X}$ or, since this is true for arbitrary $\mathbf{X}$

$$\boldsymbol{F} = \boldsymbol{R}^\dagger \boldsymbol{F} \boldsymbol{R}$$

Thus the characteristic property of the symmetry operations is that they leave invariant the form described by $\boldsymbol{F}$. The set of all such operations forms a group—isomorphic with the molecular point-group itself—and the matrices therefore form a representation, which is in fact unitary. We shall call this the " displacement representation," D, so that $R \rightarrow \boldsymbol{D}(R) = \boldsymbol{R}$. In this case $\boldsymbol{F}$ again commutes with all the matrices of the representation:

$$\boldsymbol{R}\boldsymbol{F} = \boldsymbol{F}\boldsymbol{R}$$

The last two equations are formally identical with (6.2.4) and (6.2.5), though the " rotation " here is no longer a simple physical rotation but is a hybrid operation involving both rotation and relabelling of the set of displacement vectors.

We now examine the implications of (6.2.5) when the matrices are $n$-dimensional. By suitable choice of basis, every matrix $\boldsymbol{R}$ can be brought to irreducible block form: but now each irreducible represen-

‡ For generality, we continue to use the dagger—which is equivalent to the tilde when the matrices are real.

tation will in general appear several times. Let $\boldsymbol{D}_{\alpha,\mu}(R)$ denote the $\mu$th irreducible block providing an irreducible representation $D_{\alpha,\mu} = D_\alpha$ (of dimension $g_\alpha$) and assume $\boldsymbol{F}$ partitioned similarly. Then, proceeding as in example 6.2,

$$\left.\begin{aligned} \boldsymbol{F}_{\alpha\mu,\alpha\nu} &= F^{\alpha}_{\mu\nu}\mathbf{1}_{g_\alpha} \text{ (representations } D_{\alpha,\mu}, D_{\alpha,\nu} \text{ identical)} \\ \boldsymbol{F}_{\alpha\mu,\beta\nu} &= \mathbf{0} \text{ (representations } D_{\alpha,\mu}, D_{\beta,\nu} \text{ inequivalent)} \end{aligned}\right\} \tag{6.2.6}$$

Hence, when $\boldsymbol{F}$ is referred to a symmetry adapted basis, off-diagonal elements can occur only between corresponding basis vectors of identical representations. These did not occur in examples 6.1 and 6.2 because no irreducible representation appeared more than once. Suppose, on the other hand, that $D_{1,1} = D_{1,2} = D_1$ (1-dimensional) and $D_{2,1} = D_{2,2} = D_2$ (2-dimensional). Then the reduced form of $\boldsymbol{F}$ would be

$$\begin{bmatrix} F_{11} & F_{12} & & & & \\ F_{21} & F_{22} & & & & \\ & & F_{33} & 0 & F_{34} & 0 \\ & & 0 & F_{33} & 0 & F_{34} \\ & & F_{43} & 0 & F_{44} & 0 \\ & & 0 & F_{43} & 0 & F_{44} \end{bmatrix} \tag{6.2.7}$$

It is often more convenient to rearrange the rows and columns of $\boldsymbol{F}$ so that all those referring to the first basis vector of $D_1$ appear first, then all those referring to the second basis vector, etc. . . . , then all those referring to the first basis vector of $D_2$, etc. The ordering is then (see section 5.7) according to *symmetry species*, $(\alpha, i)$, and the rearranged matrix becomes

$$\begin{bmatrix} F_{11} & F_{12} & & & & \\ F_{21} & F_{22} & & & & \\ & & F_{33} & F_{34} & & \\ & & F_{43} & F_{44} & & \\ & & & & F_{33} & F_{34} \\ & & & & F_{43} & F_{44} \end{bmatrix} \begin{matrix} \left.\vphantom{\begin{matrix}a\\a\end{matrix}}\right\} D_1\text{, first vector. Species (1, 1)} \\ \left.\vphantom{\begin{matrix}a\\a\end{matrix}}\right\} D_2\text{, first vector. Species (2, 1)} \\ \left.\vphantom{\begin{matrix}a\\a\end{matrix}}\right\} D_2\text{, second vector. Species (2, 2)} \end{matrix} \tag{6.2.8}$$

Hence there is only *one* distinct block for each irreducible representation $D_\alpha$ (the same block occurring for *each* of the $g_\alpha$ basis vectors), its dimension being the number of times that $D_\alpha$ appears. The reduction to diagonal form, which in examples 6.1 and 6.2 could evidently be achieved by symmetry considerations alone, is here incomplete: (6.2.8) has *off*-diagonal elements which (like $k_1$ and $k_2$ in example 6.2) depend upon the physical nature of the system.

It is known, however, that the block-form matrix can be brought to true diagonal form by a further unitary transformation. By choosing the axes so as to reduce the representation we are taking an important step towards this diagonal form. Many of the applications of group theory are intimately connected with the diagonalization of forms: we now examine the reduction and its significance in more detail.

### 6.3. Principal Axes. The Eigenvalue Problem

In examples 6.1 and 6.2, the choice of a symmetry adapted co-ordinate system reduced the general form (6.2.1) to

$$f = F_{11}X_1^2 + F_{22}X_2^2 + F_{33}X_3^2 \tag{6.3.1}$$

If we take a fixed value of $f$, $f = 1$ say, the compatible values of $X_1$, $X_2$, $X_3$ define points lying on a second-degree surface or *quadric*. When the $F_{ii}$ are all positive this surface is an ellipsoid and may be written in the standard form

$$\frac{X_1^2}{a_1^2} + \frac{X_2^2}{a_2^2} + \frac{X_3^2}{a_3^2} = 1 \tag{6.3.2}$$

The quantities $a_1$, $a_2$, $a_3$ are the lengths of the three *principal axes*. In these examples, the quadric gives a geometrical picture of the anisotropy of the system. Thus in example 6.1 the quadric is a sphere and indicates equality of length of the principal axes, and hence of the susceptibilities in different directions, corresponding to an isotropic situation. In example 6.2 the quadric is an ellipsoid of rotation with its axis of symmetry (to which $X_3$ is referred) along the $\mathbf{e}_3$ direction ($\bar{\mathbf{e}}_3$ Fig. 5.1)—which is the axis of symmetry of the physical system—as would be expected intuitively. The inequality of length of the principal axes indicates the inequality of the susceptibilities in different directions. Geometrically, then, the diagonalization of a matrix corresponds to a *principal axis transformation* of the quadric which it determines. For a detailed discussion of quadrics in crystal theory the reader is referred elsewhere (Ref. 1).

The need for diagonalization often appears in a different context, which may be illustrated with reference to the molecular vibration problem (Ref. 2).

EXAMPLE 6.4. If $m_1$ and $m_2$ are the masses of the atoms in the diatomic molecule of example 6.3, the kinetic energy is

$$T = \tfrac{1}{2}(m_1\dot{x}_1^2 + m_1\dot{y}_1^2 + \dots + m_2\dot{z}_2^2)$$

On introducing "mass-weighted" coordinates $q_1 = \sqrt{m_1}\, x_1$, $q_2 = \sqrt{m_1}\, y_1, \dots, q_6 = \sqrt{m_2}\, z_2$ this may be written $2T = \sum_{i=1}^{6} \dot{q}_i^2$ or, with a matrix notation,

$$2T = \dot{\boldsymbol{q}}^\dagger \dot{\boldsymbol{q}}$$

which is a quadratic form associated with the unit matrix. In terms of $\boldsymbol{q}$ the potential energy expression (example 6.3) may be written‡

$$2V = \boldsymbol{q}^\dagger \boldsymbol{F} \boldsymbol{q}$$

The last two expressions are evidently quite general. If there are $N$ atoms, as we shall henceforth assume, $\boldsymbol{q}$ and $\dot{\boldsymbol{q}}$ are columns of $3N$ components, and the equations of motion then become‡‡

$$\ddot{q}_i + \sum_{j=1}^{3N} F_{ij} q_j = 0 \qquad (i = 1, 2, \dots, 3N)$$

These may be solved by looking for "normal modes" in which each atom vibrates with the same frequency $\nu = \omega/2\pi$. This amounts to putting $q_j = v_j e^{i\omega t}$ and yields the following simultaneous equations for the "amplitudes" $v_j$:

$$\sum_{j=1}^{3N} (F_{ij} - \delta_{ij}\lambda) v_j = 0 \qquad (\lambda = \omega^2;\ i = 1, 2, \dots, 3N)$$

or in matrix form

$$\boldsymbol{F}\boldsymbol{v} = \lambda \boldsymbol{v}$$

Now it is known from section 2.11 that solution of this eigenvalue equation provides the means of diagonalizing $\boldsymbol{F}$: the eigenvectors are the columns of a matrix $\boldsymbol{V}$ such that

$$\boldsymbol{V}^\dagger \boldsymbol{F} \boldsymbol{V} = \bar{\boldsymbol{F}} = \text{diag}\,(\lambda_1, \lambda_2, \dots, \lambda_{3N})$$

At the same time they determine the relative amplitudes, in each of the $3N$ normal modes, of the coordinates describing the vibration. The corresponding eigenvalues (diagonal elements of $\bar{\boldsymbol{F}}$) determine the

‡ The 2 appears simply because of the conventional definition of force constants: thus, in one dimension, we write $V = \frac{1}{2}kx^2$, $2V = kx^2$.

‡‡ In Lagrangian form, for example, $\dfrac{\partial}{\partial t}\left(\dfrac{\partial L}{\partial \dot{q}_i}\right) - \dfrac{\partial L}{\partial q_i} = 0$ $(i = 1, 2, \dots, 3N)$ where $L = T - V$. The given equations follow on substituting for $T$ and $V$.

frequencies of the normal modes. Finally, we may use the identity

$$2V = \boldsymbol{q}^\dagger \boldsymbol{F} \boldsymbol{q} = \boldsymbol{q}^\dagger \boldsymbol{V} \boldsymbol{V}^\dagger \boldsymbol{F} \boldsymbol{V} \boldsymbol{V}^\dagger \boldsymbol{q} = \boldsymbol{Q}^\dagger \bar{\boldsymbol{F}} \boldsymbol{Q} = \sum_{i=1}^{3N} \lambda_i Q_i^2$$

to introduce the *normal coordinates*, collected in the column $\boldsymbol{Q}$:

$$\boldsymbol{Q} = \boldsymbol{V}^\dagger \boldsymbol{q}, \qquad Q_k = \boldsymbol{v}_k^\dagger \boldsymbol{q}$$

the second result following because the $k$th row of $\boldsymbol{V}^\dagger$ is $\boldsymbol{v}_k^\dagger$. In terms of normal coordinates the kinetic and potential energies both assume a simple sum-of-squares form:

$$2T = \sum_{i=1}^{3N} \dot{Q}_i^2, \qquad 2V = \sum_{i=1}^{3N} \lambda_i Q_i^2$$

We summarize the content of example 6.4 in slightly more complete form. A Hermitian form in $n$ variables $X_1, X_2, \ldots, X_n$,

$$f = \boldsymbol{X}^\dagger \boldsymbol{F} \boldsymbol{X} \tag{6.3.3}$$

involves the components $\boldsymbol{X}$ of a vector $\mathsf{X}$ in $n$-space, and a matrix $\boldsymbol{F}$. The matrix $\boldsymbol{F}$ describes an operation $\mathsf{F}$ which produces from $\mathsf{X}$ a new vector, $\mathsf{Y} = \mathsf{F}\mathsf{X}$, with components $\boldsymbol{Y} = \boldsymbol{F}\boldsymbol{X}$.

Let us denote the (orthonormal) basis, to which we refer the components $\boldsymbol{X}$, by $\mathbf{e} = (\mathrm{e}_1\ \mathrm{e}_2 \ldots, \mathrm{e}_n)$ and allow basis and components to change in the usual way (2.5.6):

$$\mathbf{e} \to \bar{\mathbf{e}} = \mathbf{e}\boldsymbol{V}, \qquad \boldsymbol{X} \to \bar{\boldsymbol{X}} = \boldsymbol{V}^{-1}\boldsymbol{X} = \boldsymbol{V}^\dagger \boldsymbol{X} \tag{6.3.4}$$

The quantity $f$, which may be regarded as the scalar product $\mathsf{X}^*(\mathsf{F}\mathsf{X})$, is then expressed in terms of $\bar{\boldsymbol{X}}$ by

$$f = \boldsymbol{X}^\dagger \boldsymbol{F} \boldsymbol{X} = \bar{\boldsymbol{X}}^\dagger \boldsymbol{V}^\dagger \boldsymbol{F} \boldsymbol{V} \bar{\boldsymbol{X}} = \bar{\boldsymbol{X}}^\dagger \bar{\boldsymbol{F}} \bar{\boldsymbol{X}} \tag{6.3.5}$$

When $\boldsymbol{F}$ is Hermitian, the solutions of the eigenvalue equation

$$\boldsymbol{F}\boldsymbol{v} = \lambda \boldsymbol{v} \tag{6.3.6}$$

may be collected as the columns of a matrix $\boldsymbol{V}$ (see section 2.11) which defines a very special basis change: it is both unitary,

$$\boldsymbol{V}^\dagger \boldsymbol{V} = \boldsymbol{V}\boldsymbol{V}^\dagger = \mathbf{1}_n \tag{6.3.7}$$

and gives the principal axis transformation with

$$\bar{\boldsymbol{F}} = \boldsymbol{V}^\dagger \boldsymbol{F} \boldsymbol{V} = \mathrm{diag}\,(\lambda_1, \lambda_2, \ldots, \lambda_n) \tag{6.3.8}$$

The eigenvectors, which by (6.3.4) are given by

$$\bar{\mathbf{e}} = \mathbf{e}\boldsymbol{V}, \qquad \bar{\mathrm{e}}_i = \mathbf{e}\boldsymbol{v}_i \tag{6.3.9}$$

then define the principal axes of the quadric. An arbitrary vector $\mathsf{X}$

has principal axis components $\bar{X}_i$ such that (again by (6.3.4))

$$\bar{\mathbf{X}} = \mathbf{V}^{\dagger}\mathbf{X}, \qquad \bar{X}_i = \mathbf{v}_i^{\dagger}\mathbf{X} \tag{6.3.10}$$

and (6.3.3) then becomes

$$f = \bar{\mathbf{X}}^{\dagger}\bar{\mathbf{F}}\bar{\mathbf{X}} = \sum_{i=1}^{n} \bar{F}_{ii} \,|\, \bar{X}_i \,|^2 \tag{6.3.11}$$

This allows us to define a standard quadric (cf. (6.3.2)) whose $i$th principal axis is of length $a_i = \bar{F}_{ii}^{-\frac{1}{2}}$ (the inverse square root of the $i$th eigenvalue).

It will be recalled that (6.3.6) is equivalent to a set of simultaneous equations (2.11.3) which are consistent only for values of $\lambda$ which satisfy the determinantal equation (2.11.4). Clearly, therefore, any preliminary reduction of $\mathbf{F}$ to block form (6.2.8), which may be achieved by symmetry considerations, will be of considerable value. For then the system of equations (2.11.3) will break into two or more separate sets, corresponding to the block form of $\mathbf{F}$ in (6.3.6), with a smaller determinant for each set. Before discussing this reduction in more detail, however, we establish a fundamental property of the eigenvectors themselves.

### 6.4. Symmetry Considerations

We consider the effect of a symmetry operation $R$ upon the vectors represented on each side of the eigenvalue equation (6.3.6). If $\mathbf{R}$ is the matrix associated with this operation and we make use of (6.2.5), we obtain

$$\mathbf{R}(\mathbf{F}\mathbf{v}) = \mathbf{F}(\mathbf{R}\mathbf{v}) = \lambda(\mathbf{R}\mathbf{v}) \tag{6.4.1}$$

The rotated eigenvectors therefore satisfy the same eigenvalue equation as they did before rotation, with the same value of $\lambda$:

$$\mathbf{F}(\mathbf{R}\mathbf{v}_i) = \lambda_i(\mathbf{R}\mathbf{v}_i) \tag{6.4.2}$$

If $\mathbf{v}_i$ is any eigenvector of $\mathbf{F}$ with eigenvalue $\lambda(=\lambda_i)$ then so also is $\mathbf{v}_i' = \mathbf{R}\mathbf{v}_i$. Consequently, $\mathbf{R}\mathbf{v}_i$ can only be a linear combination of those linearly independent eigenvectors, $\mathbf{v}_1, \ldots, \mathbf{v}_g$ say, which have the *same* eigenvalue $\lambda$. This is true for each member of the degenerate set with common eigenvalue $\lambda$ and hence these eigenvectors provide a $g$-dimensional representation, D, of the symmetry group:

$$\mathbf{v}_i' = \mathbf{R}\mathbf{v}_i = \sum_{j=1}^{g} \mathbf{v}_j D(R)_{ji} \tag{6.4.3}$$

In general, the representation D may be assumed irreducible. For if it were not we could choose new linear combinations $\bar{\mathbf{v}}_1, \ldots, \bar{\mathbf{v}}_g$ of the original eigenvectors such that under the symmetry operation two groups, $(\bar{\boldsymbol{v}}_1, \ldots, \bar{\boldsymbol{v}}_m)$ and $(\bar{\boldsymbol{v}}_{m+1}, \ldots, \bar{\boldsymbol{v}}_g)$ say, would transform separately (spanning different invariant subspaces). In this case it would no longer be necessary for the two groups to belong to the *same* eigenvalue, since no mixing is ever envisaged. Two groups of eigenvectors which are not mixed under symmetry operations but which happen to have the same eigenvalue (a somewhat rare case) are said to be *accidentally* degenerate. Our conclusions may therefore be formulated in the following way:

> Provided there are no "accidental" degeneracies, every degenerate group of eigenvectors of a matrix $\boldsymbol{F}$ provides an irreducible representation of the group of unitary transformations which leaves $\boldsymbol{F}$ invariant. (6.4.4)

The eigenvectors may therefore be classified according to symmetry species as in section 5.7.

In solving eigenvalue problems, symmetry may evidently be utilized in two ways: in reducing the eigenvalue equations to a set of lower order eigenvalue equations, by choosing a coordinate system adapted to the reduction of the representation; and in classifying the resultant solutions according to the irreducible representations which they carry. We now consider typical applications which arise in the theory of molecular vibrations.

### 6.5. Symmetry Classification of Molecular Vibrations

First, we introduce the $3N$ Cartesian components of the atomic displacements $x_1, y_1, z_1, x_2, \ldots, z_N$. Under a symmetry operation in which the displacement of atom $i$ is sent into a displacement of the equivalent atom $j$ the vector $\boldsymbol{r}_i = \begin{pmatrix} x_i \\ y_i \\ z_i \end{pmatrix}$ is rotated into $\boldsymbol{R}_c\boldsymbol{r}_i$ and then applied to atom $j$ (cf. Fig. 6.2): under the symmetry operation therefore

$$\boldsymbol{r}_j \rightarrow \boldsymbol{r}'_j = \boldsymbol{R}_c\boldsymbol{r}_i \qquad (i \rightarrow j \text{ under rotation } R)$$

In order that the kinetic energy shall appear as a sum of squares, we then introduce the mass-weighted components $\sqrt{m_i}\,\boldsymbol{r}_i$. Since $i$ and $j$ above necessarily refer to equivalent atoms, with the *same* mass, we have also $\sqrt{m_j}\,\boldsymbol{r}_j \rightarrow \sqrt{m_j}\,\boldsymbol{r}'_j = \boldsymbol{R}_c\,(\sqrt{m_i}\,\boldsymbol{r}_i)$. If we collect all the mass-weighted components in a column $\boldsymbol{q}$ the effect of a symmetry operation is

$$\boldsymbol{q} \to \boldsymbol{q}' = \boldsymbol{R}\boldsymbol{q} \tag{6.5.1}$$

where $\boldsymbol{R}$ is the partitioned matrix (cf. example 6.3) in which the $ji$-block is $\boldsymbol{R}_c$ if the symmetry operation sends the displacement of atom $i$ into a new displacement of atom $j$ but is otherwise $\boldsymbol{0}$.

EXAMPLE 6.5. In the ammonia molecule (Fig. 6.3), the point group operations permute the three equivalent hydrogen atoms. Thus under

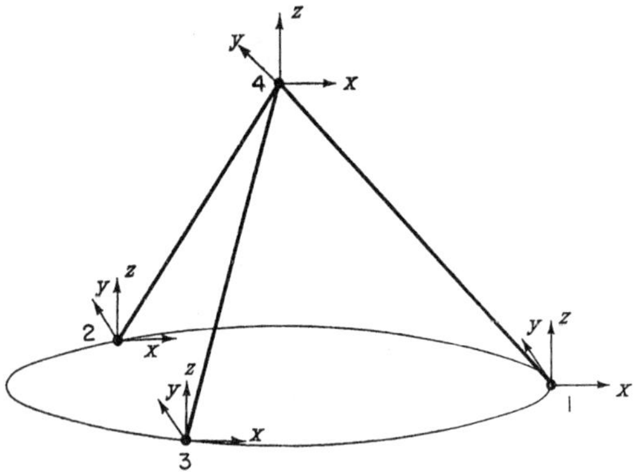


FIG. 6.3. $NH_3$ molecule. Coordinate system for displacements.

$C_3$ $1 \to 2$, $2 \to 3$, $3 \to 1$, but $4 \to 4$. The corresponding symmetry operation, applied to the displacements, sends $\boldsymbol{r}_1$ into $\boldsymbol{D}_c(C_3)\boldsymbol{r}_1$ and transfers it to atom 2, etc. Hence

$$\boldsymbol{r}_1' = \boldsymbol{D}_c(C_3)\boldsymbol{r}_3, \quad \boldsymbol{r}_2' = \boldsymbol{D}_c(C_3)\boldsymbol{r}_1, \quad \boldsymbol{r}_3' = \boldsymbol{D}_c(C_3)\boldsymbol{r}_1, \quad \boldsymbol{r}_4' = \boldsymbol{D}_c(C_3)\boldsymbol{r}_4'$$

and the matrix $\boldsymbol{D}(C_3)$ for the symmetry operation associated with $C_3$ is

$$\boldsymbol{D}(C_3) = \left(\begin{array}{c|c|c|c} \boldsymbol{0} & \boldsymbol{0} & \boldsymbol{D}_c(C_3) & \boldsymbol{0} \\ \hline \boldsymbol{D}_c(C_3) & \boldsymbol{0} & \boldsymbol{0} & \boldsymbol{0} \\ \hline \boldsymbol{0} & \boldsymbol{D}_c(C_3) & \boldsymbol{0} & \boldsymbol{0} \\ \hline \boldsymbol{0} & \boldsymbol{0} & \boldsymbol{0} & \boldsymbol{D}_c(C_3) \end{array}\right)$$

where $$\boldsymbol{D}_c(C_3) = \begin{pmatrix} -\frac{1}{2} & -\frac{1}{2}\sqrt{3} & 0 \\ \frac{1}{2}\sqrt{3} & -\frac{1}{2} & 0 \\ 0 & 0 & 1 \end{pmatrix}$$

It is a trivial matter to write down all the matrices of the group of symmetry operations. Other symmetries may be dealt with using Tables 4.1–22 : the $(xyz)$-representations are those carried by

$p_x$, $p_y$, $p_z$ in Appendix 1. The matrices determined in this way are evidently real orthogonal, $\tilde{\boldsymbol{R}}\boldsymbol{R} = \mathbf{1}$.

We now return to the vibration problem, already outlined in example 6.4. The potential energy associated with the displacements $\boldsymbol{q}$ is $2V = \boldsymbol{q}^{\dagger}\boldsymbol{F}\boldsymbol{q}$ and the normal modes of vibration, are determined, along with their frequencies, from the eigenvalue equation $\boldsymbol{F}\boldsymbol{v} = \lambda\boldsymbol{v}$. The elements of $\boldsymbol{v}_k$, the $k$th eigenvector, determine the amplitudes with which the various atoms vibrate in the $k$th normal mode. Pictorially, each triplet of elements in $\boldsymbol{v}_k$, when divided by the square root of the corresponding atomic mass (the coordinates being mass-weighted), gives the maximum displacement of an atom in the $k$th mode. To anticipate the solution, the normal modes of $NH_3$ are illustrated in this way in Fig. 6.4.

Next let us consider the symmetry classification of the normal modes using the fact that any set of eigenvectors with a common eigenvalue carries a representation $D_\alpha$ which, according to (6.4.4), we assume irreducible. It is clear from Fig. 6.4, for example, that every symmetry operation on the set of displacements represented by the first eigenvector $\boldsymbol{v}_1$ brings the set into self-coincidence, $\boldsymbol{R}\boldsymbol{v}_1 = \boldsymbol{v}_1$ (all $\boldsymbol{R}$); but a vector which is invariant under all operations of the group, carries the identity representation and this vibrational mode is therefore said to belong to the $A_1$ representation of $C_{3v}$. The second mode, associated with $\boldsymbol{v}_2$, also belongs to $A_1$. On the other hand, the degenerate pair $\boldsymbol{v}_3$ and $\boldsymbol{v}_4$ evidently transform into new linear combinations of each other under the symmetry operations: they belong, in fact, to the E representation of $C_{3v}$—as also do $\boldsymbol{v}_5$ and $\boldsymbol{v}_6$. We note also that the set of normal coordinates, $Q_1, \ldots, Q_{g_\alpha}$ (see example 6.4), formed from the set of $g_\alpha$-fold degenerate eigenvectors carrying representation $D_\alpha$, themselves transform like the basis vectors of $D_\alpha$ (assuming all quantities real). For if under operation $R$

$$\boldsymbol{v}_i \rightarrow \boldsymbol{v}'_i = \sum_{j=1}^{g_\alpha} \boldsymbol{v}_j D_\alpha(R)_{ji}$$

we have

$$\tilde{\boldsymbol{v}}_i \rightarrow \tilde{\boldsymbol{v}}'_i = \sum_{j=1}^{g_\alpha} \tilde{\boldsymbol{v}}_j D_\alpha(R)_{ji}$$

and hence, multiplying by $\boldsymbol{q}$ and using the definition $Q_i = \tilde{\boldsymbol{v}}_i\boldsymbol{q}$ (cf. (6.3.10)).

$$Q_i \rightarrow Q'_i = \sum_{j=1}^{g_\alpha} Q_j D_\alpha(R)_{ji} \qquad (6.5.2)$$

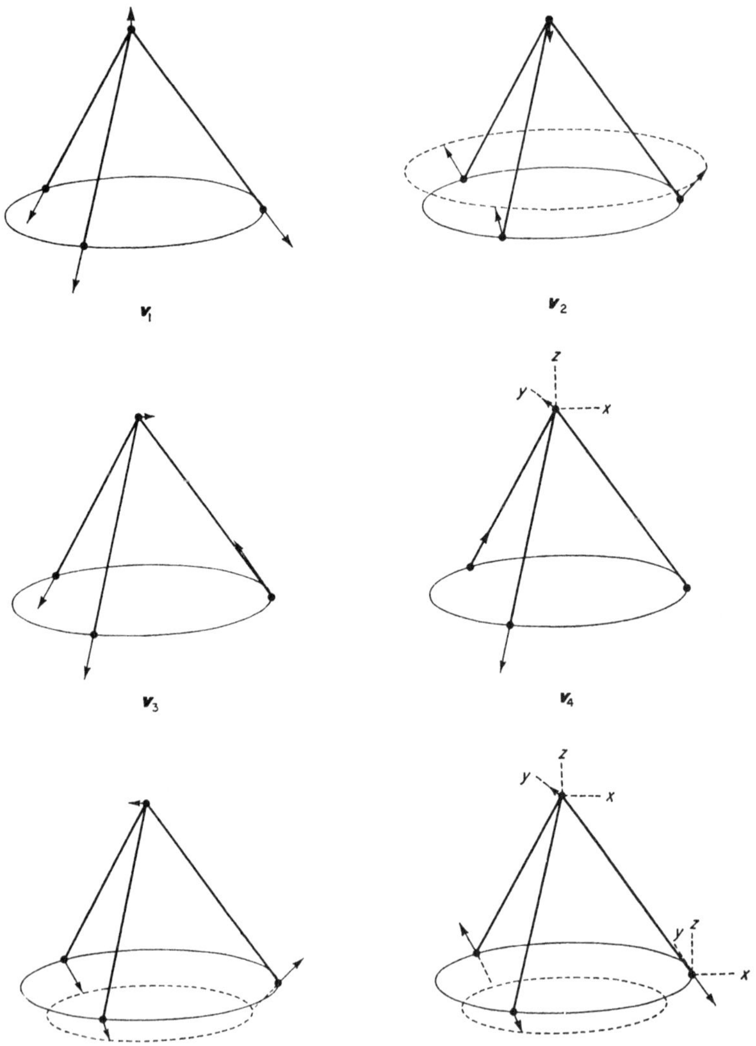


FIG. 6.4. Vibrational normal modes of $NH_3$. Each diagram indicates the displacements described by the elements of an eigenvector $\boldsymbol{v}_i$. The $\boldsymbol{v}_1$ and $\boldsymbol{v}_2$ modes are non-degenerate and of $A_1$ type. $\boldsymbol{v}_3$, $\boldsymbol{v}_4$ and $\boldsymbol{v}_5$, $\boldsymbol{v}_6$ are degenerate pairs each of E type.

Both the nature of the frequency spectrum (e.g. whether there are any doubly or triply degenerate vibrational modes) and the symmetry properties of the normal coordinates are in fact determined by molecular geometry without actually solving the eigenvalue equation, and follow immediately from the considerations of section 6.4. In $NH_3$, for example, we do not expect any *triply* degenerate modes because there are no 3-dimensional irreducible representations of $C_{3v}$. To determine whether there are eigenvectors belonging to $D_\alpha$ (i.e. whether degenerate modes of this particular type exist), we need only ask whether the $3N$-dimensional representation space has one or more subspaces carrying the $D_\alpha$ representation. But this is a familiar problem, solved in section 5.4. According to (5.4.6), the number of times $D_\alpha$ appears in the reduction of the representation D is simply $n_\alpha = \frac{1}{g}\sum_R \chi_\alpha(R)\chi(R)$ where $\chi(R)$ is the trace of the matrix $\boldsymbol{D}(R)$ associated with $R$ in the $3N$-dimensional representation D. It is therefore only necessary to know the characters in D. Reference to example 6.5 shows at once that

$$\chi(R) = U_R\,\chi_c(R)$$

where $U_R$ is the number of non-zero diagonal blocks in $\boldsymbol{R}$ and $\chi_c(R)$ is the character in the 3-dimensional Cartesian representation, namely tr $\boldsymbol{R}_c$. Now $U_R$ is simply the number of atoms left in their original places in a rotation $R$ and from (3.1.13) we know $\chi_c(R) = \pm 1 + 2\cos\theta$ for proper and improper rotations, respectively, through angle $\theta$. Hence

$$\begin{aligned} &\chi(R) = U_R(\pm 1 + 2\cos\theta) \\ &(U_R = \text{number of atoms unshifted by rotation } R, \\ &\theta = \text{angle of rotation in } R) \end{aligned} \tag{6.5.3}$$

EXAMPLE 6.6. The point group for $NH_3$ is $C_{3v}$ and has the three distinct irreducible representations given in (2.7.3). Their character systems, together with those of the ($xyz$)-representation $D_c$ and the displacement representation D (given by (6.5.3)) are tabulated below

| | $E$ | $2C_3$ | $3\sigma_v$ |
|---|---|---|---|
| $\chi_1(R)$ | 1 | 1 | 1 |
| $\chi_2(R)$ | 1 | 1 | −1 |
| $\chi_3(R)$ | 2 | −1 | 0 |
| $\chi_c(R)$ | 3 | 0 | 1 |
| $\chi(R)$ | 12 | 0 | 2 |

Hence the representation carried by the displacements is the direct sum

$$D = n_1 D_1 + n_2 D_2 + n_3 D_3$$

where $n_1 = \frac{1}{6}(12 + 0 + 0 + 2 + 2 + 2) = 3$, $n_2 = \frac{1}{6}(12 + 0 + 0 - 2 - 2 - 2) = 1$, $n_3 = \frac{1}{6}(24 + 0 + 0 + 0 + 0 + 0) = 4$. It follows from (6.4.4) that there will be three different non-degenerate symmetric modes, one mode belonging to $D_2$ (with an eigenvector which changes sign under reflection but is invariant under rotation) and four different degenerate pairs, each providing a basis for $D_3$.

*Zero-Frequency Modes*

So far, we have deliberately ignored the possibility that there may be normal modes which do not describe *vibrations* at all; but a moment's consideration shows that this is indeed so. For if all the atoms suffer a *common* displacement, $V$ will retain its equilibrium value, the restoring forces will vanish, and the equations of motion ($\ddot{q}_i = 0$) will permit *uniform translation* of the whole molecule as a solution. Formally, this non-periodic solution corresponds to $\omega^2 = \lambda = 0$ and we can find three orthogonal columns ($\boldsymbol{v}_i$) with zero eigenvalue, corresponding to uniform translation along each of the three axes. For similar reasons, we expect three modes, also with $\lambda = 0$, corresponding to uniform *rotations* about the three axes. The number of solutions of (6.3.7) which describe true vibrational modes, will in this way be reduced from $3N$ to $3N - 6$. Clearly, one aim of a group theoretical approach must be to classify and dispose of the non-vibrational modes.

The easiest way of identifying the representation to which the translational motions belong is to write down directly those linear combinations of coordinates which give the displacement of the centre of mass. These are essentially " ready-made " normal coordinates

$$\begin{aligned} X &= M^{-1}(m_1 x_1 + m_2 x_2 + \dots + m_N x_N) \\ Y &= M^{-1}(m_1 y_1 + m_2 y_2 + \dots + m_N y_N) \qquad \left(M = \sum_{i=1}^{N} m_i\right) \\ Z &= M^{-1}(m_1 z_1 + m_2 z_2 + \dots + m_N z_N) \end{aligned}$$

which are of course appropriate combinations of the mass-weighted coordinates:

$$X = M^{-1}(\sqrt{m_1}\, q_1 + \sqrt{m_2}\, q_4 + \dots + \sqrt{m_N}\, q_{3N-2}), \text{ etc.}$$

These quantities must behave under symmetry operations like the components of a single vector, the displacement of the mass-centre.

Each of the corresponding columns $\boldsymbol{v}_x, \boldsymbol{v}_y, \boldsymbol{v}_z$ represents uniform displacement along one of the axes (in keeping with the interpretation on p. 153) and must therefore give $\boldsymbol{v}_\alpha^\dagger \boldsymbol{F} \boldsymbol{v}_\alpha = 0$ ($\alpha$=x, y, z), the same value as at equilibrium. Consequently, since $\boldsymbol{v}_\alpha \neq 0$, $\boldsymbol{F}\boldsymbol{v}_\alpha = 0$ and $\boldsymbol{v}_x$, $\boldsymbol{v}_y$, $\boldsymbol{v}_z$ are eigenvectors with zero eigenvalue: they are clearly orthogonal and may be normalized, to yield true normal coordinates, by adding a factor $\sqrt{M}$. The corresponding part of the kinetic energy then takes the familiar form $\frac{1}{2}(\dot{Q}_x^2 + \dot{Q}_y^2 + \dot{Q}_z^2)$ or $\frac{1}{2}M(\dot{X}^2 + \dot{Y}^2 + \dot{Z}^2)$. Since $\boldsymbol{v}_x$, $\boldsymbol{v}_y$, $\boldsymbol{v}_z$ behave exactly like unit vectors along the three axes, these zero-frequency eigenvectors carry the representation $D_c$. The character of the representation carried by the translational modes is thus

$$\chi_{tr}(R) = \pm 1 + 2\cos\theta \quad (\theta = \text{angle of rotation in } R) \tag{6.5.4}$$

and the individual modes belong to the irreducible representations contained in $D_c$.

Just as a translation of the molecule as a whole is described by a single displacement vector, a rigid rotation or *angular* displacement is described by a single " axial " vector. The components of an axial

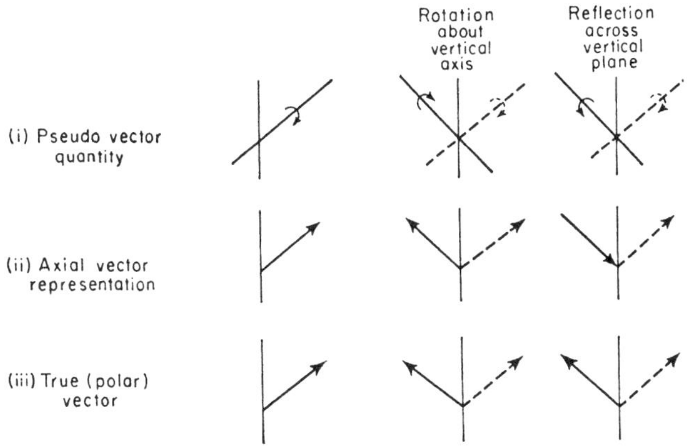


FIG. 6.5. Behaviour of a pseudo-vector under proper and improper rotations.

(i) Example of pseudo-vector—rotation about an axis. Note that the sense (curved arrow) is changed by reflection.

(ii) Pseudo-vector of (i) represented by arrow in direction of axis, magnitude indicating angle of rotation, sense corresponding to right-hand screw.

(iii) True (polar) vector changes direction but not sense on reflection: resultant vector differs in sign from that in (ii).

vector transform in the normal way under proper rotations but are multiplied by an extra factor $-1$ in an improper rotation. Entities of this kind (which are not strictly vectors at all) are best considered by tensor methods (chapter 8) but their behaviour may be inferred from a simple geometrical picture. The usual convention in elementary vector algebra is to describe an angular displacement by a line segment pointing along the direction in which a rotating corkscrew would move and of length proportional to the angle turned through, $\theta$: the association is illustrated in Fig. 6.5. Under a proper rotation of the axis the associated axial vector behaves like an ordinary vector (Fig. 6.5(a)). But an improper rotation can be regarded as a proper rotation followed by *reflection* across a plane (p. 58) and reflection reverses the sense of the angular displacement and hence of the axial vector used to represent it (Fig. 6.5(b)). Evidently an axial vector behaves like a true vector except that in an *improper* rotation it is multiplied by an extra factor of $-1$. The character of the representation carried by the components of an angular displacement is thus $(1 + 2\cos\theta)$ for a proper rotation and $-(-1 + 2\cos\theta)$ for an improper rotation. Hence

$$\boxed{\begin{array}{l}\chi_{\text{rot}}(R) = 1 \pm 2\cos\theta \\ (\theta = \text{angle of rotation in } R)\end{array}} \qquad (6.5.5)$$

We are now in a position to identify the symmetries of both translational and rotational modes and to find what pure vibrational modes remain.

EXAMPLE 6.7. The characters of the representations carried by the translational and rotational modes of $NH_3$ are

| | $E$ | $2C_3$ | $3\sigma_v$ |
|---|---|---|---|
| $\chi_{\text{tr}}(R)$ | 3 | 0 | 1 |
| $\chi_{\text{rot}}(R)$ | 3 | 0 | $-1$ |

Thus (cf. example 6.6) $D_{\text{tr}} = D_1 + D_3$ and $D_{\text{rot}} = D_2 + D_3$. Consequently, one translation belongs to $D_1$ and two to $D_3$ while one rotation belongs to $D_2$ and two to $D_3$. The remaining vibrational modes are therefore classified as follows:

| $A_1$ representation ($D_1$) | $A_2$ representation ($D_2$) | E representation ($D_3$) |
|---|---|---|
| Two non-degenerate modes | No (vibrational) modes | Two pairs of degenerate modes |

It is these modes which are indicated pictorially in Fig. 6.4.

**6.6. Symmetry Coordinates in Vibration Theory**

We now return to the problem of actually obtaining the normal coordinates by solution of the eigenvalue problem. We first make use of symmetry to reduce the matrix $\boldsymbol{F}$ to block form. This entails making a basis change which brings the matrices of all symmetry operations to irreducible block form

$$\mathbf{e} \to \hat{\mathbf{e}} = \mathbf{e}\boldsymbol{U}, \qquad \boldsymbol{q} \to \hat{\boldsymbol{Q}} = \boldsymbol{U}^{\dagger}\boldsymbol{q} \tag{6.6.1}$$

The unitary matrix $\boldsymbol{U}$ is determined purely group theoretically (as for instance in example 5.10). We can then write

$$V = \boldsymbol{q}^{\dagger}\boldsymbol{F}\boldsymbol{q} = \hat{\boldsymbol{Q}}^{\dagger}\hat{\boldsymbol{F}}\hat{\boldsymbol{Q}} \tag{6.6.2}$$

where
$$\hat{\boldsymbol{F}} = \boldsymbol{U}^{\dagger}\boldsymbol{F}\boldsymbol{U} \tag{6.6.3}$$

and from (6.2.6) $\hat{\boldsymbol{F}}$ will then have a block form related to that of the matrices of the symmetry operations (cf. (6.2.8)). The individual columns of $\boldsymbol{U}$, namely $\boldsymbol{u}_1, \boldsymbol{u}_2, \ldots, \boldsymbol{u}_{3N}$, define *symmetry vectors* and from them we obtain (using (6.3.10)) the *symmetry coordinates* which appear in the column $\hat{\boldsymbol{Q}}$:

$$\hat{Q}_i = \boldsymbol{u}_i^{\dagger}\boldsymbol{q} \tag{6.6.4}$$

These quantities are analogous to the eigenvectors and normal coordinates, of example 6.4, but appear in the reduction of $\boldsymbol{F}$ to block form rather than true diagonal form. In terms of the symmetry vectors, the individual elements of $\hat{\boldsymbol{F}}$ are given by

$$\hat{F}_{ij} = \boldsymbol{u}_i^{\dagger}\boldsymbol{F}\boldsymbol{u}_j \tag{6.6.5}$$

and vanish according to (6.2.6) *et seq.*, unless $\boldsymbol{u}_i$ and $\boldsymbol{u}_j$ are vectors of the *same species*. It is therefore convenient to arrange the rows and columns of $\hat{\boldsymbol{F}}$ so that all the elements of given species come together, as in (6.2.8), to give a diagonal block form. This merely means putting the columns of $\boldsymbol{U}$ in the desired order.

It remains only to diagonalize $\hat{\boldsymbol{F}}$ by a further basis change:

$$\hat{\mathbf{e}} \to \bar{\mathbf{e}} = \hat{\mathbf{e}}\boldsymbol{T} = \mathbf{e}\boldsymbol{U}\boldsymbol{T}, \qquad \hat{\boldsymbol{Q}} \to \boldsymbol{Q} = \boldsymbol{T}^{\dagger}\hat{\boldsymbol{Q}} = (\boldsymbol{U}\boldsymbol{T})^{\dagger}\boldsymbol{q} \tag{6.6.6}$$

We choose $\boldsymbol{T}$ so that

$$\bar{\boldsymbol{F}} = \boldsymbol{T}^{\dagger}\hat{\boldsymbol{F}}\boldsymbol{T} = (\boldsymbol{UT})^{\dagger}\boldsymbol{F}(\boldsymbol{UT}) = \text{diag}\,(\lambda_1, \lambda_2, \ldots, \lambda_{3N}) \qquad (6.6.7)$$

by taking as its columns the eigenvectors of $\hat{\boldsymbol{F}}$ in the usual way. Since $\hat{\boldsymbol{F}}$ is in block form, $\boldsymbol{T}$ will have a corresponding form, each block diagonalizing one block of $\hat{\boldsymbol{F}}$. In other words, $T_{ik} = (\boldsymbol{t}_k)_i = 0$ unless $i$ and $k$ refer to the same symmetry species. Evidently, by approaching the problem in two steps, the matrix $\boldsymbol{V}$, which in example 6.4 brought $\boldsymbol{F}$ directly to diagonal form, is factorized:

$$\boldsymbol{V} = \boldsymbol{UT} \qquad (6.6.8)$$

The eigenvectors of $\boldsymbol{F}$ are linear combinations of symmetry vectors:

$$\boldsymbol{v}_k = \sum_j (\boldsymbol{t}_k)_j \boldsymbol{u}_j \qquad (j,\ k \text{ of same species}) \qquad (6.6.9)$$

where $(\boldsymbol{t}_k)_j$ is the $j$-element of the $k$th eigenvector of $\hat{\boldsymbol{F}}$; and the normal coordinates are linear combinations of symmetry coordinates

$$Q_i = \boldsymbol{v}_i^{\dagger}\boldsymbol{q} = \sum_j (\boldsymbol{t}_k)_j \boldsymbol{u}_j^{\dagger}\boldsymbol{q} = \sum_j (\boldsymbol{t}_k)_j \hat{Q}_j \qquad (6.6.10)$$

These observations may be summarized as follows:

> An eigenvector of symmetry species ($\alpha$, $i$) is a linear combination of symmetry vectors of the same species only. A normal coordinate of symmetry species ($\alpha$, $i$) is a linear combination of symmetry coordinates of the same species only. (6.6.11)

We now use the standard method of section 5.7 to determine symmetry coordinates for the ammonia molecule. The principle of the method is simple. An arbitrary set of atomic displacements, represented by a column $\boldsymbol{u}$ may be regarded as some linear combination of the $3N$ columns defining a basis adapted to a reduction into irreducible representations. The columns may therefore be classified according to symmetry species, and we can project out of $\boldsymbol{u}$ its component of given species exactly as in example 5.11.

EXAMPLE 6.8. To determine symmetry coordinates for the $NH_3$ molecule (Fig. 6.3) let us consider the column $\boldsymbol{u}$ with elements

$$(a\ b\ c \mid 0\ 0\ 0 \mid 0\ 0\ 0 \mid d\ e\ f)$$

This describes a displacement of atom 1 through distance $a$ along the $x$-direction, $b$ along the $y$-direction, etc. Several non-zero components are taken in this instance simply because the displacements fall into separate sets which are not mixed by symmetry operations (e.g. those of the nitrogen atom and those of the hydrogens). By taking enough non-zero components in each group we make sure of being able to project out vectors of every symmetry species. The matrices $\boldsymbol{D}(R)$ associated with each rotation are easily written down by the method used in example 5.5.‡ The columns $\boldsymbol{Ru}$ which result when $\boldsymbol{u}$ is subjected to the symmetry operations are collected below

| | | $E$ | $C_3$ | $\bar{C}_3$ | $\sigma_1$ | $\sigma_2$ | $\sigma_3$ |
|---|---|---|---|---|---|---|---|
| Atom 1 | | $a$ | 0 | 0 | $a$ | 0 | 0 |
| | | $b$ | 0 | 0 | $-b$ | 0 | 0 |
| | | $c$ | 0 | 0 | $c$ | 0 | 0 |
| Atom 2 | | 0 | $-\frac{1}{2}a-\frac{1}{2}\sqrt{3}\,b$ | 0 | 0 | 0 | $-\frac{1}{2}a+\frac{1}{2}\sqrt{3}\,b$ |
| | | 0 | $\frac{1}{2}\sqrt{3}\,a-\frac{1}{2}b$ | 0 | 0 | 0 | $\frac{1}{2}\sqrt{3}\,a+\frac{1}{2}b$ |
| | | 0 | $c$ | 0 | 0 | 0 | $c$ |
| Atom 3 | | 0 | 0 | $-\frac{1}{2}a+\frac{1}{2}\sqrt{3}\,b$ | 0 | $-\frac{1}{2}a-\frac{1}{2}\sqrt{3}\,b$ | 0 |
| | | 0 | 0 | $-\frac{1}{2}\sqrt{3}\,a-\frac{1}{2}b$ | 0 | $-\frac{1}{2}\sqrt{3}\,a+\frac{1}{2}b$ | 0 |
| | | 0 | 0 | $c$ | 0 | $c$ | 0 |
| Atom 4 | | $d$ | $-\frac{1}{2}d-\frac{1}{2}\sqrt{3}\,e$ | $-\frac{1}{2}d+\frac{1}{2}\sqrt{3}\,e$ | $d$ | $-\frac{1}{2}d-\frac{1}{2}\sqrt{3}\,e$ | $-\frac{1}{2}d+\frac{1}{2}\sqrt{3}\,e$ |
| | | $e$ | $\frac{1}{2}\sqrt{3}\,d-\frac{1}{2}e$ | $-\frac{1}{2}\sqrt{3}\,d-\frac{1}{2}e$ | $-e$ | $-\frac{1}{2}\sqrt{3}\,d+\frac{1}{2}e$ | $\frac{1}{2}\sqrt{3}\,d+\frac{1}{2}e$ |
| | | $f$ | $f$ | $f$ | $f$ | $f$ | $f$ |

The symmetry vectors result when these columns are combined with the appropriate coefficients as in (5.7.8). Thus, taking all coefficients equal, the $D_1$ vector has components

$$(2a \quad 0 \quad 2c \mid -a \quad \sqrt{3}\,a \quad 2c \mid -a \quad -\sqrt{3}\,a \quad 2c \mid 0 \quad 0 \quad 6f)$$

and is a linear combination of three independent vectors, with components

$$\left.\begin{array}{l} (2 \quad 0 \quad 0 \mid -1 \quad \sqrt{3} \quad 0 \mid -1 \quad -\sqrt{3} \quad 0 \mid 0 \quad 0 \quad 0) \\ (0 \quad 0 \quad 1 \mid \phantom{-}0 \quad \phantom{\sqrt{3}}0 \quad 1 \mid \phantom{-}0 \quad \phantom{-\sqrt{3}}0 \quad 1 \mid 0 \quad 0 \quad 0) \\ (0 \quad 0 \quad 0 \mid \phantom{-}0 \quad \phantom{\sqrt{3}}0 \quad 0 \mid \phantom{-}0 \quad \phantom{-\sqrt{3}}0 \quad 0 \mid 0 \quad 0 \quad 1) \end{array}\right\} \text{Species } (1,\ 1)$$

On normalization we obtain the symmetry vectors $\boldsymbol{u}_j^{(\alpha,i)}$ defined in section 5.6. The corresponding sets of displacements are shown in Fig. 6.6. In $\boldsymbol{u}_1^{(1,1)}$ each hydrogen atom moves symmetrically outwards; the second vector, $\boldsymbol{u}_1^{(1,2)}$, gives each hydrogen atom the same upward displacement along the $z$-direction—indicated in the figure by equal open circles; while the third gives the apex atom alone a displacement, also upwards, along the $z$-direction. Evidently one linear combination of the symmetry vectors $\boldsymbol{u}_1^{(1,2)}$ and $\boldsymbol{u}_1^{(1,3)}$ describes a trans-

‡ Actually, only the matrices $\boldsymbol{D}_{\text{C}}(R)$ are needed; the atoms to which the rotated displacements refer are evident by inspection.

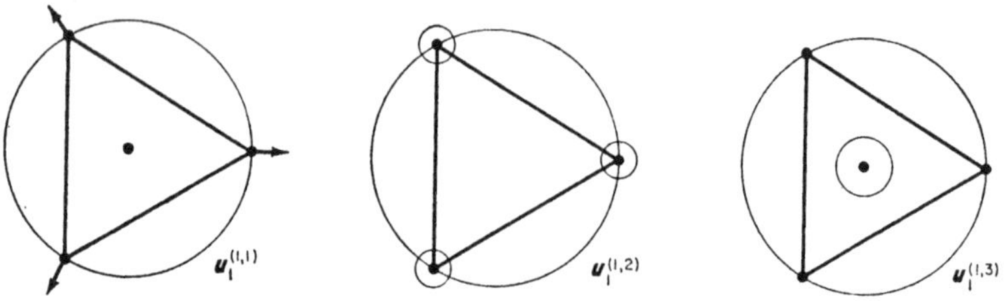


FIG. 6.6. $NH_3$ symmetry vectors for $A_1$ representation ($D_1$). Molecule viewed from above (Fig. 6.3), diameter of open circle indicating upward displacement. Displacements indicate elements of the symmetry vectors.

lation along the $z$-axis, the corresponding normal coordinate being the $z$-displacement of the mass centre. The two remaining linear combinations, which also appear as eigenvectors of the species (1, 1) block of $\hat{\boldsymbol{F}}$, describe true vibrations. They are indicated in Fig. 6.4 and describe essentially " stretching " and " bending ", respectively.

The *symmetry coordinates* belonging to the $A_1$ representation ($D_1$) follow from (6.6.4):

$$\left.\begin{aligned} \hat{Q}_1^{(1,1)} &= \tfrac{1}{3}\sqrt{3}(q_1 - \tfrac{1}{2}q_4 + \tfrac{1}{2}\sqrt{3}\, q_5 - \tfrac{1}{2}q_7 - \tfrac{1}{2}\sqrt{3}\, q_8) \\ \hat{Q}_1^{(1,2)} &= \tfrac{1}{3}\sqrt{3}(q_3 + q_6 + q_9) \\ \hat{Q}_1^{(1,3)} &= q_{12} \end{aligned}\right\} \text{Species } (1, 1)$$

These are normalized so that the sum of squares of the coefficients is unity. The matrix $\boldsymbol{U}$ which relates the symmetry coordinates to the $q$'s is then real orthogonal.

If we next make a projection using the coefficients from the $D_2$ representation, only the $b$-component in $\boldsymbol{u}$ survives. Taking $b = 1$ the normalized vector of species (2, 1) is found to be

$$(0 \;\; 1 \;\; 0 \,|\, -\tfrac{1}{2}\sqrt{3} \;\; -\tfrac{1}{2} \;\; 0 \,|\, \tfrac{1}{2}\sqrt{3} \;\; -\tfrac{1}{2} \;\; 0 \,|\, 0 \;\; 0 \;\; 0) \qquad \text{Species } (2, 1)$$

This is indicated in Fig. 6.7 and evidently describes a pure rotation

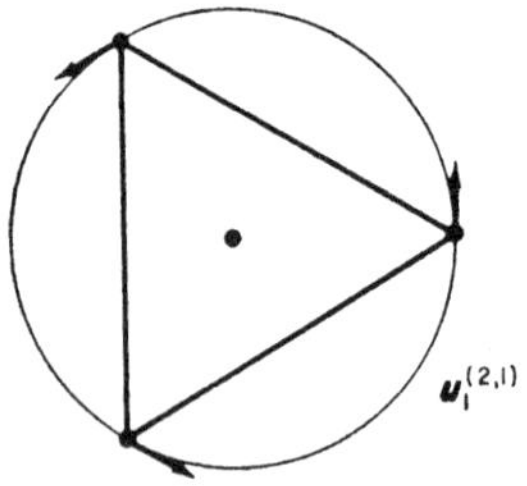


FIG. 6.7. $NH_3$ symmetry vector for $A_2$ representation ($D_2$).

about the $z$-axis. The corresponding symmetry coordinate is

$$\hat{Q}_1^{(2,1)} = (q_2 - \tfrac{1}{2}\sqrt{3}\, q_4 - \tfrac{1}{2}q_5 + \tfrac{1}{2}\sqrt{3}\, q_7 - \tfrac{1}{2}q_8)/\sqrt{3} \qquad \text{Species } (2,1)$$

Finally, we look for pairs of symmetry vectors belonging to $D_3$: these will be of species (3, 1) and (3, 2) transforming like the two basis vectors of $D_3$. Projection out of $\boldsymbol{u}$ using coefficients taken from the first column of the $D_3$ matrices gives three linearly independent vectors of each species, namely

$$\left.\begin{array}{ll}
\boldsymbol{u}_1^{(3,1)}: & \tfrac{1}{12}\sqrt{6}(4\ 0\ 0 \mid 1\ -\sqrt{3}\ \ 0 \mid 1\ \sqrt{3}\ \ 0 \mid 0\ 0\ 0) \\
\boldsymbol{u}_1^{(3,2)}: & \tfrac{1}{6}\sqrt{6}(0\ 0\ 2 \mid 0\ \ 0\ -1 \mid 0\ \ 0\ -1 \mid 0\ 0\ 0) \\
\boldsymbol{u}_1^{(3,3)}: & (0\ 0\ 0 \mid 0\ \ 0\ \ 0 \mid 0\ \ 0\ \ 0 \mid 1\ 0\ 0)
\end{array}\right\} \text{Species } (3,1)$$

$$\left.\begin{array}{ll}
\boldsymbol{u}_2^{(3,1)}: & \tfrac{1}{4}\sqrt{2}(0\ 0\ 0 \mid -1\ \sqrt{3}\ \ 0 \mid 1\ \sqrt{3}\ \ 0 \mid 0\ 0\ 0) \\
\boldsymbol{u}_2^{(3,2)}: & \tfrac{1}{2}\sqrt{2}(0\ 0\ 0 \mid 0\ \ 0\ \ 1 \mid 0\ \ 0\ -1 \mid 0\ 0\ 0) \\
\boldsymbol{u}_2^{(3,3)}: & (0\ 0\ 0 \mid 0\ \ 0\ \ 0 \mid 0\ \ 0\ \ 0 \mid 0\ 1\ 0)
\end{array}\right\} \text{Species } (3,2)$$

Two more vectors arise on using the projection operators taken from the *second* column of the $D_3$ matrices, namely,

$$\boldsymbol{u}_1^{(3,4)}: \ \tfrac{1}{4}\sqrt{2}(0\ 0\ 0 \mid \sqrt{3}\ 1\ 0 \mid \sqrt{3}\ -1\ 0 \mid 0\ 0\ 0) \qquad \text{Species } (3,1)$$

$$\boldsymbol{u}_2^{(3,4)}: \ \tfrac{1}{8}\sqrt{2}(0\ 4\ 0 \mid \sqrt{3}\ 1\ 0 \mid -\sqrt{3}\ \ 1\ 0 \mid 0\ 0\ 0) \qquad \text{Species } (3,2)$$

These normalized symmetry vectors are indicated schematically in Fig. (6.8), and corresponding symmetry coordinates, $\hat{Q}_1^{(3,\mu)}$ and $\hat{Q}_2^{(3,\mu)}$ with $\mu = 1, 2, 3, 4$, follow immediately from (6.6.4). Again, it is not difficult to identify linear combinations describing translation and rotation: suitable combinations of $\boldsymbol{u}_1^{(3,1)}$, $\boldsymbol{u}_1^{(3,3)}$, $\boldsymbol{u}_1^{(3,4)}$ and of $\boldsymbol{u}_2^{(3,1)}$, $\boldsymbol{u}_2^{(3,3)}$, $\boldsymbol{u}_2^{(3,4)}$ respectively, will describe the $x$- and $y$-translations, while other combinations of species (3, 1) and (3, 2), respectively, will describe rotations about axes in the $y$- and $x$-directions. There are two other linearly independent combinations, of each species, corresponding to the vibrational modes indicated in Fig. 6.4.

We have determined symmetry coordinates $\hat{\boldsymbol{Q}} = \boldsymbol{U}^{\dagger}\boldsymbol{q}$ such that $\hat{\boldsymbol{F}} = \boldsymbol{U}^{\dagger}\boldsymbol{F}\boldsymbol{U}$ will have block form. It is now possible to determine the normal modes and frequencies by solving the eigenvalue equation $\hat{\boldsymbol{F}}\boldsymbol{t} = \lambda\boldsymbol{t}$. At this point, it becomes apparent that the use of Cartesian coordinates, whilst permitting a simple discussion of symmetry, has certain draw-backs. In the first place, the translational and rotational modes, although identifiable, are not automatically separated from the vibrational modes, to give an equation of lower dimension. Secondly, the matrix $\boldsymbol{F}$, from which $\hat{\boldsymbol{F}}$ is obtained, is not readily available because the energy of a deformed molecule is expressed most naturally in

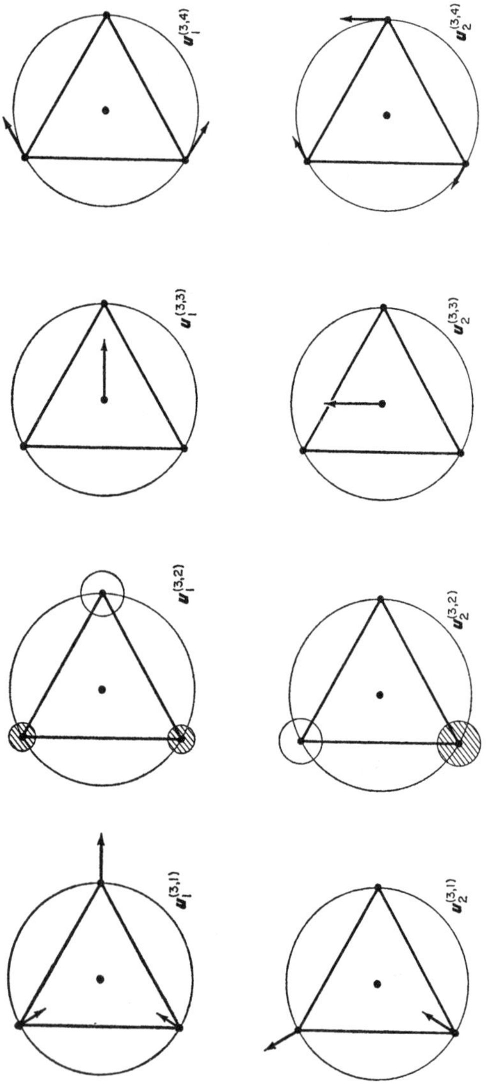


FIG. 6.8. $NH_3$ symmetry vectors for E representation ($D_3$). Top and bottom rows are of species (3, 1) and (3, 2), respectively. Open and shaded circles represent upward and downward displacements.

terms of changes in bond lengths and angles—not Cartesian displacements. For these reasons, it is more common to go over at this stage to a system of *internal* coordinates (bond lengths and angles), in terms of which the whole problem may be reformulated without ever introducing translational and rotational modes. Nevertheless, it is quite feasible to complete the calculation using the symmetry coordinates already obtained. The elements of $\boldsymbol{F}$ may be expressed in terms of the appropriate force constants (for " stretching " and " bending ") by methods due to Wilson and others (Ref. 3) : those of $\hat{\boldsymbol{F}}$ are then determined from the symmetry vectors according to (6.6.5) ; and the eigenvalue problem is then solved for each block of $\hat{\boldsymbol{F}}$ to obtain the normal modes (including those for rotation and translation). In this way we should obtain results of the kind indicated in Fig. 6.4. The more direct treatment, in which internal coordinates are employed throughout, is described by Wilson *et al.* in the work already cited (Refs. 2, 3).

## REFERENCES

1. NYE, J. F., *Physical Properties of Crystals*, chapter 1, Oxford University Press, 1957.
2. WILSON, E. B., DECIUS, J. C., and CROSS, P. C., *Molecular Vibrations*, chapter 2, McGraw-Hill, 1955.
3. WILSON, E. B., *et al.*, *Molecular Vibrations*, chapter 4, McGraw-Hill, 1955.

## BIBLIOGRAPHY

BHAGAVANTAM, S., and VENKATARAYUDU, T., *Theory of Groups and its Application to Physical Problems*, Andhra University, 1948. (Contains chapters on both crystal properties and vibration theory.)

ROSENTHAL, J., and MURPHY, G. M., *Rev. Mod. Phys.*, **8**, 317, 1936. (One of the earliest reviews of applications to molecular vibration theory.)

WILSON, E. B., DECIUS, J. C., and CROSS, P. C., *Molecular Vibrations*, McGraw-Hill, 1955. (A comprehensive standard work, making considerable use of group theory.)

CHAPTER 7

# APPLICATIONS INVOLVING FUNCTIONS AND OPERATORS

## 7.1. Transformation of Functions

In chapter 6 we have discussed the effect of symmetry operations upon algebraic forms built up from vector components. We now turn to the transformation of continuous functions of position within a physical system—the temperature, for example, or the wave function describing the electrons in a molecule. Such a function is often referred to as a *field*: if it describes a scalar quantity the field is a *scalar field*. Vector and tensor fields are also important but here it will be sufficient to discuss scalar fields. Let us first consider the temperature distribution in a solid body, writing

$$T = \psi(x, y, z) \tag{7.1.1}$$

What does this really mean? To define a function of position *within the system* we must embed a basis $\mathbf{e} = (\mathbf{e}_1\ \mathbf{e}_2\ \mathbf{e}_3)$ in the system and introduce coordinates relating every point P, with position vector $\mathbf{r}$, to the basis (Fig. 7.1(a)). We could write $\mathbf{r} = r_1\mathbf{e}_1 + r_2\mathbf{e}_2 + r_3\mathbf{e}_3$ and

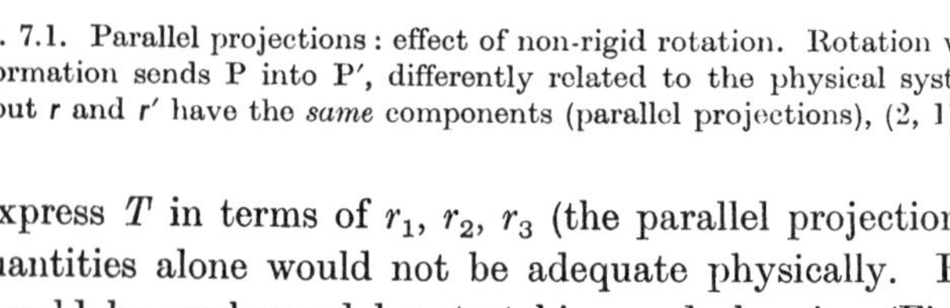

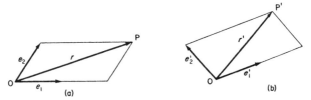

FIG. 7.1. Parallel projections: effect of non-rigid rotation. Rotation with deformation sends P into P′, differently related to the physical system, but $\mathbf{r}$ and $\mathbf{r}'$ have the *same* components (parallel projections), (2, 1).

try to express $T$ in terms of $r_1$, $r_2$, $r_3$ (the parallel projections); but these quantities alone would not be adequate physically. For their values would be unchanged by stretching and shearing (Fig. 7.1(b)) and a function $\psi(r_1, r_2, r_3)$ would therefore take the same value even when the physical relationship of the point to the rest of the system had been changed. A physical system is characterized by the *distances* and *angles* which describe the relationship of its parts, and we are interested in groups of operations which send every point to a new point in space but which preserve all distances and angles. The most suitable variables would therefore express a distance-angle, or *metrical*, relationship between a point and the basis and would keep their values *only* under rigid-body rotations (Fig. 7.2). For these reasons

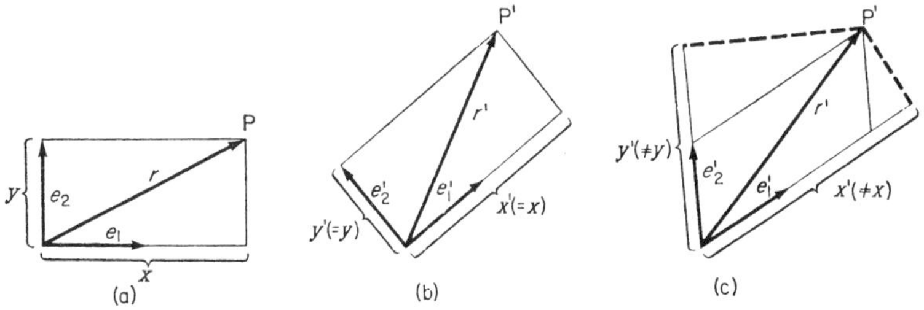


FIG. 7.2. Orthogonal projections: effect of rigid and non-rigid rotations. In a rigid rotation (b) the orthogonal projections $(x, y)$ are unchanged; so is the relationship between P and the rest of the system. When deformation occurs (c) the relationship is changed and this is revealed by the change in orthogonal projections ($x' \neq x$, $y' \neq y$).

we employ the orthogonal projections‡ $x = \mathsf{e}_1 \cdot \mathsf{r}$, $y = \mathsf{e}_2 \cdot \mathsf{r}$, $z = \mathsf{e}_3 \cdot \mathsf{r}$ which express the relationship unambiguously. To emphasize that (7.1.1) expresses a *relationship* between a field point $\mathsf{r}$ and a basis, we write it more carefully as

$$T = \psi_{\mathsf{e}}(\mathsf{r}) \tag{7.1.2}$$

remembering that the numerical quantities involved are essentially scalar products. Henceforth, it is advantageous (cf. (2.8.9)) to use an orthonormal basis: in this case we write

$$\mathsf{r} = x\mathsf{e}_1 + y\mathsf{e}_2 + z\mathsf{e}_3 \qquad (\mathsf{e}_i \cdot \mathsf{e}_j = \delta_{ij}) \tag{7.1.3}$$

where $x$, $y$, $z$ are Cartesian components. Equation (7.1.2) is then an abbreviation for a statement involving $x$, $y$, $z$.

EXAMPLE 7.1. If we use a Cartesian basis and are told that

$$T = \psi(x, y, z) = x \exp\{-a(x^2 + y^2 + z^2)\}$$

this could be regarded as

$$T = (\mathsf{e}_1 \cdot \mathsf{r}) \exp\{-a\mathsf{r} \cdot \mathsf{r}\}$$

a form which explicitly relates the field point to the basis and obviously gives the same $T$ value for all rotations which preserve the scalar

‡ In ordinary 3-space the dot is used to denote the scalar product.

product—besides being quite independent of coordinate system.

We now consider the rotation which sends $\mathbf{e}_1$ into $R\mathbf{e}_1$, etc. or, collectively, $\mathbf{e} \to \mathbf{e}' = R\mathbf{e} = \mathbf{e}\boldsymbol{R}$. The original basis now serves to define a reference frame fixed in space, against which the rotated system can be discussed. The point P, fixed in the system, is carried into a point P′ with position vector $\mathbf{r}' = R\mathbf{r} = x\mathbf{e}_1' + y\mathbf{e}_2' + z\mathbf{e}_3'$ (relating P′ to the rotated basis in exactly the way P was related to the old). The equality of $T$ at P with $T$ at P′ in the rotated system is thus expressed by

$$\left.\begin{aligned} T_{\mathrm{P}} &= T_{\mathrm{P}'}^{(\mathrm{rot})} \\ \psi_{\mathbf{e}}(\mathbf{r}) &= \psi_{\mathbf{e}'}(\mathbf{r}') = \psi_{R\mathbf{e}}(R\mathbf{r}) \end{aligned}\right\} \tag{7.1.4}$$

Here system and point are both rotated, $\mathbf{e} \to \mathbf{e}' = R\mathbf{e}$, $\mathbf{r} \to \mathbf{r}' = R\mathbf{r}$, their relationship is preserved, and the value of $T$ is unchanged. On the other hand, we may wish to describe the rotated system in terms of the fixed basis $\mathbf{e}$: in this case we should rotate the system but *not* the field point $\mathbf{r}$, so as to obtain the dependence of $T^{(\mathrm{rot})}$ upon the arbitrary point (P) with $\mathbf{r} = x\mathbf{e}_1 + y\mathbf{e}_2 + z\mathbf{e}_3$ (referred to the fixed basis). We should obtain

$$T_{\mathrm{P}}^{(\mathrm{rot})} = \psi_{R\mathbf{e}}(\mathbf{r}) = R\psi_{\mathbf{e}}(\mathbf{r}) \tag{7.1.5}$$

where $R\psi_{\mathbf{e}}$ is to be regarded as a single function symbol, defined with respect to basis $\mathbf{e}$, describing the *transformed function*. The transformation leading from the original function to the new function is said to be *induced* by the rotation of the system (cf. example 2.8).

Example 7.2. Consider the expression for $T$ in example 7.1 and suppose $R$ is a rotation through angle $\theta$ about the vector $\mathbf{e}_3$: in this case

$$R(\mathbf{e}_1\ \mathbf{e}_2\ \mathbf{e}_3) = (\mathbf{e}_1\ \mathbf{e}_2\ \mathbf{e}_3)\begin{pmatrix} \cos\theta & -\sin\theta & 0 \\ \sin\theta & \cos\theta & 0 \\ 0 & 0 & 1 \end{pmatrix}$$

The new temperature function, referred to the original basis, is thus, from (7.1.5),

$$\begin{aligned} R\psi_{\mathbf{e}}(\mathbf{r}) = \psi_{R\mathbf{e}}(\mathbf{r}) &= (R\mathbf{e}_1 \cdot \mathbf{r}) \exp\{-a\,\mathbf{r} \cdot \mathbf{r}\} \\ &= \cos\theta(\mathbf{e}_1 \cdot \mathbf{r}) \exp\{-a\mathbf{r} \cdot \mathbf{r}\} + \\ &\qquad + \sin\theta(\mathbf{e}_2 \cdot \mathbf{r}) \exp\{-a\mathbf{r} \cdot \mathbf{r}\} \end{aligned}$$

or in terms of Cartesian coordinates

$$R[xe^{-ar^2}] = \cos\theta\,[xe^{-ar^2}] + \sin\theta\,[ye^{-ar^2}]$$

Clearly, the pattern of $T$ values which the function defines is in this way rotated as a whole in just the same way as the basis (Fig. 7.3): when $r \to Rr$, $T$ in the rotated system takes exactly the value it had before rotation.

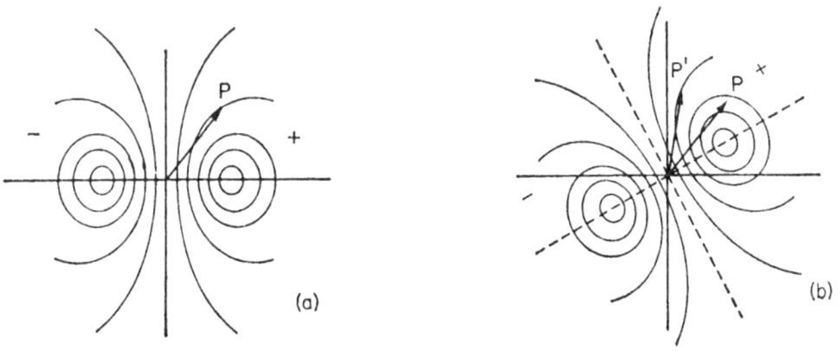


FIG. 7.3. Rotation of a field. Value of the field quantity at P′ in (b)—the image of P after rotation—is identical with that at P in (a).

Now the relationship (7.1.4) makes it possible to define the transformed function in a different way. For if $r$ is replaced by another field point $R^{-1}r$ this equation becomes

$$R\psi_e(r) = \psi_e(R^{-1}r) \tag{7.1.6}$$

The transformed function, $R\psi_e$ may therefore be obtained simply by replacing the *coordinates* in the original function by those of a *backwards rotated* field point: $r \to R^{-1}r$. There are thus two alternative rules for obtaining the transformed function:

(i) $R\psi_e(r) = \psi_{Re}(r)$. Rotate the *basis vectors,* relative to which the system is defined, and express them in terms of the original basis vectors.

(ii) $R\psi_e(r) = \psi_e(R^{-1}r)$. Rotate the field point *backwards* and use the *coordinates* of the resultant point as variables, in place of $x$, $y$, $z$.

As an example of the second procedure we again consider the function in example 7.1.

EXAMPLE 7.3. If $R \to \boldsymbol{R}$, then $R^{-1} \to \boldsymbol{R}^{-1}$ and backwards rotation gives (cf. Example 7.2)

$$\begin{pmatrix} x \\ y \\ z \end{pmatrix} \to \boldsymbol{R}^{-1} \begin{pmatrix} x \\ y \\ z \end{pmatrix} = \begin{pmatrix} \cos\theta & \sin\theta & 0 \\ -\sin\theta & \cos\theta & 0 \\ 0 & 0 & 1 \end{pmatrix} \begin{pmatrix} x \\ y \\ z \end{pmatrix}$$

since, for a Cartesian basis, $\boldsymbol{R}^{-1} = \tilde{\boldsymbol{R}}$. Hence

$$R[xe^{-ar^2}] = (x \cos\theta + y \sin\theta)\, e^{-ar^2} = \cos\theta\, [xe^{-ar^2}] + \sin\theta\, [ye^{-ar^2}]$$

which agrees with the result in example 7.2.

**7.2. Functions of Cartesian Coordinates**

Functions of $x$, $y$, $z$ may be expressed in terms of the three basic functions

$$f_1(x, y, z) = x, \quad f_2(x, y, z) = y, \quad f_3(x, y, z) = z \qquad (7.2.1)$$

In this context $f_1$, $f_2$ and $f_3$‡ (though equated with $x$, $y$, $z$, respectively, on the right) are not to be regarded as the components of a vector: $f_1(x, y, z)$ is the *function* whose level surfaces form a family of planes perpendicular to the $x$-axis. This becomes clear when we discuss the transformation properties of $f_1$, $f_2$ and $f_3$. Suppose the system rotates so that $\mathbf{e} \to R\mathbf{e} = \mathbf{e}\boldsymbol{R}$: in full, $\mathbf{e}_1 \to \mathbf{e}_1' = \mathbf{e}_1 R_{11} + \mathbf{e}_2 R_{21} + \mathbf{e}_3 R_{31}$, etc. Then $f_1 = \mathbf{e}_1 \cdot \mathbf{r} \to (\mathbf{e}_1 R_{11} + \mathbf{e}_2 R_{21} + \mathbf{e}_3 R_{31}) \cdot \mathbf{r} = f_1 R_{11} + f_2 R_{21} + f_3 R_{31}$, etc. Hence

$$R(f_1\ f_2\ f_3) = (f_1\ f_2\ f_3)\boldsymbol{R} \qquad (7.2.2)$$

and the three functions evidently transform like the *basis vectors*, $\mathbf{e}_1$, $\mathbf{e}_2$, $\mathbf{e}_3$—not like the *components* of a vector. Although it is not necessary to use a Cartesian basis, this choice does simplify the definition of the transformed function by (7.1.6). For with a Cartesian basis $\mathbf{r} \to R^{-1}\mathbf{r}$ means

$$\boldsymbol{r} \to \boldsymbol{R}^{-1}\boldsymbol{r} = \tilde{\boldsymbol{R}}\boldsymbol{r}$$

Hence, on transposing,

$$(x\ y\ z) \to (x\ y\ z)\boldsymbol{R} \qquad (7.2.3)$$

This gives exactly the transformation obtained by the first method: for on substituting these new coordinates in (7.2.1) we retrieve (7.2.2). Now that we know the basic transformation induced on the functions $f_1(=x)$, $f_2(=y)$, $f_3(=z)$, by a rotation $R$ we can easily discuss the transformations induced on any set of polynomial functions of Cartesian coordinates.

EXAMPLE 7.4. A useful illustration occurs in quantum mechanics. The wave functions for a particle in a central field (Refs. 1 and 2)

‡ Functions regarded as elements of a function space (and hence essentially as vectors) will *not* be distinguished typographically: the meaning is invariably clear from the context.

always appear in degenerate sets, the first three types being as follows:

| *s* functions |
|---|
| Function of $r$ only |
| (Non-degenerate) |

| *p* functions | $p_x$ | $p_y$ | $p_z$ |
|---|---|---|---|
| (Function of $r$) × | $x$ | $y$ | $z$ |
| (3-fold degenerate) | | | |

| *d* functions: | $d_{z^2}$ | $d_{x^2-y^2}$ | $d_{xy}$ | $d_{xy}$ | $d_{zx}$ |
|---|---|---|---|---|---|
| (Function of $r$) × | $\frac{1}{2}(3z^2-r^2)$ | $\frac{1}{2}\sqrt{3}\,(x^2-y^2)$ | $\sqrt{3}\,xy$ | $\sqrt{3}\,yz$ | $\sqrt{3}\,zx$ |
| | | (5-fold degenerate) | | | |

The functions in any particular set have a common invariant factor, only the angle dependent part being shown (with a numerical factor which allows for simultaneous normalization in the usual sense $\int |\psi|^2\,dx\,dy\,dz = 1$). The symmetries of these functions are apparent from Fig. 7.4 which indicates, purely schematically, the regions in which they take their numerically greatest values (positive or negative): in quantum chemistry the functions are described as *atomic orbitals*, replacing the Bohr orbits of early quantum theory.

Let us denote a set of $p$ functions by $p_x, p_y, p_z$. Then we can say at once that $p_x, p_y, p_z$ carry a representation ($D_p$, say) of the group $D_3$ of three dimensional rotations—for these functions behave exactly like $f_1, f_2, f_3$ above. Now the three functions span a certain subspace in a function space—the subspace comprising all solutions of the Schrödinger equation,‡ $H\psi = E\psi$, corresponding to a certain value of $E$ (the energy), all of which are obtainable as linear combinations of $p_x$, $p_y$ and $p_z$. Thus, rotations in real space induce corresponding rotations in the function space associated with the problem. The six second-degree functions ("monomials")

$$x^2, \quad y^2, \quad z^2, \quad xy, \quad yz, \quad zx \quad (\times \text{ any common invariant factor, } f(r))$$

clearly provide another representation: for if the basic functions

‡ More detailed reference to this equation is made in example 7.8.

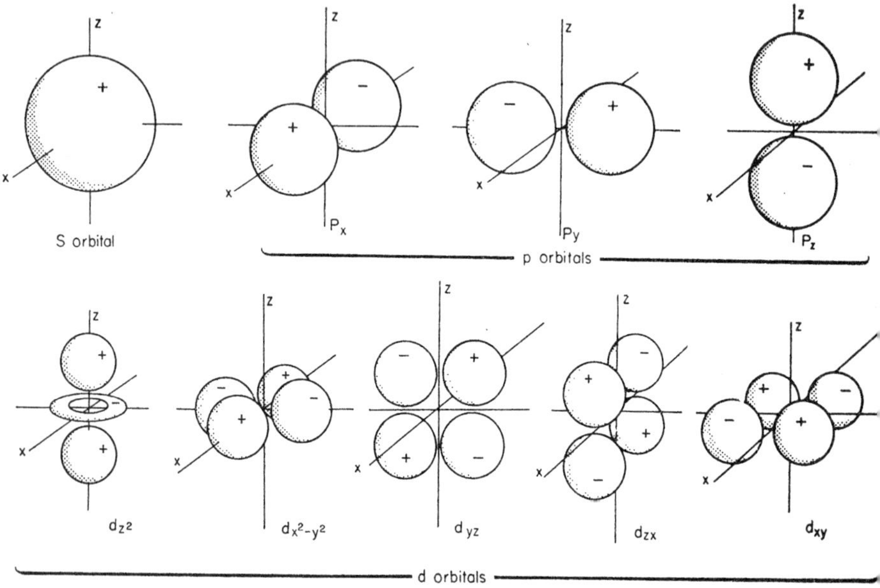


FIG. 7.4.

$f_1 = x$, etc., transform according to $(x \ y \ z) \rightarrow (x \ y \ z)\mathbf{R}$ then, for example,

$$xy \rightarrow (R_{11}x + R_{21}y + R_{31}z)(R_{12}x + R_{22}y + R_{32}z)$$

which is a linear combination of the six original functions. The $d$ functions arise when we remember that from the second degree monomials we can form one polynomial, $(x^2 + y^2 + z^2)$, which is an invariant: the linear combination, $d_0$ say, which contains the factor $(x^2 + y^2 + z^2)$ will be sent into itself under all rotations and will never be mixed with the five remaining linearly independent functions which can be set up. Consequently, $d_0$ spans an invariant subspace providing the identity representation of $\mathsf{D}_3$, while any five remaining (linearly independent) combinations will carry a five-dimensional representation. The functions listed in the table simply represent a useful orthogonal basis in this 5-dimensional subspace: the function $d_0$ is of course spherically symmetrical ($s$ type) and is therefore not included.

Example 7.4 serves to illustrate certain general features of the 3-dimensional rotation group, which is studied in detail in Ref. 3. Thus, every rotation sends the five $d$ functions into some new linear combination of themselves and the subspace which they span provides a 5-dimensional representation, $D_d$, say, of the rotation group: this is the property which characterizes the symmetry of any set of $d$ functions. Similarly the ten third-degree monomials

$$x^3,\ y^3,\ z^3,\ x^2y,\ x^2z,\ y^2x,\ y^2z,\ z^2x,\ z^2y,\ xyz \qquad (\times \text{ any common invariant factor, } f(r))$$

provide a 10-dimensional representation. But again reduction is possible if we note that the polynomials $(x^2+y^2+z^2)x$, $(x^2+y^2+z^2)y$ and $(x^2+y^2+z^2)z$ are three linear combinations which transform like $x$, $y$, $z$ and hence span an invariant subspace providing the representation $D_p$. Seven linearly independent functions remain (one particular orthogonal set corresponding to the "$f$ functions") and these provide another representation, $D_f$. What we are doing, in fact, is reducing by inspection the representations carried by monomials of given degree. The representations thus obtained can be reduced no further and the $s, p, d, f, \ldots$ functions in general carry irreducible representations of dimension 1, 3, 5, 7 ... respectively.

Since an $n$th degree polynomial in $x$, $y$, $z$ may be divided by $r^n$ to give a function of the polar angles $(\theta, \phi)$ alone, the functions in example 7.4 may all be expressed in the form $f(r)F(\theta, \phi)$. The $p$ functions, for example, become

$$f(r)\sin\theta\cos\phi, \qquad f(r)\sin\theta\sin\phi, \qquad f(r)\cos\theta$$

It is customary to normalize the radial and angular factors of such a function separately, in such a way that with volume element $dV = r^2 \sin\theta \, dr \, d\theta \, d\phi$

$$\int |Nf(r)N'F(\theta, \phi)|^2 r^2 \sin\theta \, dr \, d\theta \, d\phi = \int |Nrf(r)|^2 \, dr \int |N'F(\theta, \phi)|^2 \sin\theta \, d\theta \, d\phi = 1$$

and each integral separately has the value unity. The functions derived from example 7.4 in this way are normalized *spherical harmonics* in real form. The spherical harmonics of degree $l$ provide a $(2l+1)$-dimensional irreducible representation of the rotation group $D_3$. They appear most naturally in the *complex* form‡

$$Y_{l,m}(\theta, \phi) = (-1)^{(m+|m|)/2}\left[\left(\frac{2l+1}{4\pi}\right)\frac{(l-|m|)!}{(l+|m|)!}\right]^{\frac{1}{2}} P_l^{|m|}(\cos\theta)\, e^{im\phi} \qquad (7.2.4)$$

‡ The functions used here are discussed in Refs. 1 and 2: $Y_{l,m} = (-1)^{(m+|m|)/2}\Theta_{l,m}\Phi_m$ where $\Theta_{l,m}$ and $\Phi_m$ are given in Tables 21.1–2 of Ref. 1. The phase choice ensures that the functions carry *standard* irreducible representations of $D_3$ (Appendix A of Ref. 3) and is now almost universal.

where $P_l^{|m|}(x)$ is an *associated Legendre polynomial.* In general

$$\int Y_{l,m}(\theta, \phi)^* \, Y_{l',m'}(\theta, \phi) \sin \theta \, d\theta \, d\phi = \delta_{ll'} \delta_{mm'} \qquad (7.2.5)$$

and the functions are said to be normalized and orthogonal " with weight factor sin $\theta$ " (appropriate to integration over the solid angle). Since $e^{\pm im\phi} = \cos m\phi \pm i \sin m\phi$, *real* harmonics are easily defined:

$$Y_{l,m}^c = [(-1)^m Y_{l,m} + Y_{l,-m}]/\sqrt{2},$$
$$Y_{l,m}^s = -i[(-1^m) Y_{l,m} - Y_{l,-m}]/\sqrt{2} \quad (m > 0) \qquad (7.2.6)$$

where superscripts indicate replacement of the factor $(-1)^{(m+|m|)/2} e^{im\phi}$ in (7.2.4) by $\sqrt{2} \cos m\phi$ or $\sqrt{2} \sin m\phi$. $Y_{l,0}$ is already a real function. It is clear that the functions in Example 7.4 are indeed proportional to the real spherical harmonics with $l = 1, 2, 3$, respectively. Although the complex functions provide the standard representations of the full group $\mathsf{D}_3$, the real harmonics are often just the correct combinations to carry irreducible representations of its *sub*groups—the point groups of chapters 3 and 4. The transformation properties of the real harmonics, and hence of the $s, p, d, f, \ldots,$ functions, under the operations of the 32 crystal point groups, are of such importance as to merit detailed tabulation (appendix 1).

### 7.3. Operator Equations. Invariance

We have now discussed in some detail the transformation of functions, describing the condition of a physical system, induced by a rotation of the system. Such functions invariably satisfy some kind of differential or integral equation. Equations of this kind may be discussed most generally and systematically by introducing *operators*: we give a brief résumé of basic ideas before turning to the notion of invariance under a symmetry operation.

*Operators*

Any rule by which a function $\psi'$ may be obtained from any given function $\psi$ defines an operator: we formally write

$$\psi' = \mathsf{R}\psi$$

and observe that if the functions are regarded as vectors in a function space then $\mathsf{R}$ may be regarded as a mapping which leads from $\psi$ to its image $\psi'$: the rule for obtaining the " rotated function " in section 7.1 thus describes an operation in function space—though of a somewhat special kind. Another common operation is differentiation (cf. example 2.15). The derivative of a function may be written

$$\psi'(x) = \lim_{\delta x \to 0} \left[\frac{\psi(x + \delta x) - \psi(x)}{\delta x}\right] = \mathsf{D}\psi(x) \qquad (7.3.1)$$

Here $\mathsf{D}$ (or $\mathrm{d}/\mathrm{d}x$) is just a shorthand notation for the rule (i.e. the limiting process indicated) by which $\psi'$ is obtained from $\psi$.

Usually, the functions in which we are interested belong to some well defined " class ", satisfying specific conditions. In quantum mechanics for example, the functions are finite, continuous (except at a finite number of points), single-valued, and of " integrable square " (cf. example 2.13)—the squared modulus, integrated over all values of the variable(s), being finite. The functions may satisfy more stringent conditions (e.g. vanishing on a spherical boundary): but in physical problems the class of function is usually evident and we shall simply refer to functions " of the given class " assuming such functions finite, differentiable, etc., and in general " well behaved ". The totality of all functions of the given class then defines a corresponding function space and it is the mappings in this space, and the operators which effect them, with which we are concerned. These operators have well-defined properties, irrespective of the *particular* functions on which they work: for example, using $\mathsf{D}$ defined in (7.3.1),

$$\left.\begin{aligned} \mathsf{D}[\psi_1(x) + \psi_2(x)] &= \mathsf{D}\psi_1(x) + \mathsf{D}\psi_2(x) \\ \mathsf{D}[\psi_1(x)\psi_2(x)] &= \psi_1(x)\,\mathsf{D}\psi_2(x) + \psi_2(x)\,\mathsf{D}\psi_1(x) \end{aligned}\right\} \qquad (7.3.2)$$

for *any* two members of the class of finite, continuous and single-valued functions.

An even simpler operation is multiplication by a number, $a$, which produces a new function $a\psi(x)$: when we wish to emphasize the operational character of this process we may, as usual, write the operator in distinctive type and say

$$\mathsf{a}\psi(x) = a\psi(x) \quad (\mathsf{a} \text{ working on } \psi(x) \text{ gives the product } a\psi(x))$$

Multiplication by the variable itself may be written similarly:

$$\mathsf{x}\psi(x) = x\psi(x)$$

Once the operators of interest have been defined, they may be manipulated by themselves, without always referring to the functions on which they work. Thus we can combine operators by sequential performance (cf. the special operators describing symmetry operations). We shall agree that if

$$\left.\begin{aligned} \mathsf{R}\psi(x) &= \mathsf{S}\psi(x) \quad (\text{all } \psi \text{ of the given class}) \\ \mathsf{R} &= \mathsf{S} \end{aligned}\right\} \qquad (7.3.3)$$

Equality thus means (as in the case of symmetry operations) equality of effect. The "product" of $R$ and $S$ ($S$ operating first) is then defined by

$$(RS)\psi(x) = R(S\psi(x)) \quad \text{(all } \psi \text{ of the given class)} \qquad (7.3.4)$$

*i.e.* if $S$ followed by $R$ always gives the same result as some operator $T$, we describe the latter as the "product operator" and denote it by $RS$. Similarly, we define a sum

$$(R + S)\psi(x) = R\psi(x) + S\psi(x) \qquad \text{(all } \psi \text{ of the given class)} \qquad (7.3.5)$$

the sum being the single operator $(R + S)$ whose effect corresponds to adding together the results of the separate operations with $R$ and $S$. It should be noted that in dealing with operators in general we introduce *two* laws of combination, employing both the sum and the product. The operators are thus elements of an algebra (section 1.10).

We shall be concerned only with *linear operators* which have the characteristic property

$$R[\psi_1(x) + \psi_2(x)] = R\psi_1(x) + R\psi_2(x) \qquad (7.3.6)$$

It is clear, for example, that all operators built up by combination of (i) differentiation and (ii) multiplication by functions of $x$, will have this property: the operation of *squaring* the function will not. On the other hand, linear operators are not necessarily commutative, as the following example shows.

EXAMPLE 7.5. We consider the operators $D$ (differentiation with respect to $x$) and $x$ (multiplication by $x$). The product operator $Dx$ is defined by

$$(Dx)\psi = D(x\psi) = \psi + xD\psi = (1 + xD)\psi$$

where we use the rule for differentiating a product and write the result according to (7.3.5). Thus $Dx = 1 + xD \neq xD$. Here the $1$ is the "identity" operator (multiplication by unity) which leaves every function unchanged. The "*commutator*" $(Dx - xD) = 1$ thus has the property of leaving any function unchanged.

All the preceding considerations are readily extended to functions of several variables. In the case of differential operators, which appear most frequently in the equations governing fields of all kinds, ordinary differentiation is replaced by partial differentiation and the "Laplacian" plays a particularly important rôle. This operator, denoted by $\nabla^2$ (or sometimes $\Delta$) is defined in a Cartesian frame by

$$\nabla^2 = \frac{\partial^2}{\partial x^2} + \frac{\partial^2}{\partial y^2} + \frac{\partial^2}{\partial z^2} \tag{7.3.7}$$

differentiating any function $\psi(x, y, z)$ twice with respect to each variable and adding the results. Many of the most important equations of physics and chemistry involve the Laplacian: the following examples are typical.

EXAMPLE 7.6. When a steady state has been reached, the temperature in a uniform medium satisfies Laplace's equation

$$\nabla^2 \psi = 0$$

If $\psi$ takes prescribed values on some boundary, we require the particular solution of this equation corresponding to the given boundary conditions. When heat is generated inside the medium at a rate $\rho = \rho(x, y, z)$, $\psi$ satisfies the more general equation (Poisson's equation)

$$\nabla^2 \psi = \text{constant} \times \rho(x, y, z)$$

where the constant depends on thermal properties.

EXAMPLE 7.7. The amplitude of a plane membrane, vibrating in a normal mode (cf. example 6.4) with frequency $\omega/2\pi$, satisfies the 2-dimensional wave equation

$$\nabla^2 \psi = \omega^2 \psi \qquad (\nabla^2 = \partial^2/\partial x^2 + \partial^2/\partial y^2)$$

The equation states that $\nabla^2$ operating on the amplitude function must reproduce the same function except for a multiplicative factor $\omega^2$. The amplitude $\psi$ usually vanishes on a prescribed boundary and it turns out that functions of this class can be found only for certain definite values of the parameter $\omega$—these "eigenvalues" determining the "natural frequencies" of the vibration. This is a typical eigenvalue equation, its solutions being the *eigenfunctions* which satisfy the equation and its boundary conditions for the appropriate values of $\omega$. The equation is analogous to the *matrix* eigenvalue equation discussed in section 2.11.

EXAMPLE 7.8. Another eigenvalue equation, of immense importance in physics and chemistry, is the Schrödinger equation. The equation which determines the stationary states (and corresponding energies, $E$)

of a particle of mass $m$ in a field in which its classical potential energy would be $V = V(x, y, z)$ is‡

$$H\psi = \left(-\frac{\hbar^2}{2m}\nabla^2 + V\right)\psi = E\psi$$

The boundary conditions are usually that $\psi$ and its derivatives tend to zero at infinity; these conditions can lead to "quantization", the energy taking discrete‡‡ eigenvalues $E_1, E_2, E_3, \ldots$.

*Invariance*

We now ask what it is that distinguishes the operators describing symmetrical systems. Suppose a system (including its boundary conditions) is symmetrical, and that the field function satisfies some operator equation. Then on rotating the system into self-coincidence, the operators (and boundary conditions) which appear in the equations for the rotated system must be identical with those which applied before rotation. Symmetry therefore implies an *invariance*, under certain rotations, of the equations, and in particular the operators, which describe the system. To formalize this idea we consider the effect of rotation in general. Suppose that before rotation two functions are related by

$$\psi = H\phi \tag{7.3.8}$$

where $H$ may be, for example, a differential operator and $\phi$ is arbitrary. Then if the system is rotated

$$\phi \to \phi' = R\phi, \qquad \psi \to \psi' = R\psi \tag{7.3.9}$$

If $H'$ is an operator such that (whatever the choice of $\phi$)

$$\psi' = H'\phi' \tag{7.3.10}$$

then we say $H'$ is the "transformed operator"—which takes the place of $H$ when we deal with the rotated system. Clearly, $H'$ must be defined in the rotated frame exactly as $H$ was defined in the original frame. Since, on using (7.3.9),

$$\psi' = R\psi = RH\phi = RHR^{-1}R\phi = (RHR^{-1})\phi' \tag{7.3.11}$$

it is clear that the transformation law for $H$ may be written formally as

$$H \to H' = RHR^{-1} \tag{7.3.12}$$

‡ We use $\hbar$ to denote Planck's constant divided by $2\pi$. $H$ is the *Hamiltonian operator*.

‡‡ This is so for an attractive potential, for which "bound" states occur. Generally, $E$ may also take a continuous range of values for sufficiently high energies or in the case of repulsive potentials.

Now for rotations ($R$) which produce self-coincidence $H'$ is indistinguishable from $H$ and hence

$$\left. \begin{aligned} H' = H &= RHR^{-1} \\ \text{or} \qquad HR &= RH \end{aligned} \right\} \qquad (7.3.13)$$

Formally, then, invariant operators are those which *commute* with all symmetry operations. We may apply symmetry operations freely to operator equations provided we recognize this property. The following examples illustrate invariance.

EXAMPLE 7.9. Suppose the membrane in example 7.7 has the form of an equilateral triangle: then the system (including its boundary conditions) is brought into self-coincidence by the operations of $C_{3v}$. The operator $\nabla^2$ is invariant under these operations, as may be demon-

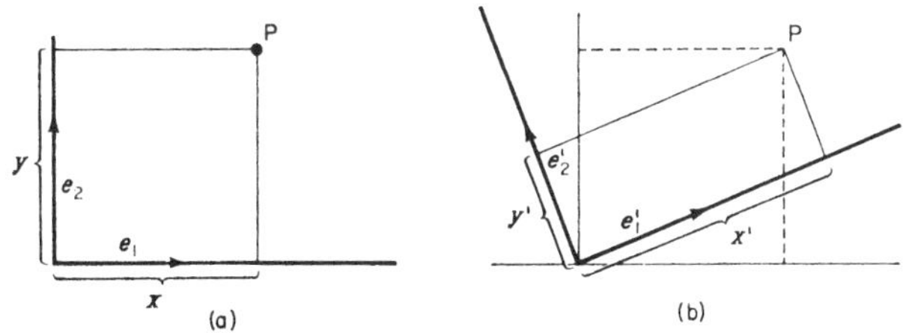


FIG. 7.5. Invariance of $\nabla^2$. Rotation of $\mathbf{e}_1$, $\mathbf{e}_2$ (with respect to which directional derivatives are defined) into $\mathbf{e}_1'$, $\mathbf{e}_2'$ leaves $\nabla^2$ (operating at any given field point P) unchanged : $\partial^2/\partial x'^2 + \partial^2/\partial y'^2 = \partial^2/\partial x^2 + \partial^2/\partial y^2$.

strated analytically by setting up the same operator for the rotated system and expressing it in terms of the original coordinates (Fig. 7.5). We readily obtain

$$\frac{\partial^2}{\partial x^2} + \frac{\partial^2}{\partial y^2} = \frac{\partial^2}{\partial x'^2} + \frac{\partial^2}{\partial y'^2}$$

A basic property of $\nabla^2$ is, in fact, its independence of the choice of directions along which the derivatives are evaluated : it is invariant not only under the operations of $C_{3v}$ but under *all* rotations of the axes.

EXAMPLE 7.10. Next consider the Schrödinger equation (example 7.8) for an electron in the field of a nucleus of charge $Ze$. Here the potential is $V = -Ze^2/r$, a function of the distance $r$ $(= \sqrt{(x^2+y^2+z^2)})$ of the field point from a nucleus at the origin. As we rotate the nucleus there is no change in potential at the field point ‡: $V$ is invariant under all rotations and the potential is spherically symmetrical. $\nabla^2$ is also invariant (cf. example 7.9) and hence $H$ is invariant under the 3-dimensional rotation group.

EXAMPLE 7.11. For a single electron in the presence of two nuclei (Fig. 7.6) the Hamiltonian operator takes the form

$$H = -\frac{\hbar^2}{2m}\nabla^2 - \frac{Z_a e^2}{r_a} - \frac{Z_b e^2}{r_b}$$

Although $\nabla^2$ is always invariant under the full rotation group, the potential at the field point will now be preserved only under a *subgroup* of operations, characterizing the symmetry of the "nuclear

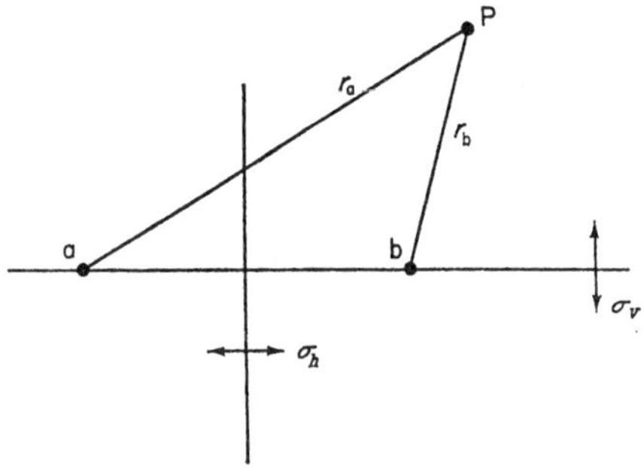


FIG. 7.6. Diatomic molecule. The Hamiltonian is invariant under $D_{\infty h}$ or $C_{\infty v}$ according as $\sigma_h$ is a symmetry operation or not.

framework" which determines the form of the problem. Any rotation of the nuclei about the axis passing through them obviously leaves $H$ invariant (subgroup $C_\infty$‡‡): so does any reflection across a plane containing the axis (element $\sigma_v$). The reflection $\sigma_h$, across a plane which is the perpendicular bisector of the axis, sends nucleus $a$ of charge $Z_a$ into position $a' = b$ and hence $r_a \to r_{a'} = r_b$. Similarly $r_b \to r_a$ and hence $\sigma_h V = -Z_a e^2/r_b - Z_b e^2/r_a$: this is the same as before reflection only if $Z_a = Z_b$. Hence for a *homonuclear* diatomic molecule there is an additional symmetry operation $\sigma_h$. The Hamiltonian operator for such a molecule is thus invariant under point group $D_{\infty h}$; in the heteronuclear case the symmetry is reduced to $C_{\infty v}$. Clearly,

‡ Here we ignore the possibility that the nucleus may have a quadrupole moment, considering only the Coulomb potential.

‡‡ The point group to which $C_n$ approaches as $n \to \infty$ (cf. example 5.16).

the group of operations under which $H$ is invariant is just the symmetry group of the nuclear framework, regarded as a geometrical object.

EXAMPLE 7.12. Finally, we consider the ammonia molecule. With the notation indicated in Fig. 7.7, the potential function is

$$V = -e^2 \left[ \frac{7}{r_0} + \frac{1}{r_1} + \frac{1}{r_2} + \frac{1}{r_3} \right]$$

and is clearly invariant under all the operations which merely permute

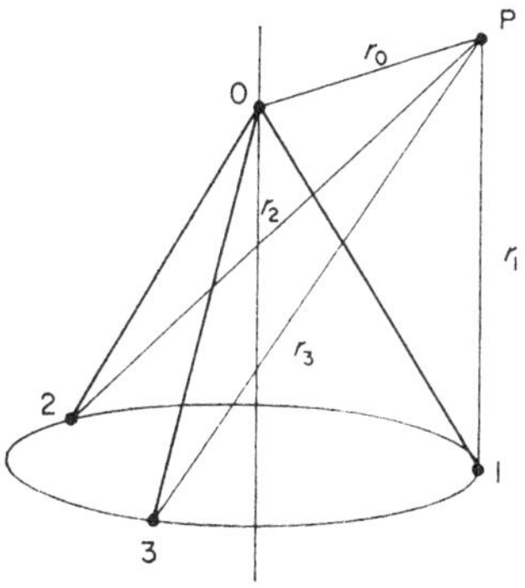


FIG. 7.7. Ammonia molecule, $NH_3$. The Hamiltonian is invariant under the operations of $C_{3v}$, which permute the hydrogens at 1, 2, 3.

the three hydrogen atoms, thereby permuting the last three terms in $V$. These operations comprise $C_{3v}$ and $H$ is therefore invariant under the group $C_{3v}$.

### 7.4. Symmetry and the Eigenvalue Problem

Eigenvalue problems, of the kind introduced in the last few examples, provide many illustrations of the power of group theoretical methods and are of such intrinsic importance that we shall discuss them fairly extensively in the present chapter. The techniques to be developed are, however, generally applicable.

Usually, it is quite impossible to obtain exact analytical solutions of eigenvalue problems—or of any equations involving, for example, partial differential operators in many variables. The aim of group theory is twofold:

(i) to infer, from whatever symmetry may be present, any general results which depend only upon the geometry of the system—results which are invariably exact and which may be obtained purely by group theoretical arguments, and

(ii) to simplify as much as possible, again using symmetry, the standard methods of obtaining *approximate* solutions.

The situation is in fact closely similar to that in chapter 6 where we first used symmetry to classify molecular vibrations and subsequently introduced symmetry *coordinates* in simplifying the actual solution of the vibration problem. In this section we shall discuss the classification and general nature of eigenfunctions, using for illustration mainly the case of a single electron in a field of given symmetry. More general problems may be dealt with by essentially similar methods and are discussed elsewhere. In later sections we discuss the use of symmetry *functions*.

We consider the eigenvalue problem

$$\mathsf{H}\psi = E\psi \tag{7.4.1}$$

where $\mathsf{H}$ and $\psi$ are the Hamiltonian operator and wave function, respectively, and in general may refer to one or many electrons. This equation resembles the *matrix* eigenvalue equation (6.3.6) and the implications of symmetry are closely similar. When the physical system is rotated we obtain

$$\mathsf{R}(\mathsf{H}\psi) = E(\mathsf{R}\psi)$$

or, by (7.3.13)

$$\mathsf{H}(\mathsf{R}\psi) = E(\mathsf{R}\psi) \tag{7.4.2}$$

Thus if $\psi$ is any eigenfunction with eigenvalue $E$, $\mathsf{R}\psi$ must also be an eigenfunction with eigenvalue $E$. If there are $g$ independent solutions, $\psi_i$ ($i = 1, 2, \ldots, g$), with this eigenvalue, we can therefore say

$$\mathsf{R}\psi = c_1\psi_1 + c_2\psi_2 + \ldots + c_g\psi_g$$

this being the most general solution with eigenvalue $E$. The functions span a $g$-dimensional *sub*space in function space and the effect of $\mathsf{R}$ is completely characterized by its effect upon each of the $g$ basis functions:

$$\begin{aligned} \mathsf{R}\psi_1 &= \psi_1 R_{11} + \psi_2 R_{21} + \ldots + \psi_g R_{g1} \\ \mathsf{R}\psi_2 &= \psi_1 R_{12} + \psi_2 R_{22} + \ldots + \psi_g R_{g2} \\ &\ldots\ldots\ldots\ldots\ldots\ldots\ldots\ldots\ldots \end{aligned} \tag{7.4.3}$$

where the coefficients may be collected into a square matrix $\boldsymbol{R}$. In accordance with section 2.6 we regard $\psi_1, \psi_2, \ldots, \psi_g$ as a set of basis vectors, collecting them into a row matrix $\boldsymbol{\psi} = (\psi_1, \psi_2, \ldots, \psi_g)$ and obtain

$$R\boldsymbol{\psi} = \boldsymbol{\psi}\boldsymbol{R} \tag{7.4.4}$$

which is the basic equation of representation theory. Since *any* solution with eigenvalue $E$ is expressible as a linear combination of $\psi_1, \psi_2, \ldots, \psi_g$ the subspace which they define is invariant under all the transformations induced by the symmetry group $\mathsf{G} = \{A, B, \ldots, R, \ldots\}$ under which $H$ is invariant: with each operation $R$ we can associate a matrix $\boldsymbol{R}$ according to (7.4.4), and the set of matrices will form a representation D, in the usual way.

The question now arises as to whether D is reducible. There are, in fact, grounds for believing that the representation will usually be *irreducible* (cf. section 6.4): for reducibility depends on finding invariant subspaces, and in this problem such subspaces are already distinguished according to eigenvalue. Reducibility would require that with a new choice of basis each matrix $\boldsymbol{R}$ would consist of diagonal blocks $\boldsymbol{R}_1$, $\boldsymbol{R}_2$, of dimension $m$ and $n$ ($m + n = g$); and in this case

$$R(\bar{\psi}_1 \ldots \bar{\psi}_m) = (\bar{\psi}_1 \ldots \bar{\psi}_m)\boldsymbol{R}_1$$

There would then be *no need* for the first $m$ and the last $n$ functions to have the same eigenvalue since they would never be mixed by the rotations. If the eigenvalues *did* coincide the degeneracy between the two sets would be termed " accidental ", arising not from the geometry of the system (the usual source of degeneracy) but from its dynamical peculiarities. Thus :

> Provided there are no " accidental " degeneracies, every degenerate group of eigenfunctions of an operator $H$ provides an irreducible representation of the group of symmetry operations which leaves $H$ invariant. (7.4.5)

This is a very general result, valid not only for point groups but for any group of operations. It is the exact counterpart of (6.4.4).

Now if the solutions of the Schrödinger equation are to form a complete set, they must include functions providing bases for *all* irreducible representations (as will be shown presently) and we can rephrase the last result :

> For every $g$-dimensional irreducible representation of the group under which $H$ is invariant we can find $g$-fold degenerate sets of eigenfunctions. Any further degeneracies would be "accidental" and expected to occur only rarely. (7.4.6)

In general, for the operators of quantum mechanics, the number of sets of each kind is infinite.

From these results, it is possible to draw some immediate conclusions about the nature of degeneracies and the transformation properties of the corresponding eigenfunctions. This is illustrated by the following examples.

Example 7.13. Let us consider a system of symmetry $C_{3v}$, the ammonia molecule, and ask what kind of eigenfunctions and degeneracies will occur. The only irreducible representations are those given in (2.7.3) and the eigenfunctions will thus be of the following forms:

(i) $\psi_1^{(1,\mu)}$ $(\mu = 1, 2, \ldots, \infty)$ transforming like the single basis vector $\mathbf{e}_1^{(1)}$ of $D_1$ (invariant under all operations)

(ii) $\psi_1^{(2,\mu)}$ $(\mu = 1, 2, \ldots, \infty)$ transforming like the single basis vector $\mathbf{e}_1^{(2)}$ of $D_2$ (invariant under rotations but changing sign under reflections)

(iii) $\psi_1^{(3,\mu)}, \psi_2^{(3,\mu)}$ $(\mu = 1, 2, \ldots, \infty)$ a degenerate pair transforming like the basis vectors $\mathbf{e}_1^{(3)}, \mathbf{e}_2^{(3)}$ of $D_3$ (or an equivalent representation)

Since (in the absence of accidental degeneracy) a $g$-fold degenerate set of eigenfunctions will provide a $g$-dimensional irreducible representation, we shall not expect any *triply* degenerate states—for there are no 3-dimensional irreducible representations of the molecular point group $C_{3v}$. Nor, for example, shall we find doubly degenerate states with wave functions which are invariant under rotation but change sign under reflection—for the $D_2$ representation is only 1-dimensional. These considerations are valid even for exact, many-electron wavefunctions, as well as for functions (molecular orbitals) describing the motion of a *single* electron in a field with the symmetry of the molecule. As a result, it is possible to classify generally, according to symmetry, both the many-electron wave functions and the molecular orbitals used in an independent-particle model. Usually, the capital letters denoting representations (chapter 4) are used in describing molecular states, while their lower case counterparts are used in labelling orbitals. The electronic states of $NH_3$ could thus be of types $A_1$, $A_2$, and E, while the molecular orbitals used to give an approximate description of the state would be of types $a_1$, $a_2$ and e.

EXAMPLE 7.14. It is frequently necessary to find the effect upon the energy levels when a symmetrical system is placed in an environment of lower symmetry. We consider the effect of a field of tetrahedral symmetry $\mathsf{T}_d$ upon a "*d* electron" (example 7.4) in the spherically symmetrical field of an atomic "core". In the absence of the tetrahedral field there are five degenerate *d* functions (i.e. five states of equal energy) providing a 5-dimensional representation of the rotation group. How far is this degeneracy likely to be resolved by the tetrahedral field?

We note from Table 4.20 that there are no 5-dimensional representations of $\mathsf{T}_d$ and that a five-fold degeneracy is not, therefore, likely to persist. However, as the field becomes infinitely weak, five eigenfunctions must pass over into the five *d* functions of example 7.4, or into linear combinations of them (i.e. into functions with the same eigenvalue). These functions, which are called the correct "zero order" combinations, will provide irreducible representations of $\mathsf{T}_d$ (i.e. the appropriate *sub*group of the rotation group). The potential degeneracies of the *d* functions are then indicated by the dimensions of the irreducible representations of $\mathsf{T}_d$, contained in the 5-dimensional representation which they carry. It is instructive to solve this problem from first principles.

We need only note the behaviour of the *d* functions under the generating operations of $\mathsf{T}_d$. The functions $x$, $y$ and $z$ transform like the three Cartesian unit vectors: so under the generating operations $\bar{S}_4^z$ and $C_3^{xyz}$ (p. 104) we have

| $\bar{S}_4^z$ | $C_3^{xyz}$ |
|---|---|
| $(x\ y\ z) \rightarrow (x'\ y'\ z') = (x\ y\ z)\begin{pmatrix} 0 & 1 & 0 \\ -1 & 0 & 0 \\ 0 & 0 & -1 \end{pmatrix}$ | $(x\ y\ z) \rightarrow (x'\ y'\ z') = (x\ y\ z)\begin{pmatrix} 0 & 0 & 1 \\ 1 & 0 & 0 \\ 0 & 1 & 0 \end{pmatrix}$ |

The behaviour of the *d* functions then follows readily as in example 7.4. For example, since $d_{z^2} = (3z^2 - r^2)f(r)$ and $d_{x^2-y^2} = \sqrt{3}\,(x^2 - y^2)f(r)$,

$$S_4^z\, d_{z^2} = (3z^2 - r^2)f(r) = d_{z^2}$$
$$C_3^{xyz}\, d_{z^2} = (3x^2 - r^2)f(r) = -\tfrac{1}{2}\, d_{z^2} + \tfrac{1}{2}\sqrt{3}\, d_{x^2-y^2}$$
$$S_4^z\, d_{x^2-y^2} = \sqrt{3}\,(y^2 - x^2)f(r) = -\, d_{x^2-y^2}$$
$$C_3^{xyz}\, d_{x^2-y} = \sqrt{3}\,(y^2 - z^2)f(r) = -\tfrac{1}{2}\sqrt{3}\, d_{z^2} - \tfrac{1}{2} d_{x^2-y^2}$$

Thus, under the generating operations, and consequently all operations of the group, $d_{z^2}$ and $d_{x^2-y^2}$ turn into new linear combinations of themselves: the basis $(d_{z^2}\ d_{x^2-y^2})$ provides, in fact, the E representation in the standard form given in Table 4.20. It is found in the same way that the basis $(d_{yz}\ d_{zx}\ d_{xy})$ carries the 3-dimensional representation $T_2$. The $d$ functions listed in example 7.4 are therefore already symmetry adapted for the group $T_d$, carrying distinct irreducible representations. In the limit of infinitely weak field, therefore, a 3-fold degenerate energy level and a 2-fold must coalesce to give the 5-fold degenerate level of the free atom: in the presence of the field there is no reason for such a coalescence and we expect a splitting into 2- and 3-fold degenerate levels. The functions which carry the 2- and 3-dimensional representations are often called the $d_\gamma$ and the $d_\varepsilon$ orbitals, respectively.

Group theory can give no information about the energy order of the resolved levels, but simple perturbation theory (e.g. Ch. 6 of Ref. 1) is often adequate for this purpose. To first order, the energy $E + \Delta E$ of the perturbed eigenfunction which passes into a correct zero order function $\psi$ of energy $E$ on removing the perturbing potential $V$ is given by

$$\Delta E = \int |\, \psi\, |^2\, V\, \mathrm{d}x\, \mathrm{d}y\, \mathrm{d}z$$

If $V$ is positive (repulsive potential) the eigenfunctions whose energies are raised highest will thus be those for which $V$ is largest in the regions where $|\, \psi\, |$ is greatest. If, in the present example, negative charges approach along tetrahedral directions (repulsive potential), it appears from Figs. 7.4 and 3.10 that the 3-fold degenerate level will lie above the 2-fold, two lobes of each $d_\varepsilon$ orbital falling on top of two negative charges.

The reader should discuss in a similar way the resolution of orbital degeneracies by fields of different symmetry. The results, which are of fundamental importance in ligand field theory (Ref. 4) follow directly from the transformation properties of the orbitals (appendix 1).

## 7.5. Approximation Methods. Symmetry Functions

The essence of most approximation methods of solving eigenvalue equations such as (7.4.1) and of dealing generally with otherwise intractible differential equations is to introduce a suitable *complete set* of functions, $\phi_1, \phi_2, \ldots, \phi_i, \ldots$, in terms of which the required solution may be expanded as

$$\psi = \sum_{i=1}^{\infty} c_i \phi_i \tag{7.5.1}$$

and to determine the first $m$ coefficients so as to get a best $m$-term approximation. If the set $\{\phi_i\}$ is complete it must certainly provide a basis for a representation of the group of rotations—and of all its finite subgroups ; and this basis can be adapted, by forming appropriate linear combinations, to a decomposition into *irreducible* representations. In dealing with a system whose symmetry group $\mathsf{G} = \{\mathsf{A}, \mathsf{B}, \ldots, \mathsf{R}, \ldots\}$ possesses irreducible representations $\mathrm{D}_1, \mathrm{D}_2, \ldots, \mathrm{D}_\alpha, \ldots$, we may therefore classify the symmetry adapted basis functions according to transformation properties. Suppose that $\mathrm{D}_\alpha$ is provided by vectors $\mathbf{e}_1^{(\alpha)}, \mathbf{e}_2^{(\alpha)}, \ldots, \mathbf{e}_{g_\alpha}^{(\alpha)}$ such that

$$\mathsf{R}\mathbf{e}_i^{(\alpha)} = \sum_j \mathbf{e}_j^{(\alpha)} D_\alpha(\mathsf{R})_{ji} \tag{7.5.2}$$

and that the set of functions $\phi_1^{(\alpha)}, \phi_2^{(\alpha)}, \ldots, \phi_{g_\alpha}^{(\alpha)}$ has identical transformation properties under the corresponding mappings $\{\mathsf{A}, \mathsf{B}, \ldots, \mathsf{R}, \ldots\}$ in function space.‡ We recall (cf. section 5.7) that $\phi_i^{(\alpha)}$ is then said to transform like the $i$th basis vector of irreducible representation $\mathrm{D}_\alpha$ or, more briefly, to belong to symmetry species $(\alpha, i)$. Generally, there will be an infinite number of such sets for each representation ; each $\mathrm{D}_\alpha$ will occur an infinite number of times. The functions of a symmetry adapted complete set therefore require a *third* label for their specification, in the usual way (cf. section 5.7). The third index simply distinguishes the different occasions on which a given representation occurs, and therefore places the functions of given species in any convenient order. The symmetry adapted basis then takes the form

$$\underbrace{\phi_1^{(1,1)}, \phi_2^{(1,1)}, \ldots, \phi_{g_1}^{(1,1)}}_{\text{First basis for } \mathrm{D}_1}\ \underbrace{\phi_1^{(1,2)}, \phi_2^{(1,2)}, \ldots, \phi_{g_1}^{(1,2)}}_{\text{Second basis for } \mathrm{D}_1} \cdots \underbrace{\phi_1^{(2,1)}, \phi_2^{(2,1)}, \ldots, \phi_{g_2}^{(2,1)}}_{\text{First basis for } \mathrm{D}_2} \cdots \tag{7.5.3}$$

We now consider the expansion of an eigenfunction in terms of the basis functions $\phi_i^{(\alpha,\mu)}$. Since the eigenfunctions provide bases for

‡ We again use the same symbol for corresponding mappings, although these now refer to a new space.

irreducible representations of G we may fix attention upon a typical solution $\psi_i^{(\alpha)}$ of species $(\alpha, i)$—one member of a $g_\alpha$-fold degenerate set providing $\mathrm{D}_\alpha$. Then, because the basis is assumed complete

$$\psi_i^{(\alpha)} = \sum_{j,\mu,\beta} c_j^{(\beta,\mu)} \phi_j^{(\beta,\mu)} \tag{7.5.4}$$

We shall now show that the only non-vanishing coefficients in such an expansion are those referring to the same symmetry species as $\psi_i^{(\alpha)}$ itself, species $(\alpha, i)$, the triple sum then becoming a single sum. Operating on (7.5.4) with $\mathsf{R}$, and using the definitive properties

$$\mathsf{R}\psi_i^{(\alpha)} = \sum_l \psi_l^{(\alpha)} D_\alpha(\mathsf{R})_{li}$$

$$\mathsf{R}\phi_j^{(\beta,\mu)} = \sum_m \phi_m^{(\beta,\mu)} D_\beta(\mathsf{R})_{mj}$$

we obtain

$$\sum_l \psi_l^{(\alpha)} D_\alpha(\mathsf{R})_{li} = \sum_{j,\mu,\beta} \sum_m c_j^{(\beta,\mu)} \phi_m^{(\beta,\mu)} D_\beta(\mathsf{R})_{mj}$$

On multiplying by $\check{D}_\alpha(\mathsf{R})_{ii}$ and summing over $\mathsf{R}$ we can make use of the orthogonality relations (5.3.14). It follows that

$$(g/g_\alpha)\,\psi_i^{(\alpha)} = \sum_{j,\mu,\beta} \sum_m c_j^{(\beta,\mu)} \phi_m^{(\beta,\mu)} (g/g_\alpha)\, \delta_{\alpha\beta}\, \delta_{im}\, \delta_{ij}$$

or

$$\psi_i^{(\alpha)} = \sum_\mu c_i^{(\alpha,\mu)} \phi_i^{(\alpha,\mu)} \tag{7.5.5}$$

Thus, when expanding a function of species $(\alpha, i)$ in terms of a complete set we may reject immediately all those functions which do not transform like the $i$th basis vector of $\mathrm{D}_\alpha$—for they would appear with zero coefficients. Maximum economy is therefore achieved by introducing a symmetry adapted basis. In words (7.5.5) becomes:

> An arbitrary function of symmetry species $(\alpha, i)$ can be expanded in terms of a complete set of basis functions of species $(\alpha, i)$ alone. (7.5.6)

It should be noted that the results established in this section again depend only upon the symmetry properties of the functions involved. Thus, in quantum mechanical applications, the functions may describe just *one* electron or may contain the coordinates of *several* electrons. To illustrate the use of symmetry functions it will be sufficient to consider the one-electron case. The Schrödinger equation of example 7.8 then arises from an "independent-particle model" in which $V$ represents an "effective field" provided by the nuclei and all other electrons. The solutions, which as we have seen may be classified according to symmetry, are then "atomic", "molecular" or "crystal" orbitals as the case may be and the symmetry adapted basis functions from which they may be constructed are then referred to as *symmetry orbitals*. The following examples indicate the nature and use of symmetry orbitals.

EXAMPLE 7.15. The water molecule (Fig. 7.8) has symmetry $C_{2v}$. In building up molecular orbitals (MO's) it is customary to employ a basis consisting of the atomic orbitals (AO's) of example 7.4 (this being the linear-combination-of-atomic-orbitals (LCAO) approximation). Let us take oxygen orbitals‡ $s^{(0)}$, $p_x^{(0)}$, $p_y^{(0)}$, $p_z^{(0)}$, $d_{z^2}^{(0)}$, . . . , and hydrogen orbitals by $s^{(1)}$, $p_x^{(1)}$, . . . and $s^{(2)}$, $p_x^{(2)}$, . . . . The irreducible

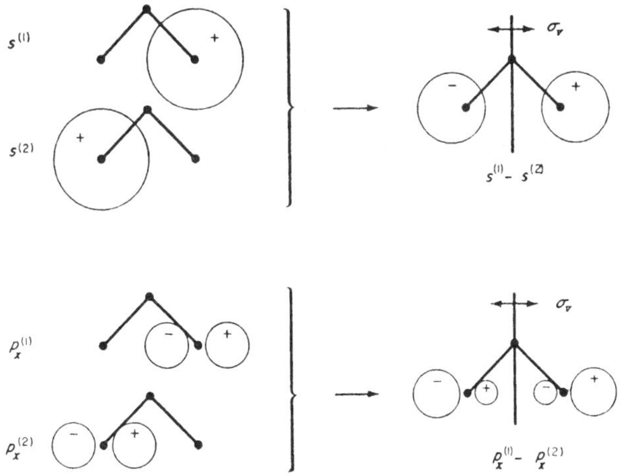


FIG. 7.8. Water molecule: construction of symmetry functions. Contours indicate (very schematically) regions in which the functions take their greatest magnitudes. Reflection $\sigma_h$ is across the plane of the paper, $C_2$ about the axis shown.

representations of $C_{2v}$ are given in Table 4.3, and the linear combinations with the corresponding symmetries are in this case obtainable by inspection. Thus, functions of species $B_1$ are symmetric under $E$ and $\sigma^{(1)}$ but antisymmetric under $C_2$ and $\sigma^{(12)}$. Clearly, from Fig. 7.8, $s^{(1)} - s^{(2)}$ (the two functions being identical but on different centres) has this property; and so has $p_x^{(1)} + p_x^{(2)}$. Some of the symmetry orbitals of the four species are shown in the table

‡ It may be necessary to consider several of each type, e.g. s orbitals $1s^{(0)}$, $2s^{(0)}$, . . . in ascending order of energy.

| $A_1$ | $A_2$ | $B_1$ | $B_2$ |
|---|---|---|---|
| $s^{(0)}$ | $p_y^{(1)} - p_y^{(2)}$ | $s^{(1)} - s^{(2)}$ | $p_y^{(0)}$ |
| $s^{(1)} + s^{(2)}$ | $d_{xy}^{(0)}$ | $p_x^{(0)}$ | $p_y^{(1)} + p_y^{(2)}$ |
| $p_z^{(0)}$ | | $p_z^{(1)} - p_z^{(2)}$ | $d_{yz}^{(0)}$ |
| $p_z^{(1)} + p_z^{(2)}$ | | $p_x^{(1)} + p_x^{(2)}$ | |
| $p_x^{(1)} - p_x^{(2)}$ | | $d_{xz}^{(0)}$ | |
| $d_{z^2}^{(0)}$ | | | |
| $d_{x^2-y^2}^{(0)}$ | | | |

Thus, if we wish to construct an approximate wave function belonging to representation $A_2$, only 2 of the 17 functions need be used.

In example 7.15 where only uni-dimensional representations occur, the symmetry functions which reduce the representation are easily selected by geometrical considerations. But the statement (7.5.5) is completely general. In order to utilize it fully we need a method of constructing symmetry functions, by suitably combining the members of an arbitrary complete set.

### 7.6. Symmetry Functions by Projection

Let us suppose that a $g_\alpha$-dimensional irreducible representation $D_\alpha$ of a finite group G is available and that we wish to obtain, from an arbitrary set of functions, sets which transform like the basis vectors of $D_\alpha$. The machinery for solving this problem has already been developed in sections 5.7–8. If we take an arbitrary function $\phi$ and expand it in terms of a symmetry adapted complete set $\{\phi_i^{(\alpha,\mu)}\}$ a similar argument shows that

$$\rho_{ij}^{(\alpha)}\,\phi = (g_\alpha/g)\sum_R \check{D}_\alpha(R)_{ij} R\phi \sim e_i^{(\alpha)} \tag{7.6.1}$$

The only difference is that the vectors of a finite-dimensional vector space are replaced by the functions of an infinite-dimensional function space and that the transformed functions $R\phi$ are understood in the sense of section 7.1. Again, we note that in the unitary case $\check{D}_\alpha(R)_{ij} = D_\alpha(R)_{ij}^*$.

EXAMPLE 7.16. Let us consider the construction of symmetry orbitals for the ammonia molecule, using atomic orbitals of types $s^{(0)}, p_x^{(0)}, p_y^{(0)}, p_z^{(0)}$ on the nitrogen atom $s^{(1)}, p_x^{(1)}, p_y^{(1)}, p_z^{(1)}$ on the first hydrogen atom, etc. It is evident that some of these orbitals are *equivalent* in the sense that they are merely permuted by symmetry operations: thus, $C_3 s^{(1)} = s^{(2)}$, etc. In order to simplify the calculation we may choose *local* axes at each atom of an equivalent set (in this case the hydrogens) so that as many of the basis orbitals as possible become equivalent or, at most, suffer only a sign change under the symmetry operations. In this case it is expedient to choose the $p_x$ and $p_y$ orbitals of each hydrogen so that they lie in " radial " and " tangential " directions (Fig. 7.9). Clearly, both the nitrogen orbitals

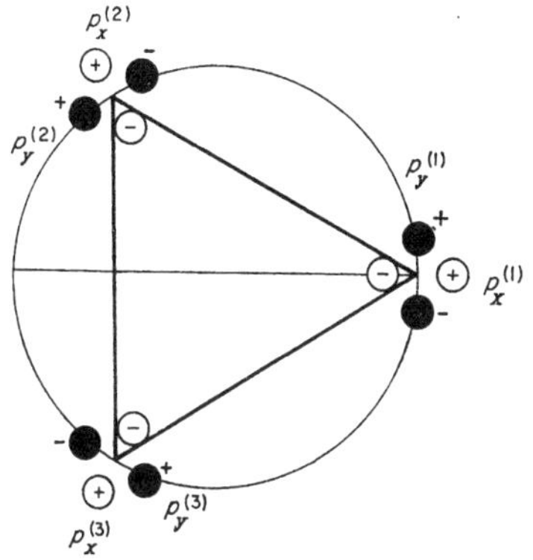


FIG. 7.9. Orientation of $p$ orbitals on three equivalent atoms. $p_x$ orbitals (open circles) and $p_y$ orbitals (shaded circles) are unmixed by symmetry operations.

and the orbitals of the equivalent set of hydrogen atoms will transform only among themselves; symmetry orbitals may therefore be constructed separately from the different groups. Taking the hydrogen group (looked at from the apex of the pyramid) we fix attention on, say, $s^{(1)}$ and ask what symmetry functions can be generated from it (Fig. 7.10). The irreducible representations are given in (2.7.3).

(i) From (7.6.1), we obtain a function transforming according to $D_1$ (i.e. an invariant) by projection with $\rho_{11}^{(1)} = \frac{1}{6} \sum_R R$ : thus

$$\sum_R R s^{(1)} = 2(s^{(1)} + s^{(2)} + s^{(3)})$$

If we normalize so that the sum of squares of the coefficients is unity,

$$\phi_1^{(1)} = (s^{(1)} + s^{(2)} + s^{(3)})/\sqrt{3}$$

This is the function indicated in Fig. 7.10(b).

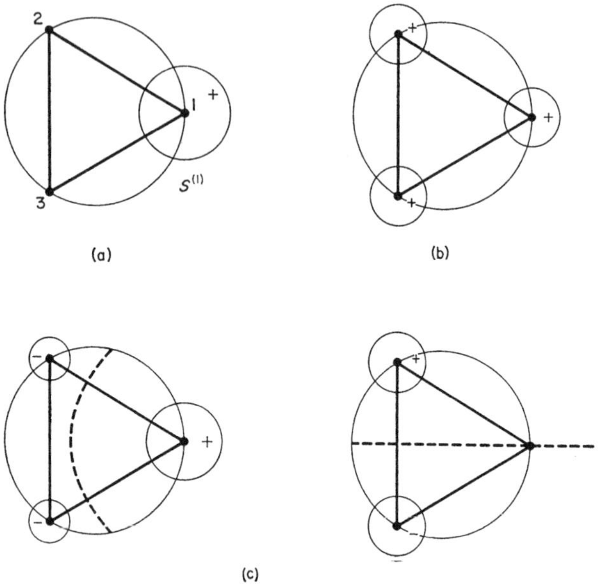


Fig. 7.10. Ammonia molecule: symmetry orbitals generated from $s^{(1)}$. (b) indicates the function belonging to $D_1$ ($A_1$ type), while (c) indicates the pair belonging to $D_3$ (E type). The broken lines indicate nodal surfaces.

(ii) If we try to obtain a function belonging to $D_2$ the result is

$$\sum_R D_2(R)_{11} R s^{(1)} = s^{(1)} + s^{(2)} + s^{(3)} - s^{(1)} - s^{(3)} - s^{(2)} = 0$$

No function belonging to $D_2$ can be constructed from the hydrogen s-orbitals.

(iii) There will be a pair of functions $\phi_1^{(3)}$ and $\phi_2^{(3)}$ belonging to $D_3$. On taking $i = 1$ in (7.6.1) we should get functions transforming like $e_1^{(3)}$—or nothing. There are two cases:

$j = 1$. $\sum_R D_3(R)^*_{11} R s^{(1)} = s^{(1)} - \frac{1}{2}s^{(2)} - \frac{1}{2}s^{(3)} + s^{(1)} - \frac{1}{2}s^{(3)} - \frac{1}{2}s^{(2)}$

giving a normalized function $\phi_1^{(3)} = (2s^{(1)} - s^{(2)} - s^{(3)})/\sqrt{6}$.

$j = 2$. In this case $\sum_R D_3(R)^*_{12} R s^{(1)} = 0$ and we cannot therefore obtain a second function transforming like $e_1^{(3)}$.

On taking $i = 2$ we obtain functions transforming like $e_2^{(3)}$.

$j = 1$. $\sum_R D_3(R)^*_{21} R s^{(1)} = \frac{1}{2}\sqrt{3}(s^{(2)} - s^{(3)} - s^{(3)} + s^{(2)})$

giving one normalized function $\phi_2^{(3)} = (s^{(2)} - s^{(3)})/\sqrt{2}$.

$j = 2$. Here $\sum_R D_3(R)^*_{22} R s^{(1)} = 0$ and again there is no second function of the same species.

The symmetry adapted linear combinations of $s^{(1)}$, $s^{(2)}$ and $s^{(3)}$ are thus

| | |
|---|---|
| Species (1, 1) | $\phi_1^{(1)} = (s^{(1)} + s^{(2)} + s^{(3)})/\sqrt{3}$ |
| Species (3, 1) | $\phi_1^{(3)} = (2s^{(1)} - s^{(2)} - s^{(3)})/\sqrt{6}$ |
| Species (3, 2) | $\phi_2^{(3)} = (s^{(2)} - s^{(3)})/\sqrt{2}$ |

This defines a basis change to new functions which carry the irreducible representations $D_1$ and $D_3$. The forms of the functions are indicated in Fig. 7.10.

Next we consider the $p$ orbitals. Because we have chosen local axes, the 9-dimensional representation which they carry is already partly reduced, the $p_x$, $p_y$ and $p_z$ types each transforming only among themselves. The $p_z$ functions, $p_z^{(1)}$, $p_z^{(2)}$, $p_z^{(3)}$, and the $p_x$ functions, $p_x^{(1)}$, $p_x^{(2)}$, $p_x^{(3)}$, behave exactly like $s^{(1)}$, $s^{(2)}$, $s^{(3)}$ under the operations of $C_{3v}$ and may therefore be disposed of immediately:

| | | |
|---|---|---|
| Species (1, 1) | $(p_z^{(1)} + p_z^{(2)} + p_z^{(3)})/\sqrt{3}$ | $(p_x^{(1)} + p_x^{(2)} + p_x^{(3)})/\sqrt{3}$ |
| Species (3, 1) | $(2p_z^{(1)} - p_z^{(2)} - p_z^{(3)})/\sqrt{6}$ | $(2p_x^{(1)} - p_x^{(2)} - p_x^{(3)})/\sqrt{6}$ |
| Species (3, 2) | $(p_z^{(2)} - p_z^{(3)})/\sqrt{2}$ | $(p_x^{(2)} - p_x^{(3)})/\sqrt{2}$ |

The $p_y$ functions, however, are not merely permuted under the symmetry operations, their signs being reversed under each reflection. The reader should verify that the symmetry functions (which now arise on taking $j = 2$) are

| | |
|---|---|
| Species (2, 1) | $(p_y^{(1)} + p_y^{(2)} + p_y^{(3)})/\sqrt{3}$ |
| Species (3, 1) | $(p_y^{(3)} - p_y^{(2)})/\sqrt{2}$ |
| Species (3, 2) | $(2p_y^{(1)} - p_y^{(2)} - p_y^{(3)})/\sqrt{6}$ |

The function belonging to $D_2$ is that indicated in Fig. 7.11(b): the pair belonging to $D_3$ are indicated in Fig. 7.11(c).

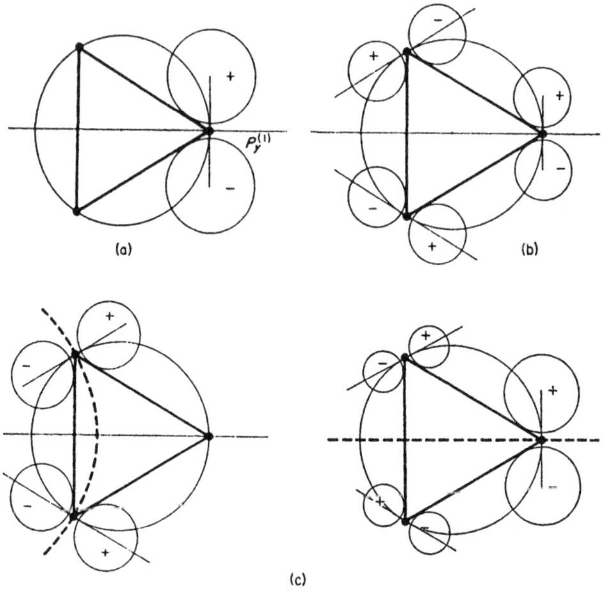


FIG. 7.11. Ammonia molecule: symmetry orbitals generated from $p_y^{(1)}$. (b) indicates the function belonging to $D_2$ ($A_2$ type), while (c) indicates the pair belonging to $D_3$ (E type). The broken lines indicate nodal surfaces.

Finally, we note that the orbitals on the nitrogen atom are already symmetry adapted. $s^{(0)}$ and $p_z^{(0)}$ each transform like $e_1^{(1)}$ while $p_x^{(0)}$ and $p_y^{(0)}$ behave like $e_1^{(3)}$, and $e_2^{(3)}$ respectively.

The symmetry orbitals are used as in example 7.15. If, for example, we wish to build up an approximation to the 2-fold degenerate eigenfunction which has a nodal plane through atoms 0 and 1 (perpendicular to the line 2–3) we should use the symmetry orbitals (species (3, 1)) $p_x^{(0)}$, $(2s^{(1)} - s^{(2)} - s^{(3)})/\sqrt{6}$, $(2p_z^{(1)} - p_z^{(2)} - p_z^{(3)})/\sqrt{6}$, $(2p_x^{(1)} - p_x^{(2)} - p_x^{(3)})/\sqrt{6}$, $(p_y^{(3)} - p_y^{(2)})/\sqrt{2}$. The inclusion of d functions may, of course, be discussed in a similar way.

EXAMPLE 7.17. As a final example of the projection method we consider a problem in solid state theory. Given a 3-dimensional lattice, how can we construct functions transforming according to the irreducible representations of the translational symmetry group? The representations, which are all uni-dimensional, were derived in section 4.3 in terms of a reciprocal lattice and appear in (4.3.11). With the translational operation $(E \mid t)$ $(t = t_1a_1 + t_2a_2 + t_3a_3)$ we associate, in representation $D_k$, the number $e^{ik.t}$, $k$ being a vector of the reciprocal space.

First we note that functions transforming like the single basis vector of representation $D_k$ must have the property

$$(E \mid t)\, \psi^{(k)}(r) = \exp\,(ik \cdot t)\, \psi^{(k)}(r)$$

If we adopt the standard interpretation (section 7.1) of a symmetry operation on a function, the left-hand side (which is the function similarly defined, but with respect to a reference frame translated through $t$) may be written as the function $\psi^{(k)}$ of the backwards-translated field point, $r - t$, i.e. as $\psi^{(k)}(r - t)$ (cf. p. 169). The symmetry functions therefore have the property (on replacing $t$ by $-t$)

$$\psi^{(k)}(r + t) = \exp\,(-ik\,.\,t)\, \psi^{(k)}(r)$$

This is in fact the basic property of crystal wave functions‡ which, since they are eigenfunctions of a Hamiltonian operator invariant under the symmetry operations, must provide bases for irreducible representations. In this context, the result was first given by Bloch.

In order to set up symmetry functions we may take a quite arbitrary function, which we shall choose for illustration as the atomic orbital $\phi_0$ of an atom at the origin, and use the projection method. A function belonging to $D_k$ is then

$$\phi^{(k)}(r) = \sum_t\,[\exp\,(ik\,.\,t)]^* \quad (E \mid t)\,\phi_0(r)$$

where the summation is over all allowed lattice translations (i.e. over the $G^3$ lattice points of a large periodic volume, p. 105). Now if $t = r_m$ sends a point at the origin to a lattice point $r_m$, we have

$(E \mid t)\, \phi_0(r) = \phi_m(r)$ (an orbital like $\phi_0$ but centred on the atom at $r_m$)

It follows that

$$\phi^{(k)}(r) = \sum_m \exp\,(-ik \cdot r_m) \quad \phi_m(r)$$

Functions of this kind are often referred to as *crystal orbitals*: again, they were first systematically investigated by Bloch.

### 7.7. Symmetry Functions and Equivalent Functions

So far, we have discussed the construction and properties of functions which transform like the basis vectors of the irreducible representations of the symmetry group. But it is sometimes useful to

‡ In much of the literature, the exponent contains the *plus* sign. This is because, following Bloch and others, the translation operator $(E|t)$ is often applied to the *variable* $r$—not to the *function*. If $[E|t]$ is used for such an operation, $\psi^{(k)}$ is defined by

$$\psi^k(r + t) = [E|t]\psi^k(r) = \exp\,(ik\,.\,t)\psi^k(r)$$

The change corresponds only to a relabelling of representations.

introduce functions which are *equivalent* except for their orientation in space and are therefore merely *permuted* by the symmetry operations: the orbitals on the three hydrogen atoms of ammonia (example 7.16) were equivalent in this sense. Equivalent functions generally carry *reducible* representations of the symmetry group, the matrices $\boldsymbol{D}(R)$ having the special form characteristic of permutation matrices (cf. example 5.1). The question which then arises is: which symmetry functions should be mixed in order to produce a given set of equivalent functions? The answer is provided simply by inspection of the character systems of the irreducible representations (provided by the symmetry functions) and the permutation representation (provided by the equivalent functions): for the permutation representation, carried by suitable linear combinations of the symmetry orbitals, must be *equivalent* to the " block-form " representation carried by the symmetry orbitals themselves. It is therefore only necessary to know *which* irreducible representations are contained in the given permutation representation, and this information is given by the methods of section 5.4. The following example makes this clear.

EXAMPLE 7.18. Is it possible to construct four equivalent functions, which are permuted like the vertices of a regular tetrahedron, from the $s$, $p$, and $d$ orbitals of a central atom? This amounts to asking which irreducible representations of $\mathsf{T}_d$ are contained in the corresponding permutation representation: for sets of functions carrying these representations can then be mixed to give the desired equivalent orbitals. The classes of $\mathsf{T}_d$ are given in Table 3.9: they are the identity, the three rotations through $2\pi/2$, the eight (positive and negative) rotations through $2\pi/3$, the six improper rotations through $2\pi/4$, and the six reflections. The character system is obtained (cf. p. 155) by counting the number of unshifted vertices in the corresponding permutations: it is compared below with the character systems for the irreducible representations given in Table 4.20 to which the $s$, $p$, and $d$ orbitals are known to belong (appendix 1).

| Representation | $E$ | $8C_3$ | $3C_2$ | $6\sigma_d$ | $6S_4$ |
|---|---|---|---|---|---|
| D (Permutation) | 4 | 1 | 0 | 2 | 0 |
| $A_1$ ($s$ function) | 1 | 1 | 1 | 1 | 1 |
| E ($d_{z^2}$, $d_{x^2-y^2}$) | 2 | −1 | 2 | 0 | 0 |
| $T_2$ $\begin{cases}(p_x\ p_y\ p_z)\\(d_{yz}\ d_{zx}\ d_{xy})\end{cases}$ | 3 | 0 | −1 | 1 | −1 |

Hence (by inspection or by using (5.4.5)) $D = A_1 + T_2$. Thus, we can construct tetrahedral equivalent orbitals from an $s$ function and any set of three $T_2$ functions. The appropriate mixtures of $s$, $p_x$, $p_y$, $p_z$ are the sp³ " hybrid " orbitals of valence theory, the mixtures of $d_{yz}$, $d_{zx}$, $d_{xy}$, $s$ are the d³s hybrids ‡: the most general tetrahedral hybrid formed from $s$, $p$, and $d$ orbitals would evidently be mixtures of

$$s, \quad \lambda p_x + \mu d_{yz}, \quad \lambda p_y + \mu d_{zx}, \quad \lambda p_z + \mu d_{xy}$$

since $(\lambda p_x + \mu d_{yz})$ is the most general function available which transforms like the first basis vector $\mathbf{e}_1$ of the $T_2$ representation; the two subsequent functions being its partners behaving like $\mathbf{e}_2$ and $\mathbf{e}_3$.

### 7.8. Determination of Equivalent Functions

By the method of section 7.7 we can establish the possibility of constructing a given set of equivalent functions from two or more sets of symmetry functions. It remains to determine the actual coefficients. One way of solving this problem is to suppose the equivalent functions given and to project from them the symmetry functions of each species as in section 7.6. This expresses the symmetry functions in terms of the equivalent functions and it is then necessary only to invert this relationship—which is trivial when the transformation matrix is unitary.

Example 7.19. Let us consider the formation of three equivalent orbitals pointing towards the base corners of a triangular pyramid, from the $s$ and $p$ atomic orbitals of a nitrogen atom at the apex: these " hybrid " orbitals might be used in a valence bond description of $NH_3$. When the axes are chosen as in Fig. 6.3 the $D_1$ ($= A_1$) and $D_3$ ($= E$) representations of $C_{3v}$ are carried by the unit vectors along the $z$-axis ($\mathbf{e}_1^{(1)}$) and along the $x$, $y$ axes ($\mathbf{e}_1^{(3)}$, $\mathbf{e}_2^{(3)}$, respectively). Since the $p$ orbitals transform like these basis vectors, we can say

$$\phi = (\lambda s + \mu p_z) \sim \mathbf{e}_1^{(1)} \qquad \text{(invariant, basis for } D_1)$$

$$\left.\begin{array}{l} p_x \sim \mathbf{e}_1^{(3)} \\ p_y \sim \mathbf{e}_2^{(3)} \end{array}\right\} \qquad \text{(basis for } D_3)$$

It is readily shown, as in example 7.18, that the three equivalent orbitals may be constructed from orbitals providing bases for $D_1$ and $D_3$, i.e. as linear combinations of $\phi$, $p_x$ and $p_y$. Let us now call the equivalent orbitals $\psi_1$, $\psi_2$, $\psi_3$ (pointing towards corners 1, 2, 3 in

‡ It is customary to list the orbitals in increasing energy order: usually $E_{3d} < E_{3s}$.

Fig. 6.3). The effect of symmetry operations on $\psi_1$ is then

| $R$ | $E$ | $C_3$ | $\bar{C}_3$ | $\sigma^{(1)}$ | $\sigma^{(2)}$ | $\sigma^{(3)}$ |
|---|---|---|---|---|---|---|
| $R\psi_1$ | $\psi_1$ | $\psi_2$ | $\psi_3$ | $\psi_1$ | $\psi_3$ | $\psi_2$ |

Exactly as in example 7.16, we may now project out of $\psi_1$ an invariant (some multiple of $\phi$), and a pair of functions behaving like $e_1^{(3)}$ and $e_2^{(3)}$ (some multiple of $p_x$ and $p_y$ respectively). Since the multiplicative factors are arbitrary, we normalize the projected functions so that the sum of the squared coefficients is unity: the two bases are then related by a unitary transformation. Thus we find $\phi = (\psi_1 + \psi_2 + \psi_3)/\sqrt{3}$, $p_x = (2\psi_1 - \psi_2 - \psi_3)/\sqrt{6}$, $p_y = (\psi_2 - \psi_3)/\sqrt{2}$

$$(\phi \ p_x \ p_y) = (\psi_1 \ \psi_2 \ \psi_3) \begin{pmatrix} 1/\sqrt{3} & 2/\sqrt{6} & 0 \\ 1/\sqrt{3} & -1/\sqrt{6} & 1/\sqrt{2} \\ 1/\sqrt{3} & -1/\sqrt{6} & -1/\sqrt{2} \end{pmatrix}$$

If the unitary matrix is denoted by $\boldsymbol{U}$ we then have $(\psi_1 \ \psi_2 \ \psi_3) = (\phi \ p_x \ p_y)\boldsymbol{U}^\dagger$ and the desired equivalent orbitals are, most generally,

$$\psi_1 = (1/\sqrt{3})\,\phi + (2/\sqrt{6})\,p_x$$
$$\psi_2 = (1/\sqrt{3})\,\phi - (1/\sqrt{6})\,p_x + (1/\sqrt{2})\,p_y$$
$$\psi_3 = (1/\sqrt{3})\,\phi - (1/\sqrt{6})\,p_x - (1/\sqrt{2})\,p_y$$

Orbitals of this kind are indicated schematically in Fig. 7.12. It is

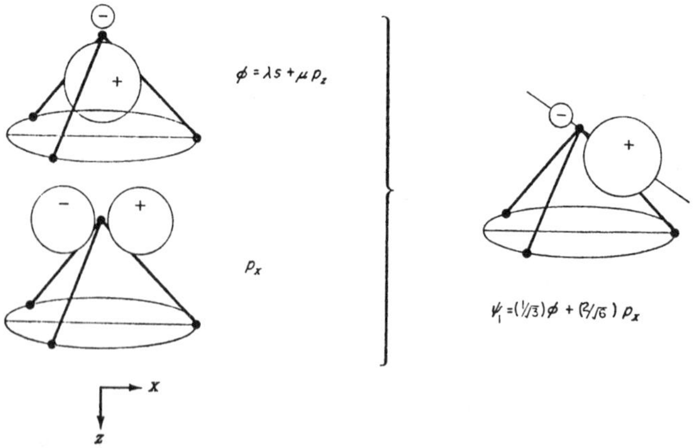


FIG. 7.12. Construction of hybrid orbital. The orbital $\psi_1$ is one of a set of three equivalent hybrids, being permuted under symmetry operations like the base corners of the pyramid.

usually most convenient to work with orthonormal functions such that

$$\int \psi_i^* \psi_j \, dx \, dy \, dz = \delta_{ij}$$

Assuming $s$, $p_x$, $p_y$, $p_z$ have this property, $\phi$ is normalized when $\lambda^2 + \mu^2 = 1$, and $\psi_1$, $\psi_2$, $\psi_3$ are orthonormal. The ratio $\lambda/\mu$ remains undetermined: it fixes the inclination of the hybrid orbitals.

We conclude by developing a slightly more direct and powerful method of generating equivalent functions. The projection method developed in section 5.7 was derived with the assumption that the representation matrices were *irreducible*. We now show that

$$\sum_R \check{D}(R)_{ij} R\phi \sim \mathbf{e}_i \qquad \text{(the } i\text{th basis vector of D)} \qquad (7.8.1)$$

even when D is a reducible representation. This means that we may project equivalent functions directly, using only the permutation representation D.

The proof follows on operating upon the function in (7.6.1) with a rotation $S$:

$$\begin{aligned} S[\sum_R \check{D}(R)_{ij} R\phi] &= \sum_R \check{D}(R)_{ij} SR\phi \\ &= \sum_T \check{D}(S^{-1}T)_{ij} T\phi \qquad (T = SR) \\ &= \sum_T \sum_k \check{D}(S^{-1})_{ik} \check{D}(T)_{kj} T\phi \\ &= \sum_k [\sum_R \check{D}(R)_{kj} R\phi] D(S)_{ki} \end{aligned}$$

where we have used the definition (5.3.13) and have replaced $T$ by $R$ in the last summation. The function on the left in (7.8.1) does, therefore, transform like the $i$th basis vector of D: $S\mathbf{e}_i = \sum_k \mathbf{e}_k D(S)_{ki}$.

An example shows the value of this result.

EXAMPLE 7.20. Suppose we wish to construct from the $s$, $p$, and $d$ orbitals of an atom at the origin, a set of six orbitals, equivalent under the operations of the group O, and pointing towards the six cube faces in Fig. 3.11. The permutation representation D is found, as in example 7.18, to be the direct sum $D = A_1 + E + T_1$. This indicates that the desired orbitals may be constructed from

| | |
|---|---|
| $s$ | (behaving like the basis vector $\mathbf{e}_1^{(A_1)}$ of $A_1$) |
| $p_x$, $p_y$, $p_z$ | (behaving like the basis vectors $\mathbf{e}_1^{(T_1)}$, $\mathbf{e}_2^{(T_1)}$, $\mathbf{e}_3^{(T_1)}$ of $T_1$) |
| $d_{z^2}$, $d_{x^2-y^2}$ | (behaving like the basis vectors $\mathbf{e}_1^{(E)}$, $\mathbf{e}_2^{(E)}$ of E) |

Let us take an arbitrary linear combination $\phi = (s\ d_{z^2}\ d_{x^2-y^2}\ p_x\ p_y\ p_z)\boldsymbol{c}$, where $\boldsymbol{c}$ is a column of arbitrary coefficients $c_1, c_2, \ldots, c_6$. Then

$$R\phi = (s\ d_{z^2}\ d_{x^2-y^2}\ p_x\ p_y\ p_z)\ \boldsymbol{Rc}$$

where $\boldsymbol{R}$ is the matrix associated with $R$ in the representation carried by the atomic orbitals: but the latter are symmetry adapted and $\boldsymbol{R}$ is therefore of block form, consisting of the $A_1$, E, and $T_1$ matrices which can be read off from Table 4.19. The coefficients $\boldsymbol{Rc}$ which determine the effect of any operation may in fact be written down by inspection: for example, the elements of $\boldsymbol{Rc}$ for $R = C_4^x$ are

$$(c_1,\ -\tfrac{1}{2}c_2 - \tfrac{1}{2}\sqrt{3}\ c_3,\ -\tfrac{1}{2}\sqrt{3}\ c_2 + \tfrac{1}{2}c_3,\ c_4,\ -c_6,\ c_5)$$

To use (7.8.1) we observe that for a permutation representation there is only one non-zero element (namely, 1), in each row and column of $\boldsymbol{D}(R)$ and that $\check{\mathrm{D}} = \mathrm{D}$ since the matrices are not only unitary but also real. The number of terms in (7.8.1) is therefore always small: in fact since $\boldsymbol{D}(R)$ must have the property

$$R\psi_j = \sum_k \psi_k D(R)_{kj}$$

it follows that $D(R)_{ij} = 0$ unless $R$ sends $\psi_j$ into $\psi_i$, in which case it is unity. We now project out, for illustration, two partners in the basis carrying the permutation representation—one, $\psi_1$, pointing along the positive $x$-axis and another, $\psi_3$, along the positive $z$-axis.

*Function* $\psi_1$

We use (7.8.1) with $j = 1$ (i.e. first column of the representation matrices): the function transforming like the first basis vector is then obtained on putting $i = 1$. But $D(R)_{11} = 1$ only for those operations which leave point 1 unmoved (Fig. 3.11) and these are $R = E, C_2^x, C_4^x, \bar{C}_4^x$. The components $\boldsymbol{Rc}$ of the rotated functions $R\phi$ are then found to be

| $R$ | Elements of $\boldsymbol{Rc}$ | | | | | |
|---|---|---|---|---|---|---|
| $E$ | $c_1$ | $c_2$ | $c_3$ | $c_4$ | $c_5$ | $c_6$ |
| $C_2^x$ | $c_1$ | $c_2$ | $c_3$ | $c_4$ | $-c_5$ | $-c_6$ |
| $C_4^x$ | $c_1$ | $(-\tfrac{1}{2}c_2 - \tfrac{1}{2}\sqrt{3}\ c_3)$ | $(-\tfrac{1}{2}\sqrt{3}\ c_2 + \tfrac{1}{2}c_3)$ | $c_4$ | $-c_6$ | $c_5$ |
| $\bar{C}_4^x$ | $c_1$ | $(-\tfrac{1}{2}c_2 - \tfrac{1}{2}\sqrt{3}\ c_3)$ | $(-\tfrac{1}{2}\sqrt{3}\ c_2 + \tfrac{1}{2}c_3)$ | $c_4$ | $c_6$ | $-c_5$ |

(7.8.2)

The components of the projected function (7.8.1) are obtained by addition and are

$$4c_1,\ \ (c_2 - \sqrt{3}\ c_3),\ \ -\sqrt{3}\ (c_2 - \sqrt{3}\ c_3),\ \ 4c_4,\ \ 0,\ \ 0$$

In other words, the function pointing along the $x$-axis is

$$\psi_1 = \lambda s + \mu(d_{z^2} - \sqrt{3}\, d_{x^2-y^2}) + \nu p_x$$

where $\lambda = 4c_1$, $\mu = (c_2 - \sqrt{3}\, c_3)$ and $\nu = 4c_4$ are arbitrary parameters.

*Function* $\psi_3$

The function which transforms like the third basis vector and is a partner to $\psi_1$ is determined by putting $i = 3$. But $D(R)_{31} = 1$ only for those operations which send point 3 into point 1 (Fig. 3.11), namely $R = \bar{C}_4^y, C_2^{zx}, \bar{C}_3^{xyz}, \bar{C}_3^{\bar{x}y\bar{z}}$. The projected function is in this case found to have components

$$4c_1, \quad -2(c_2 - \sqrt{3}\, c_3), \quad 0, \quad 0, \quad 0, \quad 4c_4$$

The function pointing along the $z$-axis is thus (with the *same* parameters as in $\psi_1$)

$$\psi_3 = \lambda s - 2\mu d_{z^2} + \nu p_z$$

It is interesting to verify directly that these functions are alike in every respect except for their orientation in space. We need only note that with

$$d_{z^2} = (2z^2 - x^2 - y^2)f(r) \text{ and } d_{x^2-y^2} = \sqrt{3}\,(x^2 - y^2)f(r)$$
$$d_{z^2} - \sqrt{3}\, d_{x^2-y^2} = (2z^2 - 4x^2 + 2y^2)f(r) = -2\, d_{x^2}$$

where $d_{x^2}$ is defined exactly like $d_{z^2}$ but with the $x$-axis taking the place of the $z$-axis. The function $\psi_1$ could thus be written in exactly the same form as $\psi_3$:

$$\psi_1 = \lambda s - 2\mu\, d_{x^2} + \nu p_x$$

The six equivalent functions follow, in fact, on replacing $x$ by $y$, $z$, $-x$, $-y$, $-z$ in turn, taking $p_{-x} = -p_x$, etc. The parameters $\lambda$, $\mu$, $\nu$ are in this case completely determined if we require the six functions to be orthonormal as in example 7.19.

## REFERENCES

1. Pauling, L., and Wilson, E. B., *Introduction to Quantum Mechanics* (chapter 5), McGraw-Hill, 1935.
2. Eyring, H., Walter, J., and Kimball, G. E., *Quantum Chemistry* (chapter 6; see also chapter 4), Wiley, 1944.
3. Wigner, E. P., *Group Theory and its Application to the Quantum Mechanics of Atomic Spectra* (section 15), Academic Press, 1959.
4. Coulson, C. A., *Valence* (chapter 10, 2nd Ed.), Oxford University Press, 1961.

## BIBLIOGRAPHY

HEINE, V., *Group Theory in Quantum Mechanics*, Pergamon Press, Oxford, 1960.

LYUBARSKII, G. YA, *The Application of Group Theory in Physics*, Pergamon Press, Oxford, 1960. (This book and that by Heine cover a vast range of detailed applications and are recommended for reference and further reading, along with the older classics by Wigner and others.)

Short accounts of group theory in quantum mechanics are available in chapter 10 of Ref. 2 and also in chapter 12 of

LANDAU, L. D., and LIFSHITZ, E. M., *Quantum Mechanics*, Pergamon Press, Oxford, 1959.

CHAPTER 8

# APPLICATIONS INVOLVING TENSORS AND TENSOR OPERATORS

## 8.1. Scalar, Vector and Tensor Properties

The two preceding chapters have been concerned primarily with the *invariance* of algebraic forms or of operators, under a group of symmetry operations, and with the simpler consequences of such invariance. On the other hand, quantities which are *not* invariant may also have an intrinsic physical importance: one example would be the set of elements of the polarizability matrix in section 6.1 which, it was remarked, defined a " tensor " property. The bearing of symmetry upon the simpler tensor properties of a system may be discussed adequately by the methods of chapter 6; but tensor properties of much greater complexity are also encountered. Thus, in the piezoelectric effect, an elastic stress (6 components) yields an electric moment (3 components); so the related quantities must be connected by a *rectangular* matrix of physical constants. In this chapter we examine more carefully the concept of a " tensor property ", and suitably generalize the analysis of chapter 6. Finally, we show how *operators* may occur in sets with tensor character and how they may be handled similarly. This prepares the way for more elaborate applications in quantum mechanics and, in particular, for a general understanding of selection rules in atomic and molecular theory.

It was shown in section 2.6 that a relationship of the form‡

$$\boldsymbol{x}' = \boldsymbol{R}\boldsymbol{x} \tag{8.1.1}$$

between two sets of vector components (columns) could be regarded either (i) as a mapping which associates an image $\mathsf{x}'$ (new vector) with every vector $\mathsf{x}$, the new basis vectors being $\mathbf{e}' = \mathbf{e}\boldsymbol{R}$, or (ii) as a transformation of components of a fixed $\mathsf{x}$ accompanying an *inverse* rotation of the basis, $\mathbf{e} \rightarrow \mathbf{e}' = \mathbf{e}\boldsymbol{R}^{-1}$. In the latter case the vector is regarded as an *invariant* and the equations

$$\mathsf{x} = \mathbf{e}\boldsymbol{x} = \mathbf{e}\boldsymbol{R}^{-1}\boldsymbol{R}\boldsymbol{x} = \mathbf{e}'\boldsymbol{x}' \tag{8.1.2}$$

‡ In this chapter we use $x_1$, $x_2$, $x_3$ for the coordinates, as is usual in tensor theory, and denote a field point by $\mathsf{x}$.

which merely relate two modes of mathematical description of the same thing, express the *contragredience* of basis vectors and components. The actual interpretation of (8.1.1) as describing a real rotation or as a change of basis is for the most part irrelevant in what follows : it is primarily the recognition of different types of quantity which will concern us.

Since, however, the second stand point is the one usually adopted in tensor theory (see, for instance, Hall 1962) we show formally that the set $\{\boldsymbol{R}\}$ of matrix transformations of components provides a representation of

(i) Actual rotations $\{R\}$ . . . of a vector $x$.
(ii) Corresponding *inverse* rotations of the reference frame *alone*, the vector $x$ remaining fixed, each rotation being defined with respect to the frame itself (not with respect to some fixed basis).

The " active " interpretation (i) has already been examined : the matrix which describes the rotation of vector $x$ is defined by $R\mathbf{e} = \mathbf{e}\boldsymbol{R}$, and if $RS = T$ it follows that $\boldsymbol{RS} = \boldsymbol{T}$. To discuss the " passive " interpretation (ii) we introduce the symbol $\hat{R}$ to describe an *inverse* rotation ($R^{-1}$) of the basis, *defined with respect to itself* (not the fixed basis as in (i)). Thus in the sequence $\hat{R}\hat{S}$ ($\hat{S}$ first), $\hat{R}$ is an inverse rotation like $R^{-1}$ but defined after first rotating by $S^{-1}$. To examine the change of components of a fixed vector in going from the original basis $\mathbf{e}$ to the final basis, everything must be referred to $\mathbf{e}$. But from (3.4.1) it is known that $\hat{R}$, the operator defined like $R^{-1}$ in a frame rotated through $S^{-1}$, is $\hat{R} = S^{-1}R^{-1}(S^{-1})^{-1}$ and hence that

$$\hat{R}\hat{S} = S^{-1}R^{-1}$$

A sequence of rotations, each defined with respect to the " floating " basis, is thus equivalent to a similar sequence *applied in reverse order* with the rotations defined in the fixed basis.‡ The corresponding change of basis vectors is now evident

$$\hat{R}\hat{S}\mathbf{e} = S^{-1}R^{-1}\mathbf{e} = \mathbf{e}\boldsymbol{S}^{-1}\boldsymbol{R}^{-1} = \mathbf{e}(\boldsymbol{RS})^{-1}$$

and the components of a fixed vector $x$ transform contragrediently :

$$\boldsymbol{x} \to \boldsymbol{x}' = \boldsymbol{RSx}$$

Hence if $\hat{R}\hat{S} = \hat{T}$, $\boldsymbol{RS} = \boldsymbol{T}$ : the same matrices $\boldsymbol{R}, \boldsymbol{S}$ . . . therefore provide a representation of the two (isomorphic) groups (i) and (ii). Henceforth, it is irrelevant which interpretation we adopt.

In subscript notation, the transformations in (8.12) are

$$x_i \to x_i' = \sum_j R_{ij}x_j \tag{8.1.3}$$

$$\mathbf{e}_i \to \mathbf{e}_i' = \sum_j \mathbf{e}_j(\boldsymbol{R}^{-1})_{ji} = \sum_j (\tilde{\boldsymbol{R}}^{-1})_{ij}\mathbf{e}_j = \sum_j \check{R}_{ij}\mathbf{e}_j \tag{8.1.4}$$

where the transposition allows us to put the set of quantities being transformed on the extreme right whilst preserving the same sequence of subscripts as in the transformation law for *components:* this is advantageous in comparing the behaviour of different sets of quantities, under a common rotation. The matrix $\check{\boldsymbol{R}} = \tilde{\boldsymbol{R}}^{-1}$ has already appeared (p. 120) in the theory of representations.

‡ This general result is well known in connection with Euler's representation of rigid body rotations (see, for example, the book by Edmonds (Ref. 1)).

The basic rotations of physical importance are those which occur in 3-dimensional space,‡ and the three components $x_1$, $x_2$, $x_3$ of an ordinary displacement vector provide the basic example of a "tensorial set".

| | |
|---|---|
| Any set of entities which, when the basis is changed, must be replaced by a new set according to $T_i \to T'_i = \sum_j R_{ij}T_j$ is a *tensorial set* transforming *co*-grediently to the coordinates $x_1$, $x_2$, $x_3$ but *contra*-grediently to the basis vectors. It is a *contravariant set.* | (8.1.5) |

From the definition $\mathsf{v} = \mathbf{e}\boldsymbol{v}$, the set of *parallel projections* of *any* given vector (e.g. the electric intensity) form a contravariant set : in order that the vector shall remain invariant its components must be changed in exactly the same way as $x_1$, $x_2$, $x_3$. Vector components are, however, very special examples of tensorial sets : later we shall find *operators* and *functions* with exactly similar transformation properties, and the interpretation will therefore be somewhat wider than that adopted in classical expositions of tensor analysis. For this reason we use the terminology of Fano and Racah (1959).

The second basic transformation law is defined by (8.1.4).

| | |
|---|---|
| Any set of entities which, when the basis is changed, must be replaced by a new set according to $T_i \to T'_i = \sum_j \breve{R}_{ij}T_j$ is a tensorial set transforming co-grediently to the basis vectors $\mathsf{e}_1$, $\mathsf{e}_2$, $\mathsf{e}_3$. It is a *covariant* set. | (8.1.6) |

It should be noted that the basic transformation law (8.1.1) with matrix $\boldsymbol{R}$, refers to the coordinates (or in tensor *calculus* to their *differentials*) and hence to a *contra*variant set. The terminology is easily generalized, as the following example shows.

EXAMPLE 8.1. Suppose the components $\boldsymbol{x}$ and $\boldsymbol{y}$ of two vectors $(\mathsf{x}, \mathsf{y})$ are related by

$$\boldsymbol{y} = \boldsymbol{T}\boldsymbol{x}$$

Then after transformation with matrix $\boldsymbol{R}$ the nine elements of $\boldsymbol{T}$ must be replaced by new elements, the relationship becoming $\boldsymbol{y}' = \boldsymbol{T}'\boldsymbol{x}'$. To obtain $\boldsymbol{T}'$ we observe that $\boldsymbol{y}' = \boldsymbol{R}\boldsymbol{y}$, $\boldsymbol{x}' = \boldsymbol{R}\boldsymbol{x}$ (vector

‡ Here we consider only non-relativistic applications.

components, by definition, having the *same* transformation law) and hence that

$$y = \boldsymbol{R}^{-1}y' = \boldsymbol{T}\boldsymbol{R}^{-1}\boldsymbol{x}' \qquad \text{or} \qquad y' = \boldsymbol{R}\boldsymbol{T}\boldsymbol{R}^{-1}\boldsymbol{x}'$$

This shows at once that after transformation

$$y' = \boldsymbol{T}'\boldsymbol{x}'$$

where

$$\boldsymbol{T}' = \boldsymbol{R}\boldsymbol{T}\boldsymbol{R}^{-1}$$

The nine elements of $\boldsymbol{T}$ therefore transform according to

$$T_{ij} \rightarrow T'_{ij} = \sum_{k,l} R_{ik}T_{kl}(\boldsymbol{R}^{-1})_{lj} = \sum_{k,l} R_{ik}\breve{R}_{jl}T_{kl}$$

These nine quantities are described as the components of a *second rank tensor*, with one degree of covariance and one of contravariance. The relationship between a tensor and its components resembles that between a vector and its components : the components depend on the manner of description but the tensor itself is an invariant entity (cf. a mapping), independent of the reference frame, which describes a certain physical relationship (e.g. between stress and strain—in which context the term "tensor" originated). The abstract relationship may be written in the usual way, $y = Tx$. The matrix $\boldsymbol{T}$, or the set of components which it contains, simply characterizes the tensor $T$ in some given coordinate system.

If a set of quantities has a transformation law of the form

$$T_{i..j...} \rightarrow T'_{i..j...} = \sum_{k,l,...} R_{ik} \ldots \breve{R}_{jl} \ldots T_{k..l...} \tag{8.1.7}$$

with $p$ $R$ factors and $q$ $\breve{R}$ factors, it is a tensorial set of rank $(p+q)$ with $p$ degrees of contravariance and $q$ of covariance. We note that tensors of rank one are usually referred to as vectors, while those of rank zero are *scalar invariants* (independent of coordinate system).

### 8.2. Significance of the Metric

It should be remembered that *when a metric exists* (section 2.8) a vector may be specified in two ways. The *parallel* projections, corresponding to resolution into components along $\mathbf{e}_1$, $\mathbf{e}_2$, $\mathbf{e}_3$, form the contravariant set. But the vector may also be described by its orthogonal projections (Fig. 7.2) :

$$\breve{x}_i = \mathbf{e}_i \cdot \mathbf{x} \tag{8.2.1}$$

and $\breve{x}_1$, $\breve{x}_2$, $\breve{x}_3$ form a covariant set: for if $\mathbf{e}_i \rightarrow \mathbf{e}'_i = \sum_j \breve{R}_{ij}\mathbf{e}_j$

$$\breve{x}_i \rightarrow \breve{x}'_i = \sum_j \breve{R}_{ij}\mathbf{e}_j \cdot \mathbf{x} = \sum_j \breve{R}_{ij}\breve{x}_j$$

the covariant law. In tensor analysis the two systems of components are distinguished by superscript (contravariant) and subscript (covariant) and it readily follows that, *for transformations in real space*, the different sets are related by

$$\check{\boldsymbol{x}} = \boldsymbol{M}\boldsymbol{x} \tag{8.2.2}$$

where $\boldsymbol{M}$ is the (real) metrical matrix (p. 44) with elements $M_{ij} = \mathbf{e}_i \cdot \mathbf{e}_j$. We note that the $k$-space components of a vector (p. 107) are nothing more than its covariant components, obtained exactly as in (8.2.2). The metric thus provides a method of " raising " or " lowering " indices. Because the assumption of real vectors is too restrictive for quantum mechanical applications, some departure from ordinary tensor theory is necessary and we shall *not* adopt the usual superscript-subscript terminology.

We now indicate the extension to Hermitian vector spaces. The transformation laws (8.1.3) and (8.1.4) were derived by considering a vector as the fundamental invariant : when a *Hermitian* metric is admitted there is a second fundamental invariant—the squared length of a vector, $|x|^2 = \boldsymbol{x}^\dagger\boldsymbol{M}\boldsymbol{x}$. If the basis is changed so that $\boldsymbol{x} \to \boldsymbol{R}\boldsymbol{x}$ and $\boldsymbol{x}^\dagger \to \boldsymbol{x}^\dagger\boldsymbol{R}^\dagger$, then (see p. 48) the metric must be changed according to

$$|x|^2 = \boldsymbol{x}^\dagger\boldsymbol{M}\boldsymbol{x} = \boldsymbol{x}'^\dagger\boldsymbol{M}'\boldsymbol{x}' \tag{8.2.3}$$

where

$$\boldsymbol{M}' = \boldsymbol{R}^{-1\dagger}\boldsymbol{M}\boldsymbol{R}^{-1} \tag{8.2.4}$$

We are therefore interested also in sets of quantities which transform according to

$$x_i^* \to x_i'^* = \sum_j R_{ij}^* x_j^* \tag{8.2.5}$$

and

$$M_{ij} \to M'_{ij} = \sum_{l,k} (\boldsymbol{R}^{-1\dagger})_{ik} M_{kl} (\boldsymbol{R}^{-1})_{lj} = \sum_{l,k} \check{R}_{ik}^* \check{R}_{jl} M_{kl} \tag{8.2.6}$$

When the factor $R_{ij}$ in the contravariant transformation law (8.1.3) is replaced by $R_{ij}^*$, as in (8.2.5), ordinary contravariance is replaced by " contravariance of the second kind " (a term due to Schouten, Ref. 2) or, more briefly, " star-contravariance ". The components of the dual vector $x^*$ (p. 43) thus form a tensorial set with one degree of star-contravariance, while the elements of the metrical matrix $\boldsymbol{M}$ form a set with one degree of covariance and one of star-covariance.

At this point, the advantages of using *unitary* bases become overwhelming : for if $\boldsymbol{R}$ is always a unitary matrix $\boldsymbol{R}^{-1} = \boldsymbol{R}^\dagger$ and star-contravariance becomes synonymous with covariance :

$$x_i^* \to x_i'^* = \sum_j R_{ij}^* x_j^* = \sum_j \check{R}_{ij} x_j^*$$

Similarly, star-covariance is identified with contravariance :

$$y_i \to y'_i = \sum_j \breve{R}^*_{ij} y_j = \sum_j R_{ij} y_j$$

In this case the two basic transformation laws (8.1.3) and (8.1.4) are adequate for discussing all quantities of interest. There is yet another reduction in the case of *real* orthogonal transformations ; for then $\boldsymbol{R}^{-1} = \boldsymbol{R}^\dagger = \tilde{\boldsymbol{R}}$ and the distinction between co- and contra-variance disappears. Tensorial sets which transform in this way are termed " Cartesian " : they provide the simplest introduction to tensor theory (see, for example, Hall 1963 : Jeffreys 1931).

### 8.3. Tensor Properties. Symmetry Restrictions

It should now be apparent that if the components of an $n$-rank tensor are regarded as components of a *vector* in a space of $3^n$-dimensions, then this space will carry a representation of the rotation group. In the case $n = 1$ the matrix $\boldsymbol{R} = \boldsymbol{D}_{\mathsf{e}}(\mathsf{R})$, the matrix representing a rotation $\mathsf{R}$ applied to the basis $\mathbf{e}$ in 3-space. For a first rank tensor, then, we can write

$$T_i \to T'_i = \sum_j D_{\mathsf{e}}(\mathsf{R})_{ij} T_j \tag{8.3.1}$$

A more general case is illustrated in the following example.

EXAMPLE 8.2. The doubly contravariant tensor with components $T_{ij}$ behaves under the rotation $\boldsymbol{x} \to \boldsymbol{x}' = \boldsymbol{R}\boldsymbol{x}$ according to

$$T_{ij} \to T'_{ij} = \sum_{k,l} R_{ik} R_{jl} T_{kl}$$

If we arrange the elements $T_{ij}$ in a single column $\boldsymbol{t}$, and adopt a double subscript notation for rows and columns, this can be written

$$\boldsymbol{t} \to \boldsymbol{t}' = \boldsymbol{Q}\boldsymbol{t} \qquad (t'_{ij} = \sum_{(kl)} Q_{ij,kl} t_{kl})$$

where $Q_{ij,kl} = R_{ik} R_{jl}$. $\boldsymbol{Q}$ is thus the outer product (p. 137) of the matrix $\boldsymbol{R}$ with itself, commonly denoted by $\boldsymbol{R} \times \boldsymbol{R} = \boldsymbol{R}^{[2]}$. Since the matrices $\boldsymbol{R}$ form a representation of the group of 3-dimensional rotations, the matrices $\boldsymbol{R}^{[2]}$—which describe the related transformation of *tensor* components—form a direct product representation. The representation carried by the nine components of a doubly contravariant tensor is thus $\mathrm{D} = \mathrm{D}_{\mathsf{e}} \times \mathrm{D}_{\mathsf{e}}$ where $\mathrm{D}_{\mathsf{e}}$ is the representation carried by the basis $\mathbf{e}$. Similarly, a second rank tensor with one degree of contravariance and one of covariance behaves according to (8.1.7) with $p = q = 1$. And since $\breve{\boldsymbol{R}}$ is the matrix associated with rotation $\mathsf{R}$

in the "contragredient" representation $\check{\mathrm{D}}_{\mathrm{e}}$ of (5.3.13) the components again carry a representation: but this time it is $\mathrm{D} = \mathrm{D}_{\mathrm{e}} \times \check{\mathrm{D}}_{\mathrm{e}}$.

In general, elements of a tensor of rank $p+q$ ($p$ degrees of contravariance and $q$ of covariance) may be regarded as components of a vector in a space of $3^{p+q}$ dimensions and this space carries a representation with matrices

$$\boldsymbol{D}(\mathsf{R}) = \boldsymbol{D}_{\mathrm{e}}(\mathsf{R}) \times \boldsymbol{D}_{\mathrm{e}}(\mathsf{R}) \times \ \ldots \ \times \check{\boldsymbol{D}}_{\mathrm{e}}(\mathsf{R}) \times \check{\boldsymbol{D}}_{\mathrm{e}}(\mathsf{R}) \times \ \ldots \qquad (8.3.2)$$

with $p$ factors of the first kind and $q$ of the second. It is convenient to condense the multiple subscripts which label the components into a single index and to write the tensor transformation as

$$T_J \to T'_J = \sum_K D(\mathsf{R})_{JK} T_K \qquad (8.3.3)$$

In the notation of example 8.2 this would take the matrix form $\boldsymbol{t} \to \boldsymbol{t}' = \boldsymbol{D}(\mathsf{R})\boldsymbol{t}$. Fortunately, it is seldom necessary to consider such representations explicitly. In discussing the effect of symmetry it is sufficient to know the *character* of the representation, and by (5.9.7) this is simply the *product* of characters of the factors in the Kronecker product.

We are now in a position to generalize the discussion of section 6.2, the transformation being regarded as the result of rotation of the system. If $\boldsymbol{R}$ describes a symmetry operation, bringing the physical system into self-coincidence, the array of quantities $\{T'_{ij\ldots}\}$ must be identical with the original $\{T_{ij\ldots}\}$. Now since the $T_{ij\ldots}$ are vector components in a space which carries a representation, D say, we can find a new basis, adapted to a decomposition into irreducible representations. The relationship

$$\boldsymbol{t} \to \boldsymbol{t}' = \boldsymbol{D}(\mathsf{R})\boldsymbol{t}$$

will then be replaced by

$$\bar{\boldsymbol{t}} \to \bar{\boldsymbol{t}}' = \bar{\boldsymbol{D}}(\mathsf{R})\bar{\boldsymbol{t}}$$

where $\bar{\boldsymbol{D}}(\mathsf{R})$ has a characteristic block form and, more fully

$$\begin{bmatrix} \bar{\boldsymbol{t}}^{(\alpha,1)'} \\ \bar{\boldsymbol{t}}^{(\alpha,2)'} \\ \vdots \\ \bar{\boldsymbol{t}}^{(\beta,1)'} \\ \\ \end{bmatrix} = \begin{bmatrix} \boxed{\boldsymbol{D}_{\alpha,1}(\mathsf{R})} & & & & \\ & \boxed{\boldsymbol{D}_{\alpha,2}(\mathsf{R})} & & & \\ & & \ddots & & \\ & & & \boxed{\boldsymbol{D}_{\beta,1}(\mathsf{R})} & \\ & & & & \ddots \end{bmatrix} \begin{bmatrix} \bar{\boldsymbol{t}}^{(\alpha,1)} \\ \bar{\boldsymbol{t}}^{(\alpha,2)} \\ \vdots \\ \bar{\boldsymbol{t}}^{(\beta,1)} \\ \vdots \end{bmatrix} \qquad (8.3.4)$$

Here $\bar{\boldsymbol{t}}^{(\alpha,\mu)}$ is a column of components referred to basis vectors which provide a basis for the $\mu$th appearance of irreducible representation

$D_\alpha$: $\bar{t}_i^{(\alpha,\mu)}$ is the component along $\mathbf{e}_i^{(\alpha,\mu)}$. Each of these columns describes a vector in an invariant subspace, in the usual way (p. 112), and under any rotation of the physical system each vector is sent into a vector of the same subspace. The meaning of invariance under a group of symmetry operations is now clear. Using a symmetry adapted basis we require $\bar{t}_i^{(\alpha,\mu)\prime} = \bar{t}_i^{(\alpha,\mu)}$, for every operation of the group and for each block $(\alpha, \mu)$ : this simply implies either that the corresponding tensor component vanishes, $\bar{t}_i^{(\alpha,\mu)\prime} = \bar{t}_i^{(\alpha,\mu)} = 0$, or that, being invariant under all operations, it carries the unidimensional identity representation $D_{\alpha,\mu}(R)_{ii} = 1$ $(D_{\alpha,\mu}(R)_{ij} = 0,\ i \neq j)$. Consequently, when referred to a symmetry adapted basis, the only non-vanishing tensor components are those which belong to the identity representation. If the identity occurs $n$ times the tensor is therefore determined by just $n$ numbers; and this restriction is a direct consequence of the symmetry of the system. This number is readily determined in the usual manner from the character system of the representation carried by the tensor components.

EXAMPLE 8.3. Let us consider the relationship $\boldsymbol{y} = \boldsymbol{T}\boldsymbol{x}$ between two three-dimensional vectors, $\boldsymbol{y}$, $\boldsymbol{x}$ being columns of contravariant components. $\boldsymbol{T}$ then behaves as in example 8.1 having one degree of co- and one of contra-variance, and the matrices of the representation carried by the tensor components are thus, by (8.3.2),

$$\boldsymbol{D}(R) = \boldsymbol{D}_e(R) \times \check{\boldsymbol{D}}_e(R)$$

The character of element $R$ is

$$\chi(R) = \chi_e(R)\check{\chi}_e(R) = |\,\chi_e(R)\,|^2$$

(since in a *unitary* representation $(\check{\boldsymbol{R}})_{ii} = (\tilde{\boldsymbol{R}}^\dagger)_{ii} = R^*_{ii}$ and the character is an invariant). Thus, according to (3.1.14).

$$\chi(R) = (\pm\, 1 + 2 \cos\, \theta)^2$$

for proper or improper rotation through $\theta$.

Let us now take two specific examples :

(i) With $C_{3v}$ symmetry, the characters are $\chi(E) = 9$, $\chi(C_3) = 0$, $\chi(\sigma_v) = 1$ and application of (5.4.5) shows that the identity representation occurs twice. The tensor can thus be characterized completely by *two* physical constants, two different principal values.

(ii) With $C_{2v}$ symmetry (e.g. orthorhombic crystal) the characters are $\chi(E) = 9$, $\chi(C_2) = \chi(\sigma) = \chi(\sigma') = 1$ and the identity representation is found to occur *three* times : in this case the tensor property is described by *three* constants—three different principal values.

(iii) In the absence of any symmetry ("symmetry group" $\{E\}$), the identity would occur nine times and the property described by $\boldsymbol{T}$ would be characterized by *nine* independent constants.‡ It should be noted that in the example above, no a priori restrictions upon the form of the *matrix* $\boldsymbol{T}$ were imposed, all nine elements being assumed independent in the absence of symmetry. In fact, however, no known physical property is described by a second rank tensor of unrestricted form: the matrix associated with a second-rank tensor property is invariably symmetric, $T_{ji} = T_{ij}$, for reasons which are physical rather than geometric (see, for example, Nye 1956). When the tensor exhibits an imposed symmetry of this kind the number of independent elements is reduced and the transformation law must be modified accordingly.

**8.4. Symmetric and Antisymmetric Tensors**

Any second rank tensor can be expressed as a sum of a symmetric tensor and an antisymmetric tensor, whose components are, respectively,

$$\left.\begin{aligned} S_{ij} &= \tfrac{1}{2}(T_{ij} + T_{ji}) = S_{ji} \\ A_{ij} &= \tfrac{1}{2}(T_{ij} - T_{ji}) = -A_{ji} \end{aligned}\right\} \tag{8.4.1}$$

Evidently, $T_{ij} = S_{ij} + A_{ij}$ is quite arbitrary. Symmetric and antisymmetric tensors are of such importance that they deserve special consideration.

In $n$ dimensions $(i,j = 1, \ldots, n)$ a symmetric tensor has $\frac{1}{2}n(n+1)$ independent components $S_{ij}(i \leqslant j)$, while an antisymmetric tensor has $\frac{1}{2}n(n-1)$ components $A_{ij}(i < j)$, zeros occurring when $i = j$. For example, from the components $a_1$, $a_2$, $a_3$ of a vector we can form nine products‡‡

$$\begin{matrix} a_1{}^2 & a_1a_2 & a_1a_3 \\ a_2a_1 & a_2{}^2 & a_2a_3 \\ a_3a_1 & a_3a_2 & a_3{}^2 \end{matrix}$$

which transform like the nine components of a second rank (doubly contravariant) tensor; but since $a_1a_2 = a_2a_1$ etc. only those along and, say, above the diagonal (namely $\frac{1}{2}n(n+1)$) are independent. These distinct products can be regarded as vector components in a

‡ At first sight, any second rank tensor can be expressed in terms of *three* principal values; but these must be defined with respect to axes which, in the absence of symmetry, are determined by the physical properties and require six more constants (two angles each) for their specification.

‡‡ There is no need to write the tensor components in a matrix array; but this form conveniently exhibits the symmetry.

space which again carries a representation of the group of all rotations but which is now only *six*-dimensional (cf. the 9-dimensional representation $D_e \times D_e$). On the other hand, we can construct an antisymmetrical tensor with components $A_{ij} = (T_{ij} - T_{ji})$ by taking for $T_{ij}$ the set of products $a_i b_j$ formed from *two* vectors with components $a_1, a_2, a_3$ and $b_1, b_2, b_3$:

$$\boldsymbol{T} = \begin{pmatrix} a_1b_1 & a_1b_2 & a_1b_3 \\ a_2b_1 & a_2b_2 & a_2b_3 \\ a_3b_1 & a_3b_2 & a_3b_3 \end{pmatrix}$$

$$\boldsymbol{A} = \begin{pmatrix} 0 & a_1b_2-a_2b_1 & a_1b_3-a_3b_1 \\ a_2b_1-a_1b_2 & 0 & a_2b_3-a_3b_2 \\ a_3b_1-a_1b_3 & a_3b_2-a_2b_3 & 0 \end{pmatrix} \qquad (8.4.2)$$

The elements of $\boldsymbol{A}$ are then the components of a second rank tensor: for if $a_i \to a'_i = \sum_k R_{ik}a_k$ and $b_j \to b'_j = \sum_l R_{jl}b_l$ then

$$(a_ib_j - a_jb_i) \to \sum_{k,l} (R_{ik}R_{jl}a_kb_l - R_{jk}R_{il}a_kb_l)$$

and since $k$ and $l$ are merely summation (" dummy ") indices we may reverse their roles in the second term to obtain

$$(a_ib_j - a_jb_i) \to \sum_{k,l} R_{ik}R_{jl}(a_kb_l - a_lb_k)$$

or

$$A_{ij} \to \sum_{k,l} R_{ik}R_{jl}A_{kl}$$

the standard law for a second rank (doubly contravariant) tensor. But there are only *three* independent elements, which we may take as those above the diagonal of $\boldsymbol{A}$. These distinct elements may be regarded as vector components in a *three*-dimensional space which again carries a representation of the rotation group. Evidently by resolving a tensor into symmetric and antisymmetric parts the representation carried by its nine independent components may be *reduced* into one carried by the six distinct symmetric combinations and another carried by the three distinct antisymmetric combinations. We now derive explicitly the transformation laws appropriate to the reduced representations carried by symmetric and antisymmetric second rank tensors.

(*i*) *Symmetric Tensors*

For a doubly contravariant tensor, the components $S_{ij}$ transform according to

$$S_{ij} \to S'_{ij} = \sum_k R_{ik}R_{jk}S_{kk} + \sum_{k<l} (R_{ik}R_{jl} + R_{il}R_{jk})S_{kl}$$

This may be written $S_{ij} \to S'_{ij} = \sum_{k \leqslant l} (\boldsymbol{R}_S^{[2]})_{ij,\,kl} S_{kl}$

or
$$\boldsymbol{s} \to \boldsymbol{s}' = \boldsymbol{R}_S^{[2]}\,\boldsymbol{s} \tag{8.4.3}$$

where the distinct tensor components ($i \leqslant j$) are collected into columns $\boldsymbol{s}$, $\boldsymbol{s}'$ and the matrix $\boldsymbol{R}_S^{[2]}$ (also with double subscript labelling of rows and columns) is defined by

$$\left.\begin{aligned} (\boldsymbol{R}_S^{[2]})_{ij,\,kk} &= R_{ik}R_{jk} \\ (\boldsymbol{R}_S^{[2]})_{ij,\,kl} &= (R_{ik}R_{jl} + R_{il}R_{jk}) \qquad (k \neq l) \end{aligned}\right\} \tag{8.4.4}$$

The matrix $\boldsymbol{R}_S^{[2]}$ is called the *symmetrized* Kronecker square of $\boldsymbol{R}$. In discussing symmetry properties, the trace of this matrix is required since the traces provide the character system of the representation carried by the independent tensor components. Clearly on putting $i = k$ and $j = l$ and summing,

$$\begin{aligned} \operatorname{tr} \boldsymbol{R}_S^{[2]} &= \sum_{i<j} (R_{ii}R_{jj} + R_{ij}R_{ji}) + \sum_i R_{ii}^2 \\ &= \tfrac{1}{2} \sum_{i,j} (R_{ii}R_{jj} + R_{ij}R_{ji}) \end{aligned}$$

Thus
$$\operatorname{tr} \boldsymbol{R}_S^{[2]} = \tfrac{1}{2}[(\operatorname{tr} \boldsymbol{R})^2 + \operatorname{tr} \boldsymbol{R}^2] \tag{8.4.5}$$

*(ii) Antisymmetric Tensors*

The transformation law for the $\frac{1}{2}n(n-1)$ components $A_{ij} (i < j)$ is very simple: using the fact that $A_{lk} = -A_{kl}$,

$$A_{ij} \to A'_{ij} = \sum_{k<l} (R_{ik}R_{jl} - R_{il}R_{jk}) A_{kl}$$

Hence
$$A_{ij} \to A'_{ij} = \sum_{k<l} (\boldsymbol{R}_A^{[2]})_{ij,\,kl} A_{kl}$$

or
$$\boldsymbol{a} \to \boldsymbol{a}' = \boldsymbol{R}_A^{[2]}\,\boldsymbol{a} \tag{8.4.6}$$

where the distinct tensor components ($i < j$) are collected into columns $\boldsymbol{a}$, $\boldsymbol{a}'$ and $\boldsymbol{R}_A^{[2]}$ is the matrix with

$$(\boldsymbol{R}_A^{[2]})_{ij,\,kl} = (R_{ik}R_{jl} - R_{il}R_{jk}) \quad (i < j,\ k < l) \tag{8.4.7}$$

It also follows that

$$\operatorname{tr} \boldsymbol{R}_A^{[2]} = \tfrac{1}{2}\,[(\operatorname{tr} \boldsymbol{R})^2 - \operatorname{tr} \boldsymbol{R}^2] \tag{8.4.8}$$

and consequently that

$$\operatorname{tr} \boldsymbol{R}_S^{[2]} + \operatorname{tr} \boldsymbol{R}_A^{[2]} = (\operatorname{tr} \boldsymbol{R})^2 = \operatorname{tr} \boldsymbol{R}^{[2]}$$

This is the expected result, because if the $\frac{1}{2}n(n+1)$ symmetric combinations of tensor components are arranged in a column $\boldsymbol{s}$, followed by the $\frac{1}{2}n(n-1)$ antisymmetric combinations $\boldsymbol{a}$, the resultant column will be transformed by the $n^2 \times n^2$ matrix

$$\left(\begin{array}{c|c} \boldsymbol{R}_S^{[2]} & \mathbf{0} \\ \hline \mathbf{0} & \boldsymbol{R}_A^{[2]} \end{array}\right)$$

But the original set of $n^2$ components carried the direct product representation with matrices $\boldsymbol{R}^{[2]}$ and the introduction of symmetric and antisymmetric combinations merely effects a partial reduction of the original representation—against which the trace is invariant.

In three dimensions an antisymmetric second rank tensor has many of the properties of a vector and the three distinct components are usually referred to as components of a " pseudo-vector ". In fact, under any proper rotation of a system the components $A_{23}$, $A_{31}$, $A_{12}$ (collected in the column $\boldsymbol{a}$) behave exactly like the components of a vector ;‡ they differ because under an *improper* rotation they are in addition multiplied by $-1$. To see how this comes about we write 8.4.6 as

$$A_\alpha \rightarrow A'_\alpha = \sum_\beta (\boldsymbol{R}_A^{[2]})_{\alpha\beta} A_\beta \qquad (\alpha, \quad \beta = 1, \quad 2, \quad 3)$$

where the Greek subscripts indicate the new labelling in which $\alpha =$ 1, 2, 3 corresponds to $ij = 23$, 31, 12 respectively. If we now choose the axis of rotation as the $z$-axis of a Cartesian system, then the rotation matrix is

$$\boldsymbol{R} = \begin{pmatrix} \cos\theta & -\sin\theta & 0 \\ \sin\theta & \cos\theta & 0 \\ 0 & 0 & \pm 1 \end{pmatrix}$$

(the lower sign in the $\pm 1$ applying in an improper rotation), and from (8.4.7)

$$(\boldsymbol{R}_A^{[2]})_{11} = R_{22}R_{33} - R_{23}R_{32} = \pm\cos\theta \qquad (ij = kl = 23)$$
$$(\boldsymbol{R}_A^{[2]})_{12} = R_{23}R_{31} - R_{21}R_{33} = \mp\sin\theta \quad \text{etc.}$$

In fact, we find $\boldsymbol{R}_A^{[2]} = \pm\boldsymbol{R}$ and accordingly

$$\boldsymbol{a} \rightarrow \boldsymbol{a}' = \pm\boldsymbol{R}\boldsymbol{a} \tag{8.4.9}$$

according as the rotation is proper or improper. This is the basic transformation law for a pseudo-vector. The rotations, conventionally represented in Fig. 6.5 by " axial " vectors, are entities of this kind : in elementary vector analysis such quantities normally occur in " vector products ". The components of the vector product $a \times b$ are in fact just the distinct elements $A_{23}$, $A_{31}$, $A_{12}$ selected from the matrix $\boldsymbol{A}$ in (8.4.2).

‡ The reason for selecting these particular components will appear presently. The choice ($i < j$) in 8.4.6–7) is then read as " $i$ before $j$, in *cyclic* order."

*Pseudo-Tensors*

In discussing the effect of geometrical symmetry upon physical systems it is important to recognize that in many instances axial vectors are involved. A very familiar example is the magnetic field " vector " which originates in a *circular current* (cf. Fig. 6.5) and is not a true (polar) vector at all. It is, however, expedient to treat such entities as pseudo-vectors, with only three components and with easily visualized transformation properties, rather than as second rank tensors (with six components); and it then becomes necessary also to introduce pseudo-tensors of higher rank.

EXAMPLE 8.4. Suppose that $\boldsymbol{y}$ and $\boldsymbol{x}$ are the components of a vector and a pseudo-vector respectively, and that they are related by

$$\boldsymbol{y} = \boldsymbol{T}\boldsymbol{x}$$

Then (cf. example 8.1) after a rotation with matrix $\boldsymbol{R}$, $\boldsymbol{y}' = \boldsymbol{R}\boldsymbol{y}$ but $\boldsymbol{x}' = \pm \boldsymbol{R}\boldsymbol{x}$. Thus $\boldsymbol{R}^{-1}\boldsymbol{y}' = \boldsymbol{T}(\pm \boldsymbol{R}^{-1})\boldsymbol{x}'$, or $\boldsymbol{y}' = \pm \boldsymbol{R}\boldsymbol{T}\boldsymbol{R}^{-1}\boldsymbol{x}'$, and the elements of $\boldsymbol{T}$ transform according to

$$T_{ij} \rightarrow T'_{ij} = \pm \sum_{k,l} R_{ik} \check{R}_{jl} T_{kl}$$

The elements of $\boldsymbol{T}$ therefore transform like the components of a second rank tensor (exactly as in example 8.1) except for *sign change* under improper rotations : in this case $\boldsymbol{T}$ describes a *pseudo-tensor.*

There is evidently no difficulty in extending the definition to the general case (8.1.7) the tensor being true or pseudo—according as it contains an even or odd number of " pseudo-factors ". Clearly, for example, if in example 8.5 were *both* pseudo-vectors, they would be related by an ordinary second-rank tensor.

The importance of these considerations becomes evident when we return to the problem of section 8.3. For an alternative to (8.3.4) now arises: in the case of a pseudo-tensor the components in a symmetry adapted basis must vanish unless they are multiplied by $\pm 1$ under the proper and improper rotations. Non-zero components therefore carry the unidimensional representation $\boldsymbol{D}(\mathsf{R}) = \pm 1$ ($\mathsf{R}$ proper or improper). To obtain the number of independent physical constants involved we therefore require the number of times the " alternating " representation occurs in the representation carried by the tensor components. We now consider examples of symmetric, antisymmetric, and pseudo-tensors and discuss the restrictions imposed by crystal symmetry.

EXAMPLE 8.5. In defining electric and magnetic susceptibilities, the relationship between field strength and induced moment per unit volume is written

$$\boldsymbol{P} = \boldsymbol{k}\boldsymbol{E} \text{ (electric)}, \qquad \boldsymbol{I} = \boldsymbol{\kappa}\boldsymbol{H} \text{ (magnetic)}$$

In the first case, $\boldsymbol{P}$ and $\boldsymbol{E}$ are components of *polar* vectors and the considerations of example 8.1 apply. If, for simplicity, we choose a Cartesian system the distinction between co- and contra-variant components disappears (since $\check{\boldsymbol{R}} = \boldsymbol{R}$) and in rotation $R$

$$k_{ij} \to k'_{ij} = \sum_{k,l} R_{ik}R_{jl}k_{kl}$$

Moreover, it is known (see for example, Nye 1957) that $k_{ji} = k_{ij}$ and the tensor is therefore symmetric.

In the second case $\boldsymbol{I}$ and $\boldsymbol{H}$ are components of *pseudo*-vectors: but because

$$\kappa_{ij} \to \kappa'_{ij} = \sum_{k,l} (\pm R_{ik})(\pm \check{R}_{jl})\kappa_{kl} = \sum_{k,l} R_{ik}\check{R}_{jl}\kappa_{kl}$$

the $\kappa_{ij}$ (like the $k_{ij}$) are *ordinary* second rank tensor components. In a Cartesian system

$$\kappa_{ij} \to \kappa'_{ij} = \sum_{k,l} R_{ik}R_{jl}\kappa_{kl}$$

And again, the tensor is symmetric ($\kappa_{ji} = \kappa_{ij}$).

To discuss the effect of crystal symmetry, allowing for the inherent symmetry of the susceptibility tensor, we need the character system of the representation carried by the six distinct tensor components. This is the symmetrized Kronecker square of that carried by the coordinates and we therefore use (8.4.5), obtaining

$$\chi(R) = \tfrac{1}{2}[(\pm 1 + 2 \cos \theta)^2 + (1 + 2 \cos 2\theta)]$$

since $R^2$ gives a rotation through $2\theta$ if $R$ is through $\theta$. Finally, then,

$$\chi(R) = \pm 2 \cos \theta + 4 \cos^2 \theta$$

For a system with $C_{3v}$ symmetry the characters are $\chi(E) = 6$, $\chi(C_3) = 0$, $\chi(\sigma_v) = 2$ and the number of independent tensor components (number of times the identity occurs) is thus 2. Recognition of the inherent symmetry of the tensor makes no difference, in this case, to the result obtained in example 8.4. In general, however, failure to recognize this symmetry could lead to over-estimation of the number of independent components. Thus, for a triclinic crystal, the present analysis gives the correct result (6) while the method of example 8.4 gives 9.

EXAMPLE 8.6. The plane of polarization of light passing through certain crystals is rotated (Fig. 8.1). The angle of rotation, per unit path length, for a beam of light with direction cosines $l_1$, $l_2$, $l_3$, is

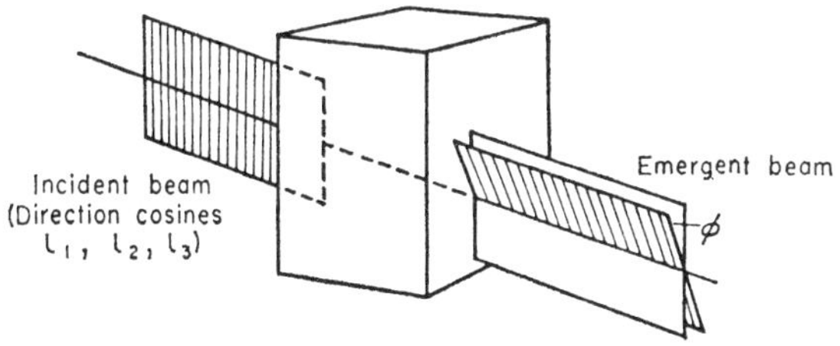


FIG. 8.1. Optical activity. The incident beam is polarized in the vertical plane ; on passage through the crystal the plane of polarization is rotated through an angle $\phi$.

proportional to a scalar quantity $G$ called the " gyration ". Now $G$ is not a *true* scalar because it involves a sense of rotation : if, for example, we reflect the whole system across a plane containing the beam (Fig. 8.1) the angle between emergent and incident planes will change from $\phi$ to $-\phi$, although under any *proper* rotation the angle is unchanged. $G$ is thus a pseudo-*scalar*. Now it is found that $G$ is related quadratically to the direction cosines of the beam :

$$G = \sum_{i,j} G_{ij} l_i l_j = \tilde{\boldsymbol{l}}\boldsymbol{G}\boldsymbol{l}$$

where $\boldsymbol{G}$ is a symmetric matrix. But $\boldsymbol{l}$ is a true vector and hence in a rotation $R$ (proper or improper)

$$\boldsymbol{l} \to \boldsymbol{l}' = \boldsymbol{R}\boldsymbol{l}, \qquad G \to G' = \pm G$$

The equation for $G$ becomes in terms of primed quantities

$$\pm G' = \tilde{\boldsymbol{l}}'\tilde{\boldsymbol{R}}^{-1}\boldsymbol{G}\boldsymbol{R}^{-1}\boldsymbol{l}'$$

and the transformed $\boldsymbol{G}$ matrix is thus

$$\boldsymbol{G}' = \pm \tilde{\boldsymbol{R}}^{-1}\boldsymbol{G}\boldsymbol{R}^{-1}$$

Taking, for simplicity, a Cartesian basis

$$G'_{ij} = \pm \sum_{k,l} \check{R}_{ik} G_{kl} \check{R}_{jl} = \pm \sum_{k,l} R_{ik} R_{jl} G_{kl}$$

and the elements of the gyration matrix $\boldsymbol{G}$ are seen to transform like the components of a second rank *pseudo*-tensor. As it happens, the character system in the representation carried by the six distinct elements of $\boldsymbol{G}$ is identical with that associated with a *true* second rank tensor (as in the preceding example): this is because the character formula (8.4.5) contains only squared terms. But now, invariance of the crystal under some symmetry group requires that the representation contains the *alternating* representation (8.5.1), the number of occurrences giving the number of independent physical constants; or that the tensor vanishes. Thus for a system with the point group $C_{3v}$, using the character system obtained in example 8.5, there are $\frac{1}{6}(6-6)$ constants. This simply means that a crystal of $C_{3v}$ class cannot be optically active, a result which is experimentally well known. On the other hand, removal of the operation $\sigma^{(1)}$ gives the group $C_3$ in which the alternating representation becomes the identity (no improper rotations); and in this case activity is possible.

## 8.5. Tensor Fields. Tensor Operators

It is now possible to extend the ideas developed in chapter 7, passing from scalar fields to tensor fields and from invariant operators to tensor operators. Emphasis will be placed upon those aspects of the subject which are important in non-relativistic quantum mechanics, not upon general tensor calculus. We bring our approach into line with that of previous chapters by using $R$ to denote an actual rotation which brings into self-coincidence the physical system considered: in this case the basis $\mathbf{e}$ (carrying the system) is rotated into $\mathbf{e}' = \mathbf{e}R$, and when we need to introduce a fixed field point its coordinates change from $\boldsymbol{x}$ to $\boldsymbol{x}' = \boldsymbol{R}^{-1}\boldsymbol{x}$. The modifications are trivial, $R$ and $\boldsymbol{R}$ in the preceding part of this chapter being replaced by $R^{-1}$ and $\boldsymbol{R}^{-1}$.

In chapter 7 we discussed the effect of rotation upon a function $\phi(x)$ describing a *scalar field*, such as the temperature or electrostatic potential. *Vector fields*, in which a vector quantity with components $\phi_1(x)$, $\phi_2(x)$, $\phi_3(x)$ is defined at every point $x$, are also familiar (e.g. the electric field or the magnetic vector potential). The distinctive feature of this kind of field is that for any new choice of directions

$$\phi_{i'}(x) = \sum_j (\boldsymbol{R}^{-1})_{ij}\phi_j(x) = \sum_j D(R^{-1})_{ij}\phi_j(x) \tag{8.5.1}$$

since when the axes are rotated the components of a vector transform contragrediently, and since the inverse matrix is the matrix of the inverse rotation. Clearly, a tensor field may be defined similarly just by relaxing the restriction to 3-component quantities: for an $n$-rank tensor, D would be a representation in $3n$ dimensions.

We now consider a rotation $R$ of a physical system (cf. Fig. 7.3) and express the fact that the $i'$ component, $\phi_{i'}^{\text{rot}}(Rx)$, of a tensor property at the image point in the rotated system has the same value as the $i$ component at point $x$ before rotation. On using (8.5.1),

$$\phi_{i'}^{\text{rot}}(Rx) = \sum_j D(R^{-1})_{ij}\phi_j^{\text{rot}}(Rx) = \phi_i(x) \tag{8.5.2}$$

Now we wish to know what new functions $\{\phi_j^{\text{rot}}\}$ are required to describe the rotated system, relative to the fixed frame, and how they are related to the $\{\phi_j\}$. We therefore solve (8.5.2) for $\phi_k^{\text{rot}}$ by multiplying by $D(R)_{ki}$ and summing over $i$. Thus

$$\sum_j \delta_{kj}\phi_j^{\text{rot}}(Rx) = \sum_i D(R)_{ki}\phi_i(x)$$

On replacing $x$ by the backwards rotated field point $R^{-1}x$, there follows:

$$\phi_k^{\text{rot}}(x) = T_R\phi_k(x) = \sum_i D(R)_{ki}\phi_i(R^{-1}x) \tag{8.5.3}$$

This equation shows how the $k$th component of a tensor field is changed by a rotation $R$ of the system, the symbol $T_R$ denoting the transformation induced in *tensor* space. The individual functions are rotated as in section 7.1; the additional *mixing* of the various component functions arises from the change of axial directions to which the components are referred.

The explicit use of tensor fields is seldom necessary in non-relativistic theory. Pauli's introduction of a 2-component wave function for the "spinning" electron marked the transition from scalar to tensor fields in quantum mechanics: but even here the use of "spin-functions" leads to a more convenient and less sophisticated development. We therefore pass to the related concept of a tensor *operator*, which is basic to an understanding of quantum mechanical selection rules, even in non-relativistic theory.

Tensor operators, working on any function $\phi(x)$ at the arbitrary field point $x$, differ from invariant operators in that their results depend on choice of the axes with respect to which they are defined. Given any operator $T_i$ of the unrotated system, we may set up an operator $T_i'$, in an exactly similar way, for the rotated system: we then try to express $T_i'$ in terms of the operators $\{T_i\}$ referred to the unrotated basis. Vector and tensor operators are then defined in strict analogy with (8.3.1), (8.3.2) and (8.3.3), which give the transformation properties of ordinary vector and tensor components. When the basis rotation is denoted by $R$ (not $R^{-1}$) we must of course replace $\boldsymbol{R}$ by $\boldsymbol{R}^{-1}$: to preserve the formal appearance of the covariant and contravariant laws it is then convenient to put $\boldsymbol{R}^{-1} = \boldsymbol{A}$. The basic contravariant law (8.1) then becomes

$$\boldsymbol{x} \to \boldsymbol{x}' = \boldsymbol{A}\boldsymbol{x} = \boldsymbol{D}_e(\mathsf{R}^{-1})\boldsymbol{x} \tag{8.5.4}$$

where $\mathrm{D}_e$ is the representation carried by the basis and $\boldsymbol{R}^{-1}$, the inverse matrix, is $\boldsymbol{D}_e(\mathsf{R}^{-1})$. If, on expressing $T_i'$ of the rotated system in terms of the $T_i$ ($i = 1, 2, 3$) we find

$$T_i' = \sum_j A_{ij}T_j = \sum_j D_e(\mathsf{R}^{-1})_{ij}T_j \tag{8.5.5}$$

we say $T_1, T_2, T_3$ are *vector operators*, forming a contravariant set.

More generally, when we encounter many-subscript operators which transform according to

$$T_{i..j...} \to T'_{i..j...} = \sum_{k,l} A_{ik} \ldots \breve{A}_{jl} \ldots T_{k..l...} \tag{8.5.6}$$

with $p$ $A$ factors and $q$ $\breve{A}$ factors we say the $T_{ij...}$ are *tensor operators*, forming a tensorial set with $p$ degrees of contravariance and $q$ of covariance. Exactly as in (8.3.3) we may relabel the components using a single subscript, and may then express the representation property as

$$T_J \to T_J' = \sum_K D(\mathsf{R}^{-1})_{JK}T_K \tag{8.5.7}$$

We now examine and classify a few typical tensor operators. In many of the most important applications, Hermitian vector spaces are involved and the tensor components may be complex: in view of this development it is expedient to employ *unitary* representations, so that the complications discussed in section 8.2 do not arise. In this case, $\boldsymbol{R}^{-1} = \boldsymbol{R}^\dagger$ and there are still only two types of transformation—contravariant and covariant: when only real quantities occur, this distinction disappears, $\boldsymbol{R}^{-1} = \tilde{\boldsymbol{R}}$.

EXAMPLE 8.7. First consider the set $\{T_i\}$ with $T_i = \mathbf{e}_i^* x$. Regarded as an operator, $T_i$ multiplies any function $\phi(x)$ by the orthogonal projection of $x$ along the $\mathbf{e}_i$ direction in the system. If the system is rotated, so that $\mathbf{e} \to \mathbf{e}' = \mathbf{e}\boldsymbol{R}$ and $\mathbf{e}'^\dagger = \boldsymbol{R}^\dagger\mathbf{e}^\dagger = \boldsymbol{A}\mathbf{e}^\dagger$, the corresponding projection along the $\mathbf{e}_i'$ direction in the rotated system is $\mathbf{e}_i'^* x$ and hence

$$T_i' = \sum_j A_{ij}\mathbf{e}_j^* x = \sum_j A_{ij}T_j$$

The three operators therefore behave like the contravariant components of a fixed vector: they are *vector operators*. In *real* space, $\boldsymbol{R}^{-1} = \tilde{\boldsymbol{R}}$ and hence $A_{ij} = R_{ji}$: the operators then also behave like the three Cartesian basis vectors $\mathbf{e}_1, \mathbf{e}_2, \mathbf{e}_3$.

EXAMPLE 8.8. The quantum mechanical angular momentum operators $L_x$, $L_y$, $L_z$, associated with the $x$, $y$, $z$ components of the angular momentum of a particle are

$$L_x = \frac{\hbar}{i}\left(y\frac{\partial}{\partial x} - z\frac{\partial}{\partial y}\right), \quad L_y = \frac{\hbar}{i}\left(z\frac{\partial}{\partial x} - x\frac{\partial}{\partial z}\right), \quad L_z = \frac{\hbar}{i}\left(x\frac{\partial}{\partial y} - y\frac{\partial}{dx}\right)$$

With the usual tensor notation $x_i$ ($i = 1, 2, 3$) for the coordinates, rotation $R$ gives $\boldsymbol{x} \to \boldsymbol{x}' = \boldsymbol{A}\boldsymbol{x}$ (basic contravariant law) and, since $\boldsymbol{x} = \boldsymbol{A}^{-1}\boldsymbol{x}'$ and hence $(\partial x_j/\partial x_i') = (\boldsymbol{A}^{-1})_{ji} = \breve{A}_{ij}$

$$\frac{\partial}{\partial x_i'} = \sum_j \frac{\partial x_j}{\partial x_i'}\frac{\partial}{\partial x_j} = \sum_j \breve{A}_{ij}\frac{\partial}{\partial x_j}$$

Also, when the basis is Cartesian, $\breve{\boldsymbol{A}} = \boldsymbol{A}$. The three angular momentum operators are then recognizable as the three distinct components of an antisymmetric second rank tensor. Exactly as in the case of the pseudo-vector (p. 214) the components $\{L_i\}$ are found to transform according to

$$L_i' = \pm\sum_j A_{ij}L_j \qquad \text{(rotation proper or improper)}$$

$\{L_x, L_y, L_z\}$ is a set of *pseudo-vector operators*: they are of such importance in quantum mechanics that their transformation properties under point group operations have also been included in appendix 1.

EXAMPLE 8.9. The operators of the previous example are essentially the components of the curl operator in vector analysis:

$$\text{curl}_{\mathsf{e}} = \sum_{(ijk)} \mathsf{e}_i\left(x_j\frac{\partial}{\partial x_k} - x_k\frac{\partial}{\partial x_j}\right) \qquad (ijk = 123,\ 231,\ 312)$$

Thus, $\text{curl}_{\mathsf{e}} = \sum_i \mathsf{e}_i l_i$, where the $l_i$ behave like pseudo-vector components, and after rotation

$$\text{curl}_{\mathsf{e}'} = \pm\sum_{i,j,k} \mathsf{e}_j R_{ji} A_{ik} l_k = \pm\,\text{curl}_{\mathsf{e}}$$

since $\sum_i R_{ji}A_{ik} = \delta_{ik}$. The curl operation is then invariant under all *proper* rotations (producing the same vector from any given vector field, irrespective of choice of axes), but changes sign when a reflection is introduced (i.e. on changing from a right-handed to a left-handed frame). The curl operator therefore has *pseudo-scalar* character. On the other hand, the gradient, divergence, and Laplacian operators all behave like ordinary scalars, being invariant under both rotations of frame and change from right- to left-handedness.

The above examples indicate how the transformation law (8.5.7) may be determined for any given set of tensor operators. Such a set carries a representation of the full rotation group and hence of all the finite point groups, which are included as subgroups. In this volume our main concern is with the finite groups, but the methods to be developed may be carried over directly to the full rotation group and form the basic mathematical tools in the quantum theory of angular momentum (Fano and Racah 1959; Edmonds 1957; Rose 1957). In order to exploit symmetry properties fully we must now introduce *irreducible* tensorial sets. To do this we merely reduce the representation carried by a set of tensor operators by setting up new linear combinations in the usual way. When this process has been carried to completion the matrices $\boldsymbol{D}(R^{-1})$ whose elements appear in (8.5.7) will appear in diagonal block form: each block is then carried by an *irreducible tensorial set* $\{T_i^{(\alpha)}\}$ such that under a basis rotation $R$

$$T_i^{(\alpha)} \rightarrow T_i^{(\alpha)\prime} = \sum_j D_\alpha(R^{-1})_{ij} T_j^{(\alpha)} \tag{8.5.8}$$

This definition is equivalent to the one given by Wigner (Ref. 3): it was derived as a natural generalization of the basic transformation (8.5.4) which introduces the representation carried by the contravariant components of a vector. On the other hand, since the aim is merely to find sets of operators carrying the irreducible representations, alternative definitions are equally acceptable. We may for instance define a standard set of irreducible tensor operators as that which behaves contragrediently to $\{T_i^{(\alpha)}\}$: if we momentarily denote the corresponding linear combinations of the original tensor operators by $\{E_i^{(\alpha)}\}$ (behaving like basis vectors in representation space and transforming contragrediently to the components $T_i^{(\alpha)}$) the definition would give

$$E_i^{(\alpha)} \rightarrow E_i^{(\alpha)\prime} = \sum_j E_j^{(\alpha)}[\boldsymbol{D}_\alpha(R^{-1})]_{ji}^{-1} = \sum_j E_j^{(\alpha)} D_\alpha(R)_{ji}$$

Most authors now employ this " covariant " definition of the irreducible tensors, in which the chosen operators behave like a row of basis functions. Although it is immaterial which standard set is introduced, the second definition offers certain advantages (e.g. the irreducible tensor operators transform precisely like sets of spherical harmonics) and will be adopted in conformity with current practice:

> If we can define operators $\{T_i^{(\alpha)}\}$ such that after a point group operation $R$ the corresponding operators of the rotated system are given by
> $$T_i^{(\alpha)\prime} = \sum_j T_j^{(\alpha)} D_\alpha(R)_{ji} \qquad (8.5.9)$$
> where $D_\alpha$ is an irreducible representation of the point group, then $\{T_i^{(\alpha)}\}$ is an irreducible tensorial set.

We now look for examples of tensor operators irreducible under the group $C_{3v}$.

EXAMPLE 8.10. Suppose the operators of example 8.7 are referred to the Cartesian basis indicated by $\bar{e}_1$, $\bar{e}_2$, $\bar{e}_3$ in Fig. 5.1. The tensorial set $\{T_i\} = \{x, y, z\}$ then breaks into two irreducible sets: the operators behave just like a set of three $p$ functions (example 7.4), or the three basis vectors, and hence $z(= T_1^{(1)})$ transforms like the basis vector of representation $D_1$ ($A_1$-type) while $x$ $(= T_1^{(3)})$ and $y$ $(= T_2^{(3)})$ transform like the two basis vectors of $D_3$ (E-type).

EXAMPLE 8.11. When the angular momentum operators of example 8.9 are referred to the same basis, $L_z$ is seen to change sign under reflections: thus $L_z(= T_1^{(2)})$ transforms like the basis vector of representation $D_2$. The operators $L_x$ and $L_y$ do not carry the representation $D_3$, because of their pseudo-vector behaviour; but they must carry a representation *equivalent* to $D_3$. To find the appropriate linear combinations we may consider their behaviour under the generators. From example 8.8 (or easily from first principles, putting $x' = cx + sy$ etc.) we obtain:

$$C_3: \; L_x' = cL_x + sL_y, \qquad L_y' = -sL_x + cL_y$$
$$\sigma^{(1)}: \; L_x' = -L_x, \qquad L_y' = L_y$$

The second result shows that $L_y$ behaves like the basis vector in the $x$ direction and it then appears that $L_y(= T_1^{(3)})$ and $-L_x(= T_2^{(3)})$ form the irreducible basis for representation $D_3$.

Clearly in more complicated instances, irreducible tensors can be projected from given tensors exactly like the symmetry *functions* of chapter 7: it is necessary only to know how the tensors behave under the symmetry operations and then to put them together with appropriate coefficients. We remark, however, that explicit reduction is not always necessary: as in previous chapters and as will be found presently, a great deal of qualitative information can be obtained just by knowing which representations a set of tensor operators carry, without actually determining the irreducible tensors.

### 8.6. Matrix Elements of Tensor Operators

In quantum mechanics it is frequently necessary to consider quantities of the form

$$T_{ij} = \langle \psi_i | T | \psi_j \rangle = \int \psi_i^*(x) T \psi_j(x)\, dx \tag{8.6.1}$$

where $T$ is an operator and $\psi_i$, $\psi_j$ are wave functions of given symmetry species. Here $x$ stands for all the variables involved and $dx$ for the corresponding volume element. $T_{ij}$ is the *matrix element* (in the Dirac notation of 2.10.5) of the operator $T$ between functions $\psi_i$ and $\psi_j$ and by using a complete set $\{\psi_i\}$ we may obtain a matrix representation of the quantum mechanical operators (cf. example 2.15). Such quantities also determine the selection rules in the theory of atomic and molecular spectra and are generally of great importance. The operator $T$ may describe a property which is invariant under rotations of the system (e.g. its energy) or may describe a vector or tensor property (e.g. a component of the electric dipole moment): in general it may be a member of a tensorial set whose elements provide a basis for an irreducible representation of the symmetry group of the system. It was the study of quantities of the form (8.6.1) which led to the general theory of irreducible tensorial sets (Fano and Racah 1959).

*Invariant Operators*

In general $\psi_i$ and $\psi_j$ may be different functions, of symmetry species $(\alpha, i)$ and $(\beta, j)$ respectively, while $T$ may be of a third species. First, however, we consider the case where $T$ is an invariant operator, writing the matrix element as

$$\langle \phi_i^{(\alpha)} | T | \psi_j^{(\beta)} \rangle = \int \phi_i^{(\alpha)*}(x) T \psi_j^{(\beta)}(x)\, dx \tag{8.6.2}$$

In many cases the symmetry functions may be *projected* from arbitrary functions $\phi$ and $\psi$ and it is instructive and useful to consider the case

$$\phi_i^{(\alpha)} = \phi_{il}^{(\alpha)} = \rho_{il}^{(\alpha)}\phi, \qquad \psi_j^{(\beta)} = \psi_{jk}^{(\beta)} = \rho_{jk}^{(\beta)}\psi \tag{8.6.3}$$

where $\rho_{il}^{(\alpha)}$ and $\rho_{jk}^{(\beta)}$ are a pair of (generalized) projection operators, as defined in section 5.8. Let us abbreviate these operators to

$$\rho_\alpha = \sum_R a_R R \qquad \rho_\beta = \sum_S b_S S \tag{8.6.4}$$

The matrix element then reduces as follows:

$$\begin{aligned} \langle \rho_\alpha \phi | T | \rho_\beta \psi \rangle &= \sum_{R,S} a_R^* b_S \langle R\phi | T | S\psi \rangle \\ &= \sum_{R,S} a_R^* b_S \langle \phi | T | R^{-1} S\psi \rangle \\ &= \langle \phi | T | \rho_\alpha^\dagger \rho_\beta \psi \rangle \end{aligned}$$

where in the first step we apply a common rotation $R^{-1}$ to all terms (leaving the integral unchanged) and in the next we introduce the "adjoint" of a projection operator:

$$\rho_\alpha = \sum_R a_R R, \qquad \rho_\alpha^\dagger = \sum_R a_R^* R^{-1} \tag{8.6.5}$$

Now the operators $\rho_{ij}^{(\alpha)}$ have the property

$$\rho_{ij}^{(\alpha)\dagger} = \sum_R D_\alpha(R)_{ij} R^{-1} = \sum_R D_\alpha(R^{-1})_{ji}^* R^{-1} = \rho_{ji}^{(\alpha)} \tag{8.6.6}$$

and the basic property expressed in equation (5.8.4), namely

$$\rho_{li}^{(\alpha)} \rho_{jk}^{(\beta)} = \rho_{lk}^{(\alpha)} \delta_{\alpha\beta} \delta_{ij}$$

then gives

$$\langle \phi_{il}^{(\alpha)} | T | \psi_{jk}^{(\beta)} \rangle = \delta_{\alpha\beta} \delta_{ij} \langle \phi | T | \rho_{lk}^{(\alpha)} \psi \rangle \tag{8.6.7}$$

This result has important consequences. The matrix element of an invariant operator between two symmetry functions must vanish unless the functions are of the same species. When they are of the same species, $(\alpha, i)$ say, we derive another basic result by supposing

$$\phi = \phi_l^{(\alpha)}, \quad \psi = \psi_k^{(\alpha)} \quad \text{(arbitrary functions of given species)}$$

and noting that (from equation (5.8.7)) $\phi_{il}^{(\alpha)} = \rho_{il}^{(\alpha)} \phi_l^{(\alpha)}$ is then a partner of $\phi_l^{(\alpha)}$ in a certain basis: the whole basis may be generated by taking different $i$ values and the resultant functions may be written $\{\phi_1^{(\alpha)}, \phi_2^{(\alpha)}, \ldots, \phi_l^{(\alpha)}, \ldots\}$. In the same way $\psi_{ik}^{(\alpha)}$ and $\rho_{lk}^{(\alpha)} \psi_k^{(\alpha)}$ are both partners of $\psi_k^{(\alpha)}$ namely $\psi_i^{(\alpha)}$ and $\psi_l^{(\alpha)}$, respectively. On putting these results in (8.6.7) we obtain

$$\langle \phi_i^{(\alpha)} | T | \psi_i^{(\alpha)} \rangle = \langle \phi_l^{(\alpha)} | T | \psi_l^{(\alpha)} \rangle$$

where the subscripts $i$, $l$ distinguish different partners in the same basis for representation $D_\alpha$. To summarize:

> Given an operator $T$, invariant under a group G, and two sets of functions $\{\phi_i^{(\alpha)}\}$, $\{\psi_j^{(\beta)}\}$ providing bases for representations $D_\alpha$ and $D_\beta$,
>
> (i) $\langle \phi_i^{(\alpha)} | T | \psi_j^{(\beta)} \rangle = 0$ unless $\alpha = \beta$ and $i = j$
>
> (ii) $\langle \phi_i^{(\alpha)} | T | \psi_i^{(\alpha)} \rangle =$ constant, $i = 1, 2, \ldots, g_\alpha$. (8.6.8)
>
> In words, the matrix element is zero unless the functions behave like the same basis vector; and then has the same value for every function of the basis.

These results have an obvious significance in quantum mechanics. They also allow us to prove a more general theorem for tensor operators. But first we consider two examples.

EXAMPLE 8.12. In the crystal field problem of example 7.14, the functions $(d_{z^2}, d_{x^2-y^2})$ provided a basis for the E representation of $T_d$. The Hamiltonian of the problem is invariant under the operations of this group. The functions are not *eigenfunctions* of $H$ unless the tetrahedral field is removed; but they are approximations, with appropriate symmetry, to the exact functions which would be rigorously degenerate. The theorem (8.6.8) shows that the *approximate* energies are also rigorously equal

$$\langle d_{z^2} | H | d_{z^2} \rangle = \langle d_{x^2-y^2} | H | d_{x^2-y^2} \rangle$$

while vanishing of the *off*-diagonal matrix elements, $\langle d_{xy} | H | d_{z^2} \rangle = 0$ etc., is simply another way of stating that functions of different species do not mix (cf. (7.5.6)).

EXAMPLE 8.13. Let us evaluate the matrix element $\langle \phi_{11}^{(3)} | H | \phi_{11}^{(3)} \rangle$ for the symmetry function of example 7.16 which is projected from the hydrogen 1*s* function $s^{(1)}$. The idempotent operator $\rho_{11}^{(3)}$ gives

$$\phi_{11}^{(3)} = \rho_{11}^{(3)} s^{(1)} = \tfrac{1}{3}(2s^{(1)} - s^{(2)} - s^{(3)})$$

On using (8.6.7), which relates specifically to projected functions, we obtain at once

$$\langle \phi_{11}^{(3)} | H | \phi_{11}^{(3)} \rangle = \tfrac{1}{3}[2\langle s^{(1)} | H | s^{(1)} \rangle - \langle s^{(1)} | H | s^{(2)} \rangle - \langle s^{(1)} | H | s^{(3)} \rangle]$$

In this case there is a further reduction because $s^{(1)}$ is itself invariant under a symmetry operation (namely $\sigma^{(1)}$): applying this operation to every term in $\langle s^{(1)} | H | s^{(3)} \rangle$, which makes no change, shows that $\langle s^{(1)} | H | s^{(3)} \rangle = \langle s^{(1)} | H | s^{(2)} \rangle$ and hence

$$\langle \phi_{11}^{(3)} | H | \phi_{11}^{(3)} \rangle = \tfrac{2}{3}[\langle s^{(1)} | H | s^{(1)} \rangle - \langle s^{(1)} | H | s^{(2)} \rangle]$$

Direct evaluation would of course involve 9 terms: in general, a projection operator contains $g$ terms and (8.6.7) reduces a sum of $g^2$ contributions to $g$.

*Tensor Operators*

We must now turn to the case in which $T$ is a tensor operator, which we also suppose transforms irreducibly. Let us consider

$$\langle \phi_i^{(\alpha)} | T_j^{(\beta)} | \psi_k^{(\gamma)} \rangle = \int \phi_i^{(\alpha)*}(x) T_j^{(\beta)} \psi_k^{(\gamma)}(x) \, dx \tag{8.6.9}$$

It is at once clear that, with Dirac notation (p. 50), this is the scalar product involving $\langle \phi_i^{(\alpha)} |$ and $T_j^{(\beta)} | \psi_k^{(\gamma)} \rangle$; and consequently that the theorem (8.6.8) may be applied if $T$ is taken as the unit operator. Unless $T_j^{(\beta)} | \psi_k^{(\gamma)} \rangle$ contains some component of symmetry $(\alpha, i)$ the scalar product (i.e. the integral (8.6.8)) must vanish. Now it is clear that such a component, if it exists, may be projected out by applying $\rho_{ii}^{(\alpha)} = (g_\alpha / g) \sum_R D_\alpha(R)_{ii}^* R$ where the rotation $R$ works on both factors and therefore, by definition, produces a linear combination of terms $T_m^{(\beta)} | \psi_n^{(\gamma)} \rangle$. Since the whole set (all values of $m$, $n$) evidently carries a direct product representation, $\mathrm{D} = \mathrm{D}_\beta \times \mathrm{D}_\gamma$, the possibility of finding a linear combination of species $(\alpha, i)$ depends simply on whether $\mathrm{D}$ contains $\mathrm{D}_\alpha$: if it does we must be able to construct basis functions for $\mathrm{D}_\alpha$ and hence at least one combination of species $(\alpha, i)$. This result in itself provides a powerful criterion for selecting non-zero matrix elements:

> The matrix element
>
> $$\langle \phi_i^{(\alpha)} | T_j^{(\beta)} | \psi_k^{(\gamma)} \rangle$$
>
> where $(\alpha, i)$ $(\beta, j)$ and $(\gamma, k)$ indicate irreducible symmetry species, vanishes identically unless $\mathrm{D}_\beta \times \mathrm{D}_\gamma$ contains $\mathrm{D}_\alpha$ at least once. (8.6.10)

It is this result which lies at the root of all the quantum mechanical selection rules. One brief example must serve to indicate a vast field of applications:

EXAMPLE 8.14. The intensity of absorption of radiation, when a system makes "dipole-induced" transitions between states $\phi_i^{(\alpha)}$ and $\psi_k^{(\gamma)}$ depends on the square of a "transition probability". For radiation polarized in the $x$ direction this is

$$\langle \phi_i^{(\alpha)} | M_x | \psi_k^{(\gamma)} \rangle$$

where $M_x$ is the $x$ component of the electric moment of the electrons. If we consider the ammonia molecule, where (example 7.13) $A_1$, $A_2$ and E states occur, the vector components $M_x$, $M_y$ (along $\bar{e}_1$, $\bar{e}_2$ of Fig. 5.1, points 1, 2, 3 denoting the three hydrogen atoms) are known from example 8.10 to form an irreducible tensorial set providing the E representation. A transition from an $A_1$ state $\phi_i^{(\alpha)}$ to an E state $\psi_k^{(\gamma)}$ is therefore "allowed" or "forbidden" according as $E \times E$ does or does not contain $A_1$. From Table 4.6 the characters of the direct product (product of characters) are seen to be $\chi(E) = 4$, $\chi(C_3) = 1$, $\chi(\sigma^{(1)}) = 0$. Hence, by inspection or from (5.4.5)

$$E \times E = A_1 + A_2 + E$$

and a transition $A_1 \rightarrow E$ under the action of $x$-polarized radiation is allowed. On the other hand, a transition $A_1 \rightarrow A_2$ is forbidden, since $E \times A_2 = E$. It must be stressed that these results are a consequence of symmetry alone: they are equally valid either in a one-electron "model" or with exact many-electron wave functions, provided the appropriate symmetry classification is used.

We now return to the argument preceding (8.6.10). In order to *evaluate* the non-zero matrix elements, we must actually find the part of $T_j^{(\beta)} | \psi_k^{(\gamma)} \rangle$ which transforms like $\phi_i^{(\alpha)}$ and therefore gives a non-vanishing contribution to the integral. This problem may be solved generally, without reference to the nature of the $\{T_j^{(\beta)}\}$ and $\{\psi_k^{(\gamma)}\}$ because we are simply asking how the products $\{e_j^{(\beta)} e_k^{(\gamma)}\}$ of two sets of entities $\{e_j^{(\beta)}\}$ and $\{e_k^{(\gamma)}\}$ can be combined to form something transforming like $e_i^{(\alpha)}$: *only the transformation properties matter.* This is the problem already recognized in section 5.9, and soluble by the methods of section 5.8. It is convenient to use the Dirac notation, writing the basis vectors as $|{}^{\beta}_{j}\rangle$, $|{}^{\gamma}_{k}\rangle$ and the products $e_j^{(\beta)} e_k^{(\gamma)}$ as $|{}^{\beta\gamma}_{jk}\rangle$. Then, the statement that the products can be combined to give a vector of species $(\alpha, i)$ may be written

$$|{}^{\alpha}_{i\mu}\rangle = \sum_{j,k} C_{jk}^{(\mu)} |{}^{\beta\gamma}_{jk}\rangle$$

where the $C^{(\mu)}_{jk}$ are suitably chosen coefficients and we recognize the fact that there may be *several* independent combinations (different $\mu$ values) of species $(\alpha, i)$. The rôle of the various indices is exposed more clearly by using a scalar product notation for the coefficients (cf. (2.10.5)). For unitary representations (orthonormal bases), formation of the scalar product $\langle{}^{\beta\gamma}_{jk}|{}^{\alpha}_{i\mu}\rangle$ reduces the right-hand side of the last equation to a single coefficient $C^{(\mu)}_{jk}$: the equation may then be written

$$|{}^{\alpha}_{i\mu}\rangle = \sum_{j,k} |{}^{\beta\gamma}_{jk}\rangle\langle{}^{\beta\gamma}_{jk}|{}^{\alpha}_{i\mu}\rangle \tag{8.6.11}$$

The number $\langle{}^{\beta\gamma}_{jk}|{}^{\alpha}_{i\mu}\rangle$ is called a "coupling coefficient" since it describes how quantities of species $(\beta, j)$ and $(\gamma, k)$ (variable $j$, $k$) may be coupled together to give one of species $(\alpha, i)$. It is a purely "geometric" coefficient, depending only on the representations $D_\beta$, $D_\gamma$ and the way in which $D_\beta \times D_\gamma$ is reduced to $D_\alpha$ (i.e. upon the "coupling scheme"): we now use them in writing down the part of $T^{(\beta)}_j|\psi^{(\gamma)}_k\rangle$ which is of species $(\alpha, i)$. First we note that (8.6.11) is a relationship between two unitary bases and, as such, possesses an inverse:

$$\bar{\mathbf{e}}_n = \sum_m \mathbf{e}_m U_{mn}, \qquad \mathbf{e}_m = \sum_n \bar{\mathbf{e}}_n U^*_{mn}$$

In other words,

$$|{}^{\beta\gamma}_{jk}\rangle = \sum_{\alpha,i,\mu} |{}^{\alpha}_{i\mu}\rangle\langle{}^{\alpha}_{i\mu}|{}^{\beta\gamma}_{jk}\rangle = \sum_{\alpha,i,\mu} |{}^{\alpha}_{i\mu}\rangle\langle{}^{\beta\gamma}_{jk}|{}^{\alpha}_{i\mu}\rangle^* \tag{8.6.12}$$

We can now express the right hand member of the scalar product (8.6.9) in terms of functions of symmetry species $(\alpha, i)$:

$$T^{(\beta)}_j|\psi^{(\gamma)}_k\rangle = \sum_{\alpha,i,\mu} |{}^{\alpha}_{i\mu}\rangle\langle{}^{\beta\gamma}_{jk}|{}^{\alpha}_{i\mu}\rangle^*$$

where $|{}^{\alpha}_{i\mu}\rangle$ now stands for the $\mu$th distinct combination of species $(\alpha, i)$. When this result is inserted in (8.6.9) the $\alpha$,$i$-summation disappears, since by (8.6.8) only one species contributes, and we obtain

$$\langle\phi^{(\alpha)}_i|T^{(\beta)}_j|\psi^{(\gamma)}_k\rangle = \sum_\mu A_\mu \langle{}^{\beta\gamma}_{jk}|{}^{\alpha}_{i\mu}\rangle^* \tag{8.6.13}$$

where

$$A_\mu = \langle\phi^{(\alpha)}_i|{}^{\alpha}_{i\mu}\rangle$$

the scalar product between $\phi^{(\alpha)}_i$ and a certain linear combination of the $T^{(\beta)}_j|\psi^{(\gamma)}_k\rangle$. Now $A_\mu$ does not depend on $j$, $k$ and is also independent of $i$ by the second part of the basic theorem (8.6.8): the $A_\mu$ are therefore *constants* in the sense that they take the same values for all the $g_\alpha g_\beta g_\gamma$ distinct matrix elements which result from different choices of $i$, $j$, $k$.

When the direct product $D_\beta \times D_\gamma$ contains $D_\alpha$ just once, the label $\mu$ may be omitted and (8.6.13) reduces to

$$\langle \phi_i^{(\alpha)} | T_j^{(\beta)} | \psi_k^{(\gamma)} \rangle = \text{constant} \times \langle {}^{\beta\gamma}_{jk} | {}^{\alpha}_{i} \rangle^* \tag{8.6.14}$$

This is the case when the symmetry group is $\mathsf{D}_3$ itself and the result is then referred to as the Wigner–Eckart theorem. Although we have not explicitly considered continuous groups, the methods used in establishing this result are of general validity and a few remarks on the full rotation group are not out of place. In the quantum mechanical central field problem, the spherical harmonics

$$Y_{J,M}\ (M = -J,\ -J+1,\ \ldots\ +J)$$

carry an irreducible representation of dimension $2J+1$ and the labels $J$, $M$ indicate the eigenvalues of the angular momentum and its $z$ component: the tensor operators may be classified similarly by comparing their behaviour with that of the spherical harmonics, and the result (8.6.14) then becomes (the co-efficients are in this case real)

$$\langle \phi_{M_1}^{(J_1)} | T_{M_2}^{(J_2)} | \psi_{M_3}^{(J_3)} \rangle = \text{constant} \times \langle {}^{J_2\,J_3}_{M_2 M_3} | {}^{J_1}_{M_1} \rangle \tag{8.6.15}$$

The coupling coefficient is then the " Clebsch–Gordan " or " Wigner " coefficient which appears when two quantum-mechanical angular momenta ($J_2$, $J_3$) are coupled to a resultant ($J_1$). Clearly, if these coefficients are known the matrix element need be evaluated only *once* (for the simplest choice of $M$'s) in order to determine the constant; all other elements then follow.

More generally, for the finite point groups, $D_\alpha$ sometimes occurs *twice* in $D_\beta \times D_\gamma$. Expansion (8.6.13) then contains two terms and the matrix element must be evaluated for two choices of $i, j, k$ in order to determine the constants: all other elements are then readily evaluated if the coupling coefficients are known. Finally, then, we express the generalized Wigner–Eckart theorem in the form:

> Given a tensorial set $\{T_j^{(\beta)}\}$ and two sets of functions $\{\phi_i^{(\alpha)}\}$, $\{\psi_k^{(\gamma)}\}$ providing bases for irreducible representations $D_\beta$, $D_\alpha$, $D_\gamma$ (respectively) of a group $\mathsf{G}$,
>
> $$\langle \phi_i^{(\alpha)} | T_j^{(\beta)} | \psi_k^{(\gamma)} \rangle = \sum_\mu A_\mu \langle {}^{\beta\gamma}_{jk} | {}^{\alpha}_{i\mu} \rangle^* \tag{8.6.16}$$
>
> Where the $A_\mu$ are independent of $i, j, k$ and the coupling coefficients express the $\mu$th basis vector of species $(\alpha, i)$ in the product space $D_\beta \times D_\gamma$ in terms of basis vector products $\mathbf{e}_j^{(\beta)}\mathbf{e}_k^{(\gamma)}$.

**8.7. Determination of Coupling Coefficients**

The determination of matrix elements of all kinds of tensor operator is a simple matter, once we know the coupling coefficients which appear in (8.6.16). It is not difficult to obtain these coefficients, once and for all time, for all pairs of representations of all the point groups. Some tables have already been given by Tanabe and others (Ref. 4), Koster and Statz (Ref. 5), and Griffith (Ref. 6). But conventions are variable and the " standard " irreducible representations suggested in chapter 4 are not in universal use. Instead of deriving extensive tables we shall therefore show how a typical table may be constructed from first principles.

We denote the basis vectors of the representations to be coupled by $|{}^{\beta}_{j}\rangle$ and $|{}^{\gamma}_{k}\rangle$ and a corresponding vector of product space by $|{}^{\beta\gamma}_{jk}\rangle$. The latter are to be combined to give a vector $|{}^{\alpha}_{i}\rangle$ transforming like the $i$th basis vector of representation $\mathrm{D}_\alpha$. One obvious method is to *project* such a vector, using an idempotent $\rho^{(\alpha)}_{ii}$, from an arbitrary vector $|{}^{\beta\gamma}_{pq}\rangle$.

The basis vectors of product space transform like the matrices of $\mathrm{D}_\beta \times \mathrm{D}_\gamma$ (outer products, which may be written down as on p. 137) and any vector of product space is represented by a column of components: if the arbitrary vector is chosen as a single basis vector the column will contain zeros except for a single unit element. Only the matrices of the generators are required, since the effect of other operations may be obtained by repeated application, and the desired symmetry vector is obtained exactly as in example 6.8. If $\mathrm{D}_\alpha$ appears twice in $\mathrm{D}_\beta \times \mathrm{D}_\gamma$ it will be possible to find two independent vectors of species $(\alpha,\ i)$ (e.g. by using also $\rho^{(\alpha)}_{ij}$ with $j \neq i$) and any two orthogonal vectors $|{}^{\alpha}_{i1}\rangle$ and $|{}^{\alpha}_{i2}\rangle$, will give suitable coefficients $\langle{}^{\beta\gamma}_{jk}|{}^{\alpha}_{i\mu}\rangle$ with $\mu = 1,\ 2$.

The projection method is straightforward, and by now so familiar that an example is unnecessary, but somewhat tedious for groups of high order. We therefore close by indicating an alternative method, in which only the generators are considered. We take one of the higher groups as an example and then show how the results may be used in an application of the Wigner–Eckart theorem of the last section.

EXAMPLE 8.15. In discussing the Zeeman effect for a transition metal ion in a tetrahedral crystal field, it is necessary to evaluate matrix elements of the angular momentum operators $L_x$, $L_y$, $L_z$ between functions which belong to the E and $T_2$ representations of the group $T_d$. The angular momentum operators form an irreducible tensorial set with pseudo-vector character, transforming like the basis vectors except for sign changes under improper operations, and consequently provide the representation $T_1$ (Table 4.20). It is therefore necessary to consider matrix elements

$$\langle {}^{E}_{i} | L^{(T_1)}_{j} | {}^{T_2}_{k} \rangle \qquad (j,\ k = 1,\ 2,\ 3;\quad i = 1,\ 2)$$

It may be shown in the usual way that $T_1 \times T_2 = A_2 + E + T_1 + T_2$, and to apply the theorem (8.6.15) we must consider the coupling of vectors $\{|{}^{T_1}_{j}\rangle\}$ and $\{|{}^{T_2}_{k}\rangle\}$ to a resultant $|{}^{E}_{i}\rangle$. The effect of the generators $C_3^{xyz}$ and $S_4^z$ upon basis vectors of the various representations is known (Table 4.20). With the ket notation the required vectors must transform as follows

$$C_3^{xyz}\,|{}^{E}_{1}\rangle = -\tfrac{1}{2}\,|{}^{E}_{1}\rangle + \tfrac{1}{2}\sqrt{3}\,|{}^{E}_{2}\rangle \qquad S_4^z\,|{}^{E}_{1}\rangle = |{}^{E}_{1}\rangle$$

$$C_3^{xyz}\,|{}^{E}_{2}\rangle = -\tfrac{1}{2}\sqrt{3}\,|{}^{E}_{1}\rangle - \tfrac{1}{2}\,|{}^{E}_{2}\rangle \qquad S_4^z\,|{}^{E}_{2}\rangle = -|{}^{E}_{2}\rangle$$

On the other hand, the effect of these rotations on the basis vectors of representations $T_1$ and $T_2$ is

| $V$ | $\lvert{}^{T_1}_{1}\rangle$ | $\lvert{}^{T_1}_{2}\rangle$ | $\lvert{}^{T_1}_{3}\rangle$ | $\lvert{}^{T_2}_{1}\rangle$ | $\lvert{}^{T_2}_{2}\rangle$ | $\lvert{}^{T_2}_{3}\rangle$ |
|---|---|---|---|---|---|---|
| $C_3^{xyz}V$ | $\lvert{}^{T_1}_{3}\rangle$ | $\lvert{}^{T_1}_{1}\rangle$ | $\lvert{}^{T_1}_{2}\rangle$ | $\lvert{}^{T_2}_{3}\rangle$ | $\lvert{}^{T_2}_{1}\rangle$ | $\lvert{}^{T_2}_{2}\rangle$ |
| $S_4^zV$ | $-\lvert{}^{T_1}_{2}\rangle$ | $\lvert{}^{T_1}_{1}\rangle$ | $-\lvert{}^{T_1}_{3}\rangle$ | $\lvert{}^{T_2}_{2}\rangle$ | $-\lvert{}^{T_2}_{1}\rangle$ | $\lvert{}^{T_2}_{3}\rangle$ |

The product kets $|{}^{T_1}_{i}{}^{T_2}_{j}\rangle$, which we abbreviate to $(ij)$, therefore behave as follows

| $V$ | (11) | (12) | (13) | (21) | (22) | (23) | (31) | (32) | (33) |
|---|---|---|---|---|---|---|---|---|---|
| $C_3^{xyz}V$ | (33) | (31) | (32) | (13) | (11) | (12) | (23) | (21) | (22) |
| $S_4^zV$ | −(22) | (21) | −(23) | (12) | (11) | (13) | −(32) | (31) | −(33) |

Let us now write the required linear combinations in product space

$$|{}^{E}_{1}\rangle = c_1(11) + c_2(12) + \ \dots\ + c_6(33)$$

$$|{}^{E}_{2}\rangle = c_1'(11) + c_2'(12) + \ \dots\ + c_6'(33)$$

and choose the coefficients so as to obtain the right transformation properties. For example, the condition $S_4^z \left|{}^{\mathrm{E}}_{1}\right\rangle = \left|{}^{\mathrm{E}}_{1}\right\rangle$ shows (using the results above) that

$$-c_1(22) + c_2(21) - c_3(23) + c_4(12) - c_5(11) + c_6(13) - c_7(32) + c_8(31) - c_9(33)$$
$$= c_1(11) + c_2(12) + c_3(13) + c_4(21) + c_5(22) + c_6(23) + c_7(31) + c_8(32) + c_9(33)$$

Hence, equating coefficients, we find

$$c_5 = -c_1, \quad c_4 = c_2, \quad c_3 = c_6 = c_7 = c_8 = c_9 = 0$$

In this way we infer

$$\left|{}^{\mathrm{E}}_{1}\right\rangle = a[(11) - (22)] + b[(12) + (21)]$$
$$\left|{}^{\mathrm{E}}_{2}\right\rangle = c[(11) + (22)] + d[(12) - (21)] + f(33)$$

The remaining coefficients are chosen so that the kets behave properly under $C_3^{xyz}$. On solving the two simultaneous equations which result, and finally normalizing, we obtain

$$\left|{}^{\mathrm{E}}_{1}\right\rangle = \left|{}^{\mathrm{T_1T_2}}_{1\ 1}\right\rangle\left(\tfrac{1}{\sqrt{2}}\right) + \left|{}^{\mathrm{T_1T_2}}_{2\ 2}\right\rangle\left(-\tfrac{1}{\sqrt{2}}\right)$$
$$\left|{}^{\mathrm{E}}_{2}\right\rangle = \left|{}^{\mathrm{T_1T_2}}_{1\ 1}\right\rangle\left(-\tfrac{1}{\sqrt{6}}\right) + \left|{}^{\mathrm{T_1T_2}}_{2\ 2}\right\rangle\left(-\tfrac{1}{\sqrt{6}}\right) + \left|{}^{\mathrm{T_1T_2}}_{3\ 3}\right\rangle\left(\sqrt{\tfrac{2}{3}}\right)$$

The numerical coefficients are the required coupling constants $\left\langle{}^{\mathrm{T_1T_2}}_{i\ j}\middle|{}^{\mathrm{E}}_{k}\right\rangle$.

EXAMPLE 8.16. Application of the Wigner–Eckart theorem (8.6.15) to the problem which introduced the last example is now straightforward. In order to make the results more immediately significant we employ the appropriate $d$ functions of example 7.14. With an obvious notation, the E functions are $|z^2\rangle$ and $|x^2-y^2\rangle$, while the $\mathrm{T}_2$ functions are $|yz\rangle$, $|zx\rangle$ and $|xy\rangle$. Hence

$$\langle z^2|L_x|yz\rangle = \left\langle{}^{\mathrm{E}}_{1}\middle|L_1^{(\mathrm{T_1})}\middle|{}^{\mathrm{T_2}}_{1}\right\rangle = \text{constant} \times \left\langle{}^{\mathrm{T_1T_2}}_{1\ 1}\middle|{}^{\mathrm{E}}_{1}\right\rangle^* = A/\sqrt{2}$$
$$\langle z^2|L_y|zx\rangle = \left\langle{}^{\mathrm{E}}_{1}\middle|L_2^{(\mathrm{T_1})}\middle|{}^{\mathrm{T_2}}_{2}\right\rangle = \text{constant} \times \left\langle{}^{\mathrm{T_1T_2}}_{2\ 2}\middle|{}^{\mathrm{E}}_{1}\right\rangle^* = -A/\sqrt{2}$$

where the coefficients are taken from the expression for $\left|{}^{\mathrm{E}}_{1}\right\rangle$ in the last example. The remaining coefficients, appearing in $\left|{}^{\mathrm{E}}_{2}\right\rangle$, give

$$\langle x^2-y^2|L_x|yz\rangle = \langle x^2-y^2|L_y|zx\rangle = -A/\sqrt{6},$$
$$\langle x^2-y^2|L_z|xy\rangle = 2A/\sqrt{6}$$

All other elements between an E function and a $\mathrm{T}_2$ function must vanish.

On the other hand, $\mathrm{T}_1 \times \mathrm{T}_2$ contains $\mathrm{T}_2$ as well as E and there will thus be other non-zero elements when both functions are of $\mathrm{T}_2$ type:

these will involve the coupling coefficients $\left\langle \begin{matrix} T_1 & T_2 \\ i & j \end{matrix} \middle| \begin{matrix} T_2 \\ k \end{matrix} \right\rangle$. It should be noted that the constant which then appears, $A'$ say, will in general differ from $A$, depending on all three representations. When the functions are *pure d* functions (spherical harmonic angle dependence) the two constants become identical, for the E and $T_2$ functions then belong to the *same* representation of the full rotation group; but when the functions are modified by the tetrahedral field there is no need for the constants to take the same value. The modified functions might, for instance, be molecular orbitals (example 7.15) obtained by combining the $d$ orbitals of the transition metal ion with appropriate symmetry orbitals on the surrounding ligands.

## REFERENCES

1. EDMONDS, A. R. *Angular Momentum in Quantum Mechanics* (Section 1.3), Princeton University Press, 1957.
2. SCHOUTON, J. A., *Tensor Analysis for Physicists* (p. 240 *et seq.*), Oxford University Press, 1951.
3. WIGNER, E. P., *Group Theory and its Application to the Quantum Mechanics of Atomic Spectra* (p. 244), Academic Press, 1959.
4. TANABE, Y., and SUGANO, S., *J. Phys. Soc. Japan*, 1954, **9**, 753; TANABE, Y., and KAMIMURA, H., *ibid.*, 1958, **13**, 394.
5. STATZ, H., and KOSTER, G. F., *Phys. Rev.*, 1959, **115**, 1568.
6. GRIFFITH, J. S., *The Theory of Transition Metal Ions* (Appendix 2), Cambridge University Press, 1961.

## BIBLIOGRAPHY

EDMONDS, A. R., *Angular Momentum in Quantum Mechanics*, Princeton University Press, 1957.

FANO, U., and RACAH, G., *Irreducible Tensorial Sets*, Academic Press, 1959.

GRIFFITH, J. S., *The Theory of Transition Metal Ions*, Cambridge University Press, 1961.

HALL, G. G., *Matrices and Tensors*, Pergamon Press, Oxford, 1962.

HEINE, V., *Group Theory in Quantum Mechanics*, Pergamon Press, Oxford, 1960.

JEFFREYS, H. J., *Cartesian Tensors*, Cambridge University Press, 1931.

LYUBARSKII, G. YA., *The Application of Group Theory in Physics*, Pergamon Press, Oxford, 1960.

NYE, J. F., *Physical Properties of Crystals*, Oxford University Press, 1957.

ROSE, M. E., *Elementary Theory of Angular Momentum*, Wiley, 1957.

WIGNER, E. P., *Group Theory and its Application to the Quantum Mechanics of Atomic Spectra*, Academic Press, 1959.

APPENDIX 1

# REPRESENTATIONS CARRIED BY HARMONIC FUNCTIONS

THE $s$, $p$, $d$, and $f$ functions (example 7.4) have angle dependent factors constructed from spherical harmonics of degree 0, 1, 2, 3, respectively, the $2l+1$ independent functions of degree $l$ carrying a representation of the rotation group $\mathsf{D}_3$. Each set of functions also carries a representation of each of the 32 point groups discussed in chapters 3 and 4. We now tabulate the harmonics, or appropriate linear combinations, which carry the standard irreducible representations defined in chapter 4. Appropriate sets of (real) $s$, $p$, $d$, $f$ functions are defined below in terms of (i) Cartesian coordinates, and (ii) real spherical harmonics.

TABLE A1.1

*Cubic harmonics*

$s$ functions

| | | |
|---|---|---|
| $s$ | $N_0$ | $Y_{0,0}$ |

$p$ functions

| | | | |
|---|---|---|---|
| $p_x$ | $N_1r^{-1}$ | $x$ | $Y^c_{1,1}$ |
| $p_y$ | ,, | $y$ | $Y^s_{1,1}$ |
| $p_z$ | ,, | $z$ | $Y_{1,0}$ |

$d$ functions

| | | | |
|---|---|---|---|
| $d_{z^2}$ | $N_2r^{-2}$ | $\frac{1}{2}(3z^2-r^2)$ | $Y_{2,0}$ |
| $d_{x^2-y^2}$ | ,, | $\frac{1}{2}\sqrt{3}\,(x^2-y^2)$ | $Y^c_{2,2}$ |
| $d_{yz}$ | ,, | $\sqrt{3}\,yz$ | $Y^s_{2,1}$ |
| $d_{zx}$ | ,, | $\sqrt{3}\,zx$ | $Y^c_{2,1}$ |
| $d_{xy}$ | ,, | $\sqrt{3}\,xy$ | $Y^s_{2,2}$ |

*f* functions

| | | | |
|---|---|---|---|
| $f_x$ | $N_3 r^{-3}$ | $\frac{1}{2}(5x^2 - 3r^2)x$ | $-\frac{1}{4}(\sqrt{6}\, Y^c_{3,1} - \sqrt{10}\, Y^c_{3,3})$ |
| $f_y$ | ,, | $\frac{1}{2}(5y^2 - 3r^2)y$ | $-\frac{1}{4}(\sqrt{6}\, Y^s_{3,1} + \sqrt{10}\, Y^s_{3,3})$ |
| $f_z$ | ,, | $\frac{1}{2}(5z^2 - 3r^2)z$ | $Y_{3,0}$ |
| $f_x'$ | ,, | $\frac{1}{2}\sqrt{15}\,(y^2 - z^2)x$ | $-\frac{1}{4}(\sqrt{10}\, Y^c_{3,1} + \sqrt{6}\, Y^c_{3,3})$ |
| $f_y'$ | ,, | $\frac{1}{2}\sqrt{15}\,(z^2 - x^2)y$ | $\frac{1}{4}(\sqrt{10}\, Y^s_{3,1} - \sqrt{6}\, Y^s_{3,3})$ |
| $f_z'$ | ,, | $\frac{1}{2}\sqrt{15}\,(x^2 - y^2)z$ | $Y^c_{3,2}$ |
| $f_{xyz}$ | ,, | $\sqrt{15}\, xyz$ | $Y^s_{3,2}$ |

*Note.* The normalizing factor which appears in the Cartesian expressions is given by $N_l = \sqrt{[(2l+1)/4\pi]}$. For definition of the real spherical harmonics, see p. 174.

The *f* functions here defined have simple transformation properties under the operations of the triclinic, monoclinic, orthorhombic, tetragonal and cubic groups, the first three transforming like *p* functions and the second three similarly except for sign changes: they are in fact "cubic" harmonics. For trigonal and hexagonal groups, the ordinary harmonics are adequate: the Cartesian forms, and relationship to the cubic harmonics, appear in Table A1.2 for those cases not already given.

## Table A1.2

*Real spherical harmonics (cubic forms)*

| | | | |
|---|---|---|---|
| $Y^c_{3,1}$ | $N_3 r^{-3}$ | $\frac{1}{4}\sqrt{6}\,(5z^2 - r^2)x$ | $-\frac{1}{4}(\sqrt{6} f_x + \sqrt{10} f_x')$ |
| $Y^s_{3,1}$ | ,, | $\frac{1}{4}\sqrt{6}\,(5z^2 - r^2)y$ | $-\frac{1}{4}(\sqrt{6} f_y - \sqrt{10} f_y')$ |
| $Y^c_{3,3}$ | ,, | $\frac{1}{4}\sqrt{10}\,(x^2 - 3y^2)x$ | $\frac{1}{4}(\sqrt{10} f_x - \sqrt{6} f_x')$ |
| $Y^s_{3,3}$ | ,, | $\frac{1}{4}\sqrt{10}\,(3x^2 - y^2)y$ | $-\frac{1}{4}(\sqrt{10} f_y + \sqrt{6} f_y')$ |

See Table A1.1 for normalizing factor and for other harmonics. For definitions, see p. 174.

We now list the bases which carry the standard irreducible representations (Tables 4.1–4.22) of all the crystal point groups, starting with each holohedry and showing how the classification changes as the symmetry is reduced. The notation $(A_{1g}, B_{1g}) \rightarrow A_1$, for example, will indicate that on restriction to a subgroup the functions belonging to representations $A_{1g}$ and $B_{1g}$ *all* provide the subgroup representation $A_1$. Also, where complex representations occur, the functions have been chosen to give the standard *real* forms to which they are equivalent (cf. example 4.1). Finally, the properties of the angular momentum operators $L_x$, $L_y$, $L_z$, which provide bases according to (8.5.9), have been included in view of their special importance in quantum mechanics: other sets of quantities with a similar pseudo-vector character (e.g. rotations about the three coordinate axes—often denoted by curved arrows) have identical properties.

TABLE A1.3

*Triclinic bases*

| Group $S_2$ | | |
|---|---|---|
| | $A_g$ | $s$; $d_{z^2}$, $d_{x^2-y^2}$, $d_{yz}$, $d_{zx}$, $d_{xy}$; $L_x$, $L_y$, $L_z$ |
| | $A_u$ | $p_x$, $p_y$, $p_z$; $f_x$, $f_y$, $f_z$, $f_x'$, $f_y'$, $f_z'$, $f_{xyz}$ |

TABLE A1.4

*Monoclinic bases*

| Group $C_{2h}$ | | | | |
|---|---|---|---|---|
| $A_g$ | $s$; $d_{z^2}$, $d_{x^2-y^2}$, $d_{xy}$, $L_z$ | $A_u$ | $p_z$; $f_z$, $f_z'$, $f_{xyz}$ |
| $B_g$ | $d_{yz}$, $d_{zx}$; $L_x$, $L_y$ | $B_u$ | $p_x$, $p_y$; $f_x$, $f_y$, $f_x'$, $f_y'$ |

Group $C_2$ $(A_g, A_u) \rightarrow A$ $(B_g, B_u) \rightarrow B$

Group $C_{1h}$ $(A_g, B_u) \rightarrow A'$ $(A_u, B_g) \rightarrow A''$

TABLE A1.5

*Orthorhombic bases*

| Group $D_{2h}$ | | | |
|---|---|---|---|
| $A_g$ | $s$; $d_{z^2}$, $d_{x^2-y^2}$ | $A_u$ | $f_{xyz}$ |
| $B_{1g}$ | $d_{xy}$; $L_z$ | $B_{1u}$ | $p_z$; $f_z$, $f_z'$ |
| $B_{2g}$ | $d_{zx}$; $L_y$ | $B_{2u}$ | $p_y$; $f_y$, $f_y'$ |
| $B_{3g}$ | $d_{yz}$; $L_x$ | $B_{3u}$ | $p_x$; $f_x$, $f_x'$ |

Group $D_2$ $(A_g, A_u) \rightarrow A$, $(B_{1g}, B_{1u}) \rightarrow B_1$, $(B_{2g}, B_{2u}) \rightarrow B_2$, $(B_{3g}, B_{3u}) \rightarrow B_3$

Group $C_{2v}$ $(A_g, B_{1u}) \rightarrow A_1$, $(A_u, B_{1g}) \rightarrow A_2$, $(B_{3g}, B_{2u}) \rightarrow B_2$, $(B_{2g}, B_{3u}) \rightarrow B_1$

## Table A1.6

*Trigonal bases*

Group $D_{3d}$

| | |
|---|---|
| $A_{1g}$ | $s$; $d_{z^2}$ |
| $A_{2g}$ | $L_z$ |
| $E_g$ | $(d_{x^2-y^2}, \; -d_{xy})$, $(d_{yz}, \; -d_{zx})$; $(L_x, \; L_y)$ |
| $A_{1u}$ | $Y^c_{3,3}$ |
| $A_{2u}$ | $p_z$; $f_z$; $Y^s_{3,3}$ |
| $E_u$ | $(p_x, \; p_y)$; $(f_{xyz}, \; f_z')$; $(Y^c_{3,1}, \; Y^s_{3,1})$ |

*Note.* The phases of the functions (+ or — signs) are chosen so that the E representations appear in standard form, the order of the functions being that of the standard basis vectors (Note that for $C_n$ groups the real E representations are reducible, *cf.* p. 92). Any changes necessary, in passing to the subgroups, are indicated.

Group $D_3$ $(A_{1g}, \; A_{1u}) \rightarrow A_1$, $(A_{2g}, \; A_{2u}) \rightarrow A_2$, $(E_g, \; E_u) \rightarrow E$

Group $C_{3v}$ $(A_{1g}, \; A_{2u}) \rightarrow A_1$, $(A_{2g}, \; A_{1u}) \rightarrow A_2$, $(E_g, \; E_u) \rightarrow E$
Order and phase changes: $(d_{yz}, \; -d_{zx}) \rightarrow (d_{zx}, \; d_{yz})$, $(f_{xyz}, \; f_z') \rightarrow (f_z', \; -f_{xyz})$, $(L_x, \; L_y) \rightarrow (L_y, \; -L_x)$
New functions differ from original by 60° rotation about $z$ axis.

Group $C_3$ $(A_{1g}, \; A_{1u}, \; A_{2g}, \; A_{2u}) \rightarrow A$, $(E_g, \; E_u) \rightarrow E$

Group $S_6$ $(A_{1g}, \; A_{2g}) \rightarrow A_g$, $(A_{1u}, \; A_{2u}) \rightarrow A_u$

## Table A1.7

*Hexagonal bases*

Group $D_{6h}$

| | | | |
|---|---|---|---|
| $A_{1g}$ | $s$; $d_z{}^2$ | $A_{1u}$ | |
| $A_{2g}$ | $L_z$ | $A_{2u}$ | $p_z$; $f_z$ |
| $B_{1g}$ | | $B_{1u}$ | $Y^c_{3,3}$ |
| $B_{2g}$ | | $B_{2u}$ | $Y^s_{3,3}$ |
| $E_{1g}$ | $(d_{yz}, \; -d_{zx})$; $(L_x, \; L_y)$ | $E_{1u}$ | $(p_x, p_y)$; $(Y^c_{3,1}, Y^s_{3,1})$ |
| $E_{2g}$ | $(d_{x^2-y^2}, \; -d_{xy})$ | $E_{2u}$ | $(f_{xyz}, \; f_z')$ |

*Note.* Order and phases are chosen to give standard E representations (see Table A1.6).

Group $D_{3h}$ $(A_{1g}, \; B_{1u}) \to A_1'$, $(B_{1g}, \; A_{1u}) \to A_1''$, $(A_{2g}, \; B_{2u}) \to A_2'$,
$(B_{2g}, \; A_{2u}) \to A_2''$, $(E_{2g}, \; E_{1u}) \to E'$, $(E_{1g}, \; E_{2u}) \to E''$

Group $D_6$ $(A_{1g}, \; A_{1u}) \to A_1$, $(A_{2g}, \; A_{2u}) \to A_2$, $(B_{1g}, \; B_{1u}) \to B_1$,
$(B_{2g}, \; B_{2u}) \to B_2$, $(E_{1g}, \; E_{1u}) \to E_1$, $(E_{2g}, \; E_{2u}) \to E_2$

Group $C_{6v}$ $(A_{1g}, \; A_{2u}) \to A_1$, $(A_{2g}, \; A_{1u}) \to A_2$, $(B_{1g}, \; B_{2u}) \to B_2$,
$(B_{2g}, \; B_{1u}) \to B_1$, $(E_{1g}, \; E_{1u}) \to E_1$, $(E_{2g}, \; E_{2u}) \to E_2$
Order and phase changes: $(d_{yz}, \; -d_{zx}) \to (d_{zx}, \; d_{yz})$,
$(f_{xyz}, \; f_z') \to (f_z', \; -f_{xyz})$, $(L_x, \; L_y) \to (L_y, \; -L_x)$

Group $C_{6h}$ $(A_{1g}, \; A_{2g}) \to A_g$, $(A_{1u}, \; A_{2u}) \to A_u$, $(B_{1g}, \; B_{2g}) \to B_g$,
$(B_{1u}, \; B_{2u}) \to B_u$

Group $C_{3h}$ $(A_{1g}, \; A_{2g}, \; B_{1u}, \; B_{2u}) \to A'$, $(B_{1g}, \; B_{2g}, \; A_{1u}, \; A_{2u}) \to A''$
$(E_{1g}, \; E_{2u}) \to E''$, $(E_{2g}, \; E_{1u}) \to E'$

Group $C_6$ $(A_{1g}, \; A_{2g}, \; A_{1u}, \; A_{2u}) \to A$, $(B_{1g}, \; B_{2g}, \; B_{1u}, \; B_{2u}) \to B$,
$(E_{1g}, \; E_{1u}) \to E_1$, $(E_{2g}, \; E_{2u}) \to E_2$

## TABLE A1.8

### *Tetragonal bases*

Group $D_{4h}$

| | | | |
|---|---|---|---|
| $A_{1g}$ | $s$; $d_{z^2}$ | $A_{1u}$ | |
| $A_{2g}$ | $L_z$ | $A_{2u}$ | $p_z$; $f_z$ |
| $B_{1g}$ | $d_{x^2-y^2}$ | $B_{1u}$ | $f_{xyz}$ |
| $B_{2g}$ | $d_{xy}$ | $B_{2u}$ | $f_z'$ |
| $E_g$ | $(d_{yz}, \; -d_{zx})$; $(L_x, \; L_y)$ | $E_u$ | $(p_x, \; p_y)$; $(f_x, \; f_y)$, $(f_x', \; -f_y')$ |

*Note.* Order and phases are chosen to give standard E representations (see Table A1.6).

Group $D_{2d}$ $(A_{1g}, \; B_{1u}) \to A_1$, $(B_{1g}, \; A_{1u}) \to B_1$, $(A_{2g}, \; B_{2u}) \to A_2$,
$(B_{2g}, \; A_{2u}) \to B_2$, $(E_g, \; E_u) \to E$
Phase change: $(d_{yz}, \; -d_{zx}) \to (d_{yz}, \; d_{zx})$

Group $D_4$ $(A_{1g}, \; A_{1u}) \to A_1$, $(B_{1g}, \; B_{1u}) \to B_1$, $(A_{2g}, \; A_{2u}) \to A_2$,
$(B_{2g}, \; B_{2u}) \to B_2$, $(E_g, \; E_u) \to E$

Group $S_4$ $(A_{1g}, A_{2g}, B_{1u}, B_{2u}) \to A$, $(B_{1g}, B_{2g}, A_{1u}, A_{2u}) \to B$, $(E_g, E_u) \to E$
Phase change: $(d_{yz}, \; -d_{zx}) \to (d_{yz}, \; d_{zx})$

Group $C_{4h}$ $(A_{1g}, A_{2g}) \to A_g$, $(A_{1u}, A_{2u}) \to A_u$, $(B_{1g}, B_{2g}) \to B_g$, $(B_{1u}, B_{2u}) \to B_u$

Group $C_{4v}$ $(A_{1g}, \; A_{2u}) \to A_1$, $(A_{2g}, \; A_{1u}) \to A_2$, $(B_{1g}, \; B_{2u}) \to B_1$,
$(B_{2g}, \; B_{1u}) \to B_2$, $(E_g, \; E_u) \to E$
Order and phase changes: $(d_{yz}, \; -d_{zx}) \to (d_{zx}, \; d_{yz})$,
$(L_x, \; L_y) \to (L_y, \; -L_x)$

Group $C_4$ $(A_{1g}, A_{2g}, A_{1u}, A_{2u}) \to A$, $(B_{1g}, B_{2g}, B_{1u}, B_{2u}) \to B$, $(E_g, E_u) \to E$

## TABLE A1.9

*Cubic bases*

Group $O_h$

| | | | |
|---|---|---|---|
| $A_{1g}$ | $s$ | $A_{1u}$ | |
| $A_{2g}$ | | $A_{2u}$ | $f_{xyz}$ |
| $E_g$ | $(d_{z^2},\ d_{x^2-y^2})$ | $E_u$ | |
| $T_{1g}$ | $(L_x,\ L_y,\ L_z)$ | $T_{1u}$ | $(p_x,\ p_y,\ p_z)$; $(f_x,\ f_y,\ f_z)$ |
| $T_{2g}$ | $(d_{yz},\ d_{zx},\ d_{xy})$ | $T_{2u}$ | $(f_x',\ f_y',\ f_z')$ |

*Note.* Order and phases are chosen to give standard E and T representations (see Table A1.6).

Group O $(A_{1g},\ A_{1u}) \to A_1$, $(A_{2g},\ A_{2u}) \to A_2$, $(E_g,\ E_u) \to E$,
$(T_{1g},\ T_{1u}) \to T_1$, $(T_{2g},\ T_{2u}) \to T_2$.

Group $T_d$ $(A_{1g},\ A_{2u}) \to A_1$, $(A_{2g},\ A_{1u}) \to A_2$, $(E_g,\ E_u) \to E$,
$(T_{1g},\ T_{2u}) \to T_1$, $(T_{2g},\ T_{1u}) \to T_2$.

Group $T_h$ $(A_{1g},\ A_{2g}) \to A_g$, $(A_{1u},\ A_{2u}) \to A_u$, $(T_{1g},\ T_{2g}) \to T_g$, $(T_{1u},\ T_{2u}) \to T_u$

Group T $(A_{1g},\ A_{2g},\ A_{1u},\ A_{2u}) \to A$, $(E_g,\ E_u) \to E$, $(T_{1g},\ T_{2g},\ T_{1u},\ T_{2u}) \to T$

APPENDIX 2

# ALTERNATIVE BASES FOR CUBIC GROUPS

IN the subgroup reductions of Table A1.9, the $z$ axis remains the principal axis throughout. Since, however, the cube possesses a variety of equivalent axes, it is often necessary to consider distortions in which some other axis becomes the principal axis of a subgroup: and it is then most useful to form linear combinations of the functions of Table A1.1 which are adapted to the corresponding reduction. The principal types of distortion which occur are illustrated in Fig. A2.1.

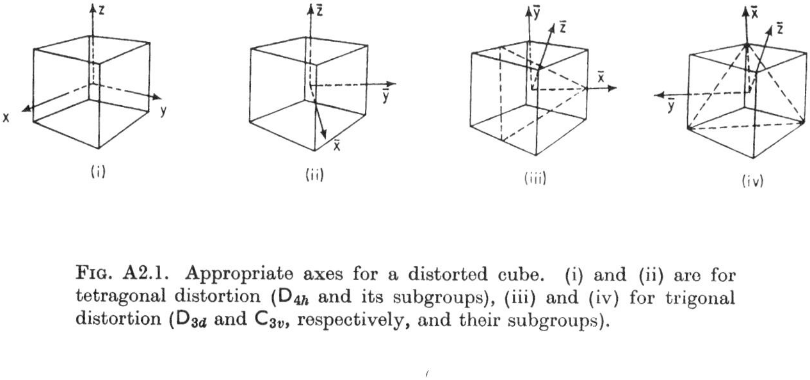


FIG. A2.1. Appropriate axes for a distorted cube. (i) and (ii) are for tetragonal distortion ($D_{4h}$ and its subgroups), (iii) and (iv) for trigonal distortion ($D_{3d}$ and $C_{3v}$, respectively, and their subgroups).

In (i) the $z$ axis remains a 4-fold axis, as for example in compression of the cube along the $z$ direction, the original axes remain appropriate and the bases for the tetragonal groups may be read from Table A1.8.

In (ii) the $x$ and $y$ axes are rotated so as to be appropriate for further tetragonal distortion in which the cube section normal to the $z$ axis becomes diamond shaped: the bases for tetragonal groups defined with respect to these axes may be read from Table A1.8, provided the functions used are the rotated functions $\bar{p}_x$, $\bar{p}_y$, . . . , $\bar{f}_{xyz}$ given in Table A2.1.

## Table A2.1

*Cubic bases for* $D_{4h}$ *reduction*

| $D_{4h}$ | $\bar{p}_x$ | $\bar{p}_y$ | $\bar{p}_z$ | |
|---|---|---|---|---|
| | $\sqrt{2}/2$ | $-\sqrt{2}/2$ | 0 | $p_x$ |
| | $\sqrt{2}/2$ | $\sqrt{2}/2$ | 0 | $p_y$ |
| | 0 | 0 | 1 | $p_z$ |

| $D_{4h}$ | $\bar{d}_{z^2}$ | $\bar{d}_{x^2-y^2}$ | $\bar{d}_{yz}$ | $\bar{d}_{zx}$ | $\bar{d}_{xy}$ | |
|---|---|---|---|---|---|---|
| | 1 | 0 | 0 | 0 | 0 | $d_{z^2}$ |
| | 0 | 0 | 0 | 0 | $-1$ | $d_{x^2-y^2}$ |
| | 0 | 0 | $\sqrt{2}/2$ | $\sqrt{2}/2$ | 0 | $d_{yz}$ |
| | 0 | 0 | $-\sqrt{2}/2$ | $\sqrt{2}/2$ | 0 | $d_{zx}$ |
| | 0 | 1 | 0 | 0 | 0 | $d_{xy}$ |

| $D_{4h}$ | $\bar{f}_x$ | $\bar{f}_y$ | $\bar{f}_z$ | $\bar{f}_x'$ | $\bar{f}_y'$ | $\bar{f}_z'$ | $\bar{f}_{xyz}$ | |
|---|---|---|---|---|---|---|---|---|
| | $-\sqrt{2}/8$ | $\sqrt{2}/8$ | 0 | $\sqrt{30}/8$ | $\sqrt{30}/8$ | 0 | 0 | $f_x$ |
| | $-\sqrt{2}/8$ | $-\sqrt{2}/8$ | 0 | $\sqrt{30}/8$ | $-\sqrt{30}/8$ | 0 | 0 | $f_y$ |
| | 0 | 0 | 1 | 0 | 0 | 0 | 0 | $f_z$ |
| | $\sqrt{30}/8$ | $-\sqrt{30}/8$ | 0 | $\sqrt{2}/8$ | $\sqrt{2}/8$ | 0 | 0 | $f_x'$ |
| | $-\sqrt{30}/8$ | $-\sqrt{30}/8$ | 0 | $-\sqrt{2}/8$ | $\sqrt{2}/8$ | 0 | 0 | $f_y'$ |
| | 0 | 0 | 0 | 0 | 0 | 0 | $-1$ | $f_z'$ |
| | 0 | 0 | 0 | 0 | 0 | 1 | 0 | $f_{xyz}$ |

*Note.* The axes employed in defining the $D_{4h}$ functions (barred) are indicated in Fig. A2.1 (ii).

In distortions which produce a system with trigonal symmetry, it is a cube diagonal which persists as the 3-fold principal axis, and in order to take over the results of Table A1.6 we take the diagonal through point (1, 1, 1) as a new $z$ axis. There are then two useful choices of orientation for the $x$ and $y$ axes, these being shown in Fig. A2.1 (iii) and (iv): the first is appropriate to compression of the cube along the diagonal, to symmetry $D_{3d}$, and would correspond to "flattening" of the octahedron with vertices at the cube face centres;

while the second would be appropriate to a tetrahedron in which the (1, 1, 1) vertex was distinguished from the others. The appropriate bases for $D_{3d}$ (and subgroups) and $C_{3v}$ (and subgroups), with axes as in (iii) and (iv) respectively, are listed in Table A2.2: the functions in Table A1.6 should be interpreted as these rotated functions.

**Table A2.2**

*Cubic bases for $D_{3d}$ and $C_{3v}$ reduction*

| | | | | |
|---|---|---|---|---|
| $C_{3v}$ | $-\bar{p}_y$ | $\bar{p}_x$ | $\bar{p}_z$ | |
| $D_{3d}$ | $\bar{p}_x$ | $\bar{p}_y$ | $\bar{p}_z$ | |
| | $-\sqrt{2}/2$ | $-\sqrt{6}/6$ | $\sqrt{3}/3$ | $p_x$ |
| | $\sqrt{2}/2$ | $-\sqrt{6}/6$ | $\sqrt{3}/3$ | $p_y$ |
| | 0 | $\sqrt{6}/3$ | $\sqrt{3}/3$ | $p_z$ |

| | | | | | | |
|---|---|---|---|---|---|---|
| $C_{3v}$ | $\bar{d}_{z^2}$ | $-\bar{d}_{x^2-y^2}$ | $\bar{d}_{zx}$ | $-\bar{d}_{yz}$ | $-\bar{d}_{xy}$ | |
| $D_{3d}$ | $\bar{d}_{z^2}$ | $\bar{d}_{x^2-y^2}$ | $\bar{d}_{yz}$ | $\bar{d}_{zx}$ | $\bar{d}_{xy}$ | |
| | 0 | $-\sqrt{3}/3$ | $\sqrt{6}/3$ | 0 | 0 | $d_{z^2}$ |
| | 0 | 0 | 0 | $-\sqrt{6}/3$ | $\sqrt{3}/3$ | $d_{x^2-y^2}$ |
| | $\sqrt{3}/3$ | 1/3 | $\sqrt{2}/6$ | $\sqrt{6}/6$ | $\sqrt{3}/3$ | $d_{yz}$ |
| | $\sqrt{3}/3$ | 1/3 | $\sqrt{2}/6$ | $-\sqrt{6}/6$ | $-\sqrt{3}/3$ | $d_{zx}$ |
| | $\sqrt{3}/3$ | $-2/3$ | $-\sqrt{2}/3$ | 0 | 0 | $d_{xy}$ |

| | | | | | | | | |
|---|---|---|---|---|---|---|---|---|
| $C_{3v}$ | $\bar{f}_z$ | $-\bar{f}_{xyz}$ | $-\bar{f}_z'$ | $\bar{Y}^s_{3,3}$ | $-\bar{Y}^c_{3,3}$ | $-\bar{Y}^s_{3,1}$ | $\bar{Y}^c_{3,1}$ | |
| $D_{3d}$ | $\bar{f}_z$ | $\bar{f}_{xyz}$ | $\bar{f}_z'$ | $\bar{Y}^c_{3,3}$ | $\bar{Y}^s_{3,3}$ | $\bar{Y}^c_{3,1}$ | $\bar{Y}^s_{3,1}$ | |
| | $-2\sqrt{3}/9$ | $\sqrt{15}/6$ | $\sqrt{5}/6$ | 0 | $-\sqrt{15}/9$ | $-\sqrt{3}/6$ | $-1/6$ | $f_x$ |
| | $-2\sqrt{3}/9$ | $-\sqrt{15}/6$ | $\sqrt{5}/6$ | 0 | $-\sqrt{15}/9$ | $\sqrt{3}/6$ | $-1/6$ | $f_y$ |
| | $-2\sqrt{3}/9$ | 0 | $-\sqrt{5}/3$ | 0 | $-\sqrt{15}/9$ | 0 | 1/3 | $f_z$ |
| | 0 | 1/6 | $-\sqrt{3}/6$ | $-\sqrt{3}/3$ | 0 | $\sqrt{5}/6$ | $-\sqrt{15}/6$ | $f_x'$ |
| | 0 | 1/6 | $\sqrt{3}/6$ | $-\sqrt{3}/3$ | 0 | $\sqrt{5}/6$ | $\sqrt{15}/6$ | $f_y'$ |
| | 0 | $-1/3$ | 0 | $-\sqrt{3}/3$ | 0 | $-\sqrt{5}/3$ | 0 | $f_z'$ |
| | $\sqrt{5}/3$ | 0 | 0 | 0 | $-2/3$ | 0 | 0 | $f_{xyz}$ |

# INDEX


Algebra, 19–21
Angular momentum, 221, 223, 231–3
Associative law, 3, 20, 26
Axioms
  group, 6
  vector space, 27
  metric, 42, 47

Basal units, 21
Basis, 30
  change of, 31–3, 40–1, 241–3
  types of, 42, 44–8
Basis for representation, 34–41
  symmetry adapted, 127, 146–7, 187–8, 209–10
Bloch theorem, 195
Bravais lattices, 78, 81–5

Central field problem, 170–4
Characters, 121–6
  in direct product, 138
Classes, 12–13, 122, 125–6
Clebsch-Gordan coefficients, 230
Closure property, 3, 21
Collection, 2
Combination law, 2
  symmetry operations, 5
  complexes, 13
  elements of algebra, 20
  vectors, 22, 26
Commutative law, 4, 20, 26
Completeness, 50, 132, 187
Complex, 13
Conjugate elements, 12
  subgroups, 13
Contragredience, 33, 120, 204
Coset, 11–12
Coupling coefficients, 229–33
Covariance and contravariance, 205
  of second kind, 207
Crystal (ligand) field, 185–6, 226, 231, 233–4
Crystal properties, 142–3, 203, 208–11, 216–8
Crystal symmetry, 78–80, 85–90
  classes, 80, 81–4
  systems, 81–5

Degeneracy, 150, 183–4
  accidental, 151, 183
  resolution of, 185
Diagonalization of matrix, 52, 147–50
Dimensionality, 28–9
Dirac notation, 50–1, 224
Direct product
  of groups, 18
  of spaces, 134–5
  of matrices, 136–7
  of representations, 137, 208–9
Direct sum, 109

Eigenfunctions, 177
  symmetry properties, 183–4
Eigenvalue equation,
  matrix, 51–2, 147–9
  operator, 177–8, 182
  approximate solution of, 187–90
Eigenvectors, 51–2, 148–9
  symmetry properties, 150–1
Equivalent functions, 195–201
Equivalent points, 5, 59
Equivalent representations, 41, 109–12
  test for, 122

Factor group, 13
Field, 19, 166 (scalar)
  vector and tensor, 218
  transformation of, 168–9, 218–9
Fourier resolution, 46–7
Functions
  of integrable square, 46
  orthogonal, 47, 173–4
  of a matrix, 112
  of Cartesian coordinates, 170–4
  *See also* Eigenfunction, Fields, Harmonic functions, Symmetry functions
Function space, 29, 38, 46, 49–50, 134

Generators, 9–11
  of translation groups, 23–6
  of point groups, 59–60, 71–5
Group algebra, 20–21, 131–3

Group
definition, 3, 6
Abelian, 4
finite, 7
abstract, 8
cyclic, 10, 23, 58, 105
translation, 24
space, 25, 85
point, 54
rotation, 171–4, 208–9, 222
*See also* Point groups, Translation groups
Group element(s), 2, 6
powers and products, 9–11
conjugate, 12

Hamiltonian operator, 178, 180–2
Harmonic functions
spherical, 173, 230
cubic, 235–6, 241–3
Hermitian scalar product, 44
operator, 52
form, 141, 149
Holohedry, 82
Homomorphism, 14–16
Hybridization, 197–201
Hypercomplex number, 21

Idempotent, 131–3
Image, 15, 33, 59
*See also* Mappings
Invariance
of subgroup, 13, 76, 79
of metric, 81–3
of subspace, 112–4, 118, 151, 183
of matrices and forms, 141–2
of operators, 178–9, 224
of tensor, 209–10
Inverse, 6, 39
of matrix, 33
Inversion, 58, 62, 95
Irreducible matrices, 117
Irreducible representations, 110
orthogonality relations, 116–20
test for, 123
other properties, 123–6
*See also* Reducibility, Point groups, Translation groups
Irreducible tensorial set, 222–4
Isomorphism, 16

Kronecker delta, 45
Kronecker product, 137, 209
symmetrized, 213
k-space, 107

Lattices, 22–6
Bravais, 78, 81–5
point group of, 82
primitive and centred, 85
reciprocal, 106
Linear dependence, 28–9
Linear operators, 176
Linear vector space, 27–9

Mappings, 15–18, 33–41
unitary, 48
Matrix, 17
notation, 31–2
inverse, 33
Hermitian transpose (adjoint), 44
metrical, 44, 48
positive definite, 47, 49
unitary, 48
eigenvalues of, 51–2
Hermitian symmetric, 52
real orthogonal, 56
Matrix elements (as scalar products), 49–50
Dirac notation, 50–1
of invariant operators, 224–6
of tensor operators, 227–30
Metric, 41–4, 47–8
of lattices, 81–3
in tensor theory, 206–7
Molecules, 64, 142–4, 148–55, 160–5, 180–1, 184–5, 189–94, 197–201
*See also* Hybridization, Symmetry co-ordinates, Symmetry functions, Vibrations
Multiplication table, 7–8
of group $C_{3v}$, 9

Normal co-ordinates, 149, 153
Normal modes, 37, 148, 153–4
of zero frequency, 156–8

Operator(s), 34, 50, 174–8
Hermitian, 52
idempotent, 131
projection, 131
" shift ", 133
Laplacian, 176–7
invariant, 178–9, 224
vector and tensor, 219–24, 227
Optical activity, 217–8
Orbitals
atomic, 171–4, 189
molecular, 189
crystal, 189, 195
symmetry, 189–94
hybrid, 197–201

Orthogonality
of vectors, 45, 52
of functions, 47, 173
of character vectors, 125–6
Orthogonality relations, 116–7, 119–20, 122

Partners in a representation, 129, 132–3, 225
Point groups, 54
projection diagrams, 60
generators, 60–5, 71–5
tables, 66–9 (axial), 72–5 (cubic)
irreducible representations, 91–6, 96–105 (tables)
Positive definite property, 47
Principal axes, 147, 149–50
Product
of group elements, 3
*See also* Combination law, Direct product, Scalar product
Projection, 128–31
operators, 131–3, 160–3, 190–5, 198–201, 224–7
Pseudo-vectors and -tensors, 157, 214–8, 221

Quadratic forms, 140–5
and eigenvalue equation, 148–50
Quadric, 147, 150
Quantum mechanical applications, 170–4, 179–201, 221–34

Rearrangement theorem, 9
Reciprocal lattice, 106–7
Reducibility
meaning of, 109–10, 112–14
theorems on, 110–16
test for, 123
Reduction
methods of, 126–31
of direct product, 138, 208–14, 228–33
Regular representation, 124–5
Representation
concept of, 16–18
example ($C_{3v}$), 16–17, 40
equivalence, 40–1, 109–12
real and complex forms, 92
unitary, 110–11
contragredient, 120
regular, 124
*See also* Irreducible representation, Reducibility, Point groups, Translation groups

Representation
carried by basis vectors, 39, 91
by eigenvectors, 150
by functions of Cartesian coordinates, 170
by spherical harmonics, 173, 235
by eigenfunctions, 183
by tensor components, 208
by tensor operators, 220–3
by harmonic functions, 235–43
Resolution of the identity, 132
Rotation(s), 2, 5, 33
proper and improper, 56
of field, 169, 218
of group $D_3$, 171–2, 235
passive interpretation, 204
*See also* Transformation of . . .

Scalar product, 41–2, 45–7, 50
Schrodinger's equation, 171, 177–8, 188
*See also* Eigenvalue equation
Schur's lemma, 117, 143
corollary, 118
Selection rules, 227–8
Semi-direct product, 75
Space group, 25, 85–90
Spectral set, 132
Spherical basis, 42, 46
Spherical harmonic, 173–4, 230
Subgroup, 11
invariant (normal), 13
Susceptibility
electric and magnetic, 216
Symmetry
coordinates and vectors, 159–65
functions (orbitals), 189–95
Symmetry element, 59, 80, 87
Symmetry operations, 4–6, 24, 54–6, 87–9
nomenclature, 61–2, 70–1
compatibility of, 75–81
in molecular vibration theory, 143–5
Symmetry species, 127
Symmetry-adapted basis, 127, 146–7, 187–8, 209–10
Syngony, 83

Tensor (tensorial set), 205
components of, 206
rank of, 206
Cartesian, 208
symmetric and antisymmetric, 211–4
pseudo-, 214–8
Tensor operator, 219–22
irreducible, 222–3
matrix elements of, 224–30

Transformation, 33–8
  similarity, 41
  unitary, 47–8
  orthogonal, 56
  congruent, 81
  active and passive, 204
Transformation
  of vector components, 33
  of fields, 166–70 (scalar), 218–9 (tensor)
  of operators, 178, 219–23
  of tensor components, 204–6
Translation groups, 22–6, 79
  irreducible representation of, 105–8

Unitary basis, 45
  transformation, 47–8
  representation, 110–11

Vector(s), 22
  addition of, 22, 26
Vector(s)—*contd.*
  dual, 43
  bra- and ket-, 50
  axial and polar, 157
  pseudo-, 157, 214
Vector components, 30–31
  as scalar products, 46, 51
  covariant and contravariant, 206–7
Vector space, 27–9
  Hermitian, 44
Vibrations
  molecular, 148–9
  classification, 151
  *See also* Normal modes, Symmetry operations, Symmetry coordinates

Wigner coefficients, 230
Wigner-Eckart theorem, 230, 233

Zeeman effect, 231

A CATALOG OF SELECTED

# DOVER BOOKS

IN SCIENCE AND MATHEMATICS

# Astronomy

CHARIOTS FOR APOLLO: The NASA History of Manned Lunar Spacecraft to 1969, Courtney G. Brooks, James M. Grimwood, and Loyd S. Swenson, Jr. This illustrated history by a trio of experts is the definitive reference on the Apollo spacecraft and lunar modules. It traces the vehicles' design, development, and operation in space. More than 100 photographs and illustrations. 576pp. 6 3/4 x 9 1/4. 0-486-46756-2

EXPLORING THE MOON THROUGH BINOCULARS AND SMALL TELESCOPES, Ernest H. Cherrington, Jr. Informative, profusely illustrated guide to locating and identifying craters, rills, seas, mountains, other lunar features. Newly revised and updated with special section of new photos. Over 100 photos and diagrams. 240pp. 8 1/4 x 11. 0-486-24491-1

WHERE NO MAN HAS GONE BEFORE: A History of NASA's Apollo Lunar Expeditions, William David Compton. Introduction by Paul Dickson. This official NASA history traces behind-the-scenes conflicts and cooperation between scientists and engineers. The first half concerns preparations for the Moon landings, and the second half documents the flights that followed Apollo 11. 1989 edition. 432pp. 7 x 10. 0-486-47888-2

APOLLO EXPEDITIONS TO THE MOON: The NASA History, Edited by Edgar M. Cortright. Official NASA publication marks the 40th anniversary of the first lunar landing and features essays by project participants recalling engineering and administrative challenges. Accessible, jargon-free accounts, highlighted by numerous illustrations. 336pp. 8 3/8 x 10 7/8. 0-486-47175-6

ON MARS: Exploration of the Red Planet, 1958-1978--The NASA History, Edward Clinton Ezell and Linda Neuman Ezell. NASA's official history chronicles the start of our explorations of our planetary neighbor. It recounts cooperation among government, industry, and academia, and it features dozens of photos from Viking cameras. 560pp. 6 3/4 x 9 1/4. 0-486-46757-0

ARISTARCHUS OF SAMOS: The Ancient Copernicus, Sir Thomas Heath. Heath's history of astronomy ranges from Homer and Hesiod to Aristarchus and includes quotes from numerous thinkers, compilers, and scholasticists from Thales and Anaximander through Pythagoras, Plato, Aristotle, and Heraclides. 34 figures. 448pp. 5 3/8 x 8 1/2. 0-486-43886-4

AN INTRODUCTION TO CELESTIAL MECHANICS, Forest Ray Moulton. Classic text still unsurpassed in presentation of fundamental principles. Covers rectilinear motion, central forces, problems of two and three bodies, much more. Includes over 200 problems, some with answers. 437pp. 5 3/8 x 8 1/2. 0-486-64687-4

BEYOND THE ATMOSPHERE: Early Years of Space Science, Homer E. Newell. This exciting survey is the work of a top NASA administrator who chronicles technological advances, the relationship of space science to general science, and the space program's social, political, and economic contexts. 528pp. 6 3/4 x 9 1/4. 0-486-47464-X

STAR LORE: Myths, Legends, and Facts, William Tyler Olcott. Captivating retellings of the origins and histories of ancient star groups include Pegasus, Ursa Major, Pleiades, signs of the zodiac, and other constellations. "Classic." – *Sky & Telescope.* 58 illustrations. 544pp. 5 3/8 x 8 1/2. 0-486-43581-4

A COMPLETE MANUAL OF AMATEUR ASTRONOMY: Tools and Techniques for Astronomical Observations, P. Clay Sherrod with Thomas L. Koed. Concise, highly readable book discusses the selection, set-up, and maintenance of a telescope; amateur studies of the sun; lunar topography and occultations; and more. 124 figures. 26 halftones. 37 tables. 335pp. 6 1/2 x 9 1/4. 0-486-42820-6

# Chemistry

MOLECULAR COLLISION THEORY, M. S. Child. This high-level monograph offers an analytical treatment of classical scattering by a central force, quantum scattering by a central force, elastic scattering phase shifts, and semi-classical elastic scattering. 1974 edition. 310pp. 5 3/8 x 8 1/2. 0-486-69437-2

HANDBOOK OF COMPUTATIONAL QUANTUM CHEMISTRY, David B. Cook. This comprehensive text provides upper-level undergraduates and graduate students with an accessible introduction to the implementation of quantum ideas in molecular modeling, exploring practical applications alongside theoretical explanations. 1998 edition. 832pp. 5 3/8 x 8 1/2. 0-486-44307-8

RADIOACTIVE SUBSTANCES, Marie Curie. The celebrated scientist's thesis, which directly preceded her 1903 Nobel Prize, discusses establishing atomic character of radioactivity; extraction from pitchblende of polonium and radium; isolation of pure radium chloride; more. 96pp. 5 3/8 x 8 1/2. 0-486-42550-9

CHEMICAL MAGIC, Leonard A. Ford. Classic guide provides intriguing entertainment while elucidating sound scientific principles, with more than 100 unusual stunts: cold fire, dust explosions, a nylon rope trick, a disappearing beaker, much more. 128pp. 5 3/8 x 8 1/2. 0-486-67628-5

ALCHEMY, E. J. Holmyard. Classic study by noted authority covers 2,000 years of alchemical history: religious, mystical overtones; apparatus; signs, symbols, and secret terms; advent of scientific method, much more. Illustrated. 320pp. 5 3/8 x 8 1/2. 0-486-26298-7

CHEMICAL KINETICS AND REACTION DYNAMICS, Paul L. Houston. This text teaches the principles underlying modern chemical kinetics in a clear, direct fashion, using several examples to enhance basic understanding. Solutions to selected problems. 2001 edition. 352pp. 8 3/8 x 11. 0-486-45334-0

PROBLEMS AND SOLUTIONS IN QUANTUM CHEMISTRY AND PHYSICS, Charles S. Johnson and Lee G. Pedersen. Unusually varied problems, with detailed solutions, cover of quantum mechanics, wave mechanics, angular momentum, molecular spectroscopy, scattering theory, more. 280 problems, plus 139 supplementary exercises. 430pp. 6 1/2 x 9 1/4. 0-486-65236-X

ELEMENTS OF CHEMISTRY, Antoine Lavoisier. Monumental classic by the founder of modern chemistry features first explicit statement of law of conservation of matter in chemical change, and more. Facsimile reprint of original (1790) Kerr translation. 539pp. 5 3/8 x 8 1/2. 0-486-64624-6

MAGNETISM AND TRANSITION METAL COMPLEXES, F. E. Mabbs and D. J. Machin. A detailed view of the calculation methods involved in the magnetic properties of transition metal complexes, this volume offers sufficient background for original work in the field. 1973 edition. 240pp. 5 3/8 x 8 1/2. 0-486-46284-6

GENERAL CHEMISTRY, Linus Pauling. Revised third edition of classic first-year text by Nobel laureate. Atomic and molecular structure, quantum mechanics, statistical mechanics, thermodynamics correlated with descriptive chemistry. Problems. 992pp. 5 3/8 x 8 1/2. 0-486-65622-5

ELECTROLYTE SOLUTIONS: Second Revised Edition, R. A. Robinson and R. H. Stokes. Classic text deals primarily with measurement, interpretation of conductance, chemical potential, and diffusion in electrolyte solutions. Detailed theoretical interpretations, plus extensive tables of thermodynamic and transport properties. 1970 edition. 590pp. 5 3/8 x 8 1/2. 0-486-42225-9

# Engineering

FUNDAMENTALS OF ASTRODYNAMICS, Roger R. Bate, Donald D. Mueller, and Jerry E. White. Teaching text developed by U.S. Air Force Academy develops the basic two-body and n-body equations of motion; orbit determination; classical orbital elements, coordinate transformations; differential correction; more. 1971 edition. 455pp. 5 3/8 x 8 1/2. 0-486-60061-0

INTRODUCTION TO CONTINUUM MECHANICS FOR ENGINEERS: Revised Edition, Ray M. Bowen. This self-contained text introduces classical continuum models within a modern framework. Its numerous exercises illustrate the governing principles, linearizations, and other approximations that constitute classical continuum models. 2007 edition. 320pp. 6 1/8 x 9 1/4. 0-486-47460-7

ENGINEERING MECHANICS FOR STRUCTURES, Louis L. Bucciarelli. This text explores the mechanics of solids and statics as well as the strength of materials and elasticity theory. Its many design exercises encourage creative initiative and systems thinking. 2009 edition. 320pp. 6 1/8 x 9 1/4. 0-486-46855-0

FEEDBACK CONTROL THEORY, John C. Doyle, Bruce A. Francis and Allen R. Tannenbaum. This excellent introduction to feedback control system design offers a theoretical approach that captures the essential issues and can be applied to a wide range of practical problems. 1992 edition. 224pp. 6 1/2 x 9 1/4. 0-486-46933-6

THE FORCES OF MATTER, Michael Faraday. These lectures by a famous inventor offer an easy-to-understand introduction to the interactions of the universe's physical forces. Six essays explore gravitation, cohesion, chemical affinity, heat, magnetism, and electricity. 1993 edition. 96pp. 5 3/8 x 8 1/2. 0-486-47482-8

DYNAMICS, Lawrence E. Goodman and William H. Warner. Beginning engineering text introduces calculus of vectors, particle motion, dynamics of particle systems and plane rigid bodies, technical applications in plane motions, and more. Exercises and answers in every chapter. 619pp. 5 3/8 x 8 1/2. 0-486-42006-X

ADAPTIVE FILTERING PREDICTION AND CONTROL, Graham C. Goodwin and Kwai Sang Sin. This unified survey focuses on linear discrete-time systems and explores natural extensions to nonlinear systems. It emphasizes discrete-time systems, summarizing theoretical and practical aspects of a large class of adaptive algorithms. 1984 edition. 560pp. 6 1/2 x 9 1/4. 0-486-46932-8

INDUCTANCE CALCULATIONS, Frederick W. Grover. This authoritative reference enables the design of virtually every type of inductor. It features a single simple formula for each type of inductor, together with tables containing essential numerical factors. 1946 edition. 304pp. 5 3/8 x 8 1/2. 0-486-47440-2

THERMODYNAMICS: Foundations and Applications, Elias P. Gyftopoulos and Gian Paolo Beretta. Designed by two MIT professors, this authoritative text discusses basic concepts and applications in detail, emphasizing generality, definitions, and logical consistency. More than 300 solved problems cover realistic energy systems and processes. 800pp. 6 1/8 x 9 1/4. 0-486-43932-1

THE FINITE ELEMENT METHOD: Linear Static and Dynamic Finite Element Analysis, Thomas J. R. Hughes. Text for students without in-depth mathematical training, this text includes a comprehensive presentation and analysis of algorithms of time-dependent phenomena plus beam, plate, and shell theories. Solution guide available upon request. 672pp. 6 1/2 x 9 1/4. 0-486-41181-8

## Mathematics–Logic and Problem Solving

PERPLEXING PUZZLES AND TANTALIZING TEASERS, Martin Gardner. Ninety-three riddles, mazes, illusions, tricky questions, word and picture puzzles, and other challenges offer hours of entertainment for youngsters. Filled with rib-tickling drawings. Solutions. 224pp. 5 3/8 x 8 1/2. 0-486-25637-5

MY BEST MATHEMATICAL AND LOGIC PUZZLES, Martin Gardner. The noted expert selects 70 of his favorite "short" puzzles. Includes The Returning Explorer, The Mutilated Chessboard, Scrambled Box Tops, and dozens more. Complete solutions included. 96pp. 5 3/8 x 8 1/2. 0-486-28152-3

THE LADY OR THE TIGER?: and Other Logic Puzzles, Raymond M. Smullyan. Created by a renowned puzzle master, these whimsically themed challenges involve paradoxes about probability, time, and change; metapuzzles; and self-referentiality. Nineteen chapters advance in difficulty from relatively simple to highly complex. 1982 edition. 240pp. 5 3/8 x 8 1/2. 0-486-47027-X

SATAN, CANTOR AND INFINITY: Mind-Boggling Puzzles, Raymond M. Smullyan. A renowned mathematician tells stories of knights and knaves in an entertaining look at the logical precepts behind infinity, probability, time, and change. Requires a strong background in mathematics. Complete solutions. 288pp. 5 3/8 x 8 1/2. 0-486-47036-9

THE RED BOOK OF MATHEMATICAL PROBLEMS, Kenneth S. Williams and Kenneth Hardy. Handy compilation of 100 practice problems, hints and solutions indispensable for students preparing for the William Lowell Putnam and other mathematical competitions. Preface to the First Edition. Sources. 1988 edition. 192pp. 5 3/8 x 8 1/2. 0-486-69415-1

KING ARTHUR IN SEARCH OF HIS DOG AND OTHER CURIOUS PUZZLES, Raymond M. Smullyan. This fanciful, original collection for readers of all ages features arithmetic puzzles, logic problems related to crime detection, and logic and arithmetic puzzles involving King Arthur and his Dogs of the Round Table. 160pp. 5 3/8 x 8 1/2. 0-486-47435-6

UNDECIDABLE THEORIES: Studies in Logic and the Foundation of Mathematics, Alfred Tarski in collaboration with Andrzej Mostowski and Raphael M. Robinson. This well-known book by the famed logician consists of three treatises: "A General Method in Proofs of Undecidability," "Undecidability and Essential Undecidability in Mathematics," and "Undecidability of the Elementary Theory of Groups." 1953 edition. 112pp. 5 3/8 x 8 1/2. 0-486-47703-7

LOGIC FOR MATHEMATICIANS, J. Barkley Rosser. Examination of essential topics and theorems assumes no background in logic. "Undoubtedly a major addition to the literature of mathematical logic." – *Bulletin of the American Mathematical Society.* 1978 edition. 592pp. 6 1/8 x 9 1/4. 0-486-46898-4

INTRODUCTION TO PROOF IN ABSTRACT MATHEMATICS, Andrew Wohlgemuth. This undergraduate text teaches students what constitutes an acceptable proof, and it develops their ability to do proofs of routine problems as well as those requiring creative insights. 1990 edition. 384pp. 6 1/2 x 9 1/4. 0-486-47854-8

FIRST COURSE IN MATHEMATICAL LOGIC, Patrick Suppes and Shirley Hill. Rigorous introduction is simple enough in presentation and context for wide range of students. Symbolizing sentences; logical inference; truth and validity; truth tables; terms, predicates, universal quantifiers; universal specification and laws of identity; more. 288pp. 5 3/8 x 8 1/2. 0-486-42259-3

# Mathematics–Algebra and Calculus

VECTOR CALCULUS, Peter Baxandall and Hans Liebeck. This introductory text offers a rigorous, comprehensive treatment. Classical theorems of vector calculus are amply illustrated with figures, worked examples, physical applications, and exercises with hints and answers. 1986 edition. 560pp. 5 3/8 x 8 1/2. 0-486-46620-5

ADVANCED CALCULUS: An Introduction to Classical Analysis, Louis Brand. A course in analysis that focuses on the functions of a real variable, this text introduces the basic concepts in their simplest setting and illustrates its teachings with numerous examples, theorems, and proofs. 1955 edition. 592pp. 5 3/8 x 8 1/2. 0-486-44548-8

ADVANCED CALCULUS, Avner Friedman. Intended for students who have already completed a one-year course in elementary calculus, this two-part treatment advances from functions of one variable to those of several variables. Solutions. 1971 edition. 432pp. 5 3/8 x 8 1/2. 0-486-45795-8

METHODS OF MATHEMATICS APPLIED TO CALCULUS, PROBABILITY, AND STATISTICS, Richard W. Hamming. This 4-part treatment begins with algebra and analytic geometry and proceeds to an exploration of the calculus of algebraic functions and transcendental functions and applications. 1985 edition. Includes 310 figures and 18 tables. 880pp. 6 1/2 x 9 1/4. 0-486-43945-3

BASIC ALGEBRA I: Second Edition, Nathan Jacobson. A classic text and standard reference for a generation, this volume covers all undergraduate algebra topics, including groups, rings, modules, Galois theory, polynomials, linear algebra, and associative algebra. 1985 edition. 528pp. 6 1/8 x 9 1/4. 0-486-47189-6

BASIC ALGEBRA II: Second Edition, Nathan Jacobson. This classic text and standard reference comprises all subjects of a first-year graduate-level course, including in-depth coverage of groups and polynomials and extensive use of categories and functors. 1989 edition. 704pp. 6 1/8 x 9 1/4. 0-486-47187-X

CALCULUS: An Intuitive and Physical Approach (Second Edition), Morris Kline. Application-oriented introduction relates the subject as closely as possible to science with explorations of the derivative; differentiation and integration of the powers of x; theorems on differentiation, antidifferentiation; the chain rule; trigonometric functions; more. Examples. 1967 edition. 960pp. 6 1/2 x 9 1/4. 0-486-40453-6

ABSTRACT ALGEBRA AND SOLUTION BY RADICALS, John E. Maxfield and Margaret W. Maxfield. Accessible advanced undergraduate-level text starts with groups, rings, fields, and polynomials and advances to Galois theory, radicals and roots of unity, and solution by radicals. Numerous examples, illustrations, exercises, appendixes. 1971 edition. 224pp. 6 1/8 x 9 1/4. 0-486-47723-1

AN INTRODUCTION TO THE THEORY OF LINEAR SPACES, Georgi E. Shilov. Translated by Richard A. Silverman. Introductory treatment offers a clear exposition of algebra, geometry, and analysis as parts of an integrated whole rather than separate subjects. Numerous examples illustrate many different fields, and problems include hints or answers. 1961 edition. 320pp. 5 3/8 x 8 1/2. 0-486-63070-6

LINEAR ALGEBRA, Georgi E. Shilov. Covers determinants, linear spaces, systems of linear equations, linear functions of a vector argument, coordinate transformations, the canonical form of the matrix of a linear operator, bilinear and quadratic forms, and more. 387pp. 5 3/8 x 8 1/2. 0-486-63518-X

# Mathematics-Probability and Statistics

BASIC PROBABILITY THEORY, Robert B. Ash. This text emphasizes the probabilistic way of thinking, rather than measure-theoretic concepts. Geared toward advanced undergraduates and graduate students, it features solutions to some of the problems. 1970 edition. 352pp. 5 3/8 x 8 1/2. 0-486-46628-0

PRINCIPLES OF STATISTICS, M. G. Bulmer. Concise description of classical statistics, from basic dice probabilities to modern regression analysis. Equal stress on theory and applications. Moderate difficulty; only basic calculus required. Includes problems with answers. 252pp. 5 5/8 x 8 1/4. 0-486-63760-3

OUTLINE OF BASIC STATISTICS: Dictionary and Formulas, John E. Freund and Frank J. Williams. Handy guide includes a 70-page outline of essential statistical formulas covering grouped and ungrouped data, finite populations, probability, and more, plus over 1,000 clear, concise definitions of statistical terms. 1966 edition. 208pp. 5 3/8 x 8 1/2. 0-486-47769-X

GOOD THINKING: The Foundations of Probability and Its Applications, Irving J. Good. This in-depth treatment of probability theory by a famous British statistician explores Keynesian principles and surveys such topics as Bayesian rationality, corroboration, hypothesis testing, and mathematical tools for induction and simplicity. 1983 edition. 352pp. 5 3/8 x 8 1/2. 0-486-47438-0

INTRODUCTION TO PROBABILITY THEORY WITH CONTEMPORARY APPLICATIONS, Lester L. Helms. Extensive discussions and clear examples, written in plain language, expose students to the rules and methods of probability. Exercises foster problem-solving skills, and all problems feature step-by-step solutions. 1997 edition. 368pp. 6 1/2 x 9 1/4. 0-486-47418-6

CHANCE, LUCK, AND STATISTICS, Horace C. Levinson. In simple, non-technical language, this volume explores the fundamentals governing chance and applies them to sports, government, and business. "Clear and lively ... remarkably accurate." – *Scientific Monthly*. 384pp. 5 3/8 x 8 1/2. 0-486-41997-5

FIFTY CHALLENGING PROBLEMS IN PROBABILITY WITH SOLUTIONS, Frederick Mosteller. Remarkable puzzlers, graded in difficulty, illustrate elementary and advanced aspects of probability. These problems were selected for originality, general interest, or because they demonstrate valuable techniques. Also includes detailed solutions. 88pp. 5 3/8 x 8 1/2. 0-486-65355-2

EXPERIMENTAL STATISTICS, Mary Gibbons Natrella. A handbook for those seeking engineering information and quantitative data for designing, developing, constructing, and testing equipment. Covers the planning of experiments, the analyzing of extreme-value data; and more. 1966 edition. Index. Includes 52 figures and 76 tables. 560pp. 8 3/8 x 11. 0-486-43937-2

STOCHASTIC MODELING: Analysis and Simulation, Barry L. Nelson. Coherent introduction to techniques also offers a guide to the mathematical, numerical, and simulation tools of systems analysis. Includes formulation of models, analysis, and interpretation of results. 1995 edition. 336pp. 6 1/8 x 9 1/4. 0-486-47770-3

INTRODUCTION TO BIOSTATISTICS: Second Edition, Robert R. Sokal and F. James Rohlf. Suitable for undergraduates with a minimal background in mathematics, this introduction ranges from descriptive statistics to fundamental distributions and the testing of hypotheses. Includes numerous worked-out problems and examples. 1987 edition. 384pp. 6 1/8 x 9 1/4. 0-486-46961-1

# Mathematics–Geometry and Topology

PROBLEMS AND SOLUTIONS IN EUCLIDEAN GEOMETRY, M. N. Aref and William Wernick. Based on classical principles, this book is intended for a second course in Euclidean geometry and can be used as a refresher. More than 200 problems include hints and solutions. 1968 edition. 272pp. 5 3/8 x 8 1/2. 0-486-47720-7

TOPOLOGY OF 3-MANIFOLDS AND RELATED TOPICS, Edited by M. K. Fort, Jr. With a New Introduction by Daniel Silver. Summaries and full reports from a 1961 conference discuss decompositions and subsets of 3-space; n-manifolds; knot theory; the Poincaré conjecture; and periodic maps and isotopies. Familiarity with algebraic topology required. 1962 edition. 272pp. 6 1/8 x 9 1/4. 0-486-47753-3

POINT SET TOPOLOGY, Steven A. Gaal. Suitable for a complete course in topology, this text also functions as a self-contained treatment for independent study. Additional enrichment materials make it equally valuable as a reference. 1964 edition. 336pp. 5 3/8 x 8 1/2. 0-486-47222-1

INVITATION TO GEOMETRY, Z. A. Melzak. Intended for students of many different backgrounds with only a modest knowledge of mathematics, this text features self-contained chapters that can be adapted to several types of geometry courses. 1983 edition. 240pp. 5 3/8 x 8 1/2. 0-486-46626-4

TOPOLOGY AND GEOMETRY FOR PHYSICISTS, Charles Nash and Siddhartha Sen. Written by physicists for physics students, this text assumes no detailed background in topology or geometry. Topics include differential forms, homotopy, homology, cohomology, fiber bundles, connection and covariant derivatives, and Morse theory. 1983 edition. 320pp. 5 3/8 x 8 1/2. 0-486-47852-1

BEYOND GEOMETRY: Classic Papers from Riemann to Einstein, Edited with an Introduction and Notes by Peter Pesic. This is the only English-language collection of these 8 accessible essays. They trace seminal ideas about the foundations of geometry that led to Einstein's general theory of relativity. 224pp. 6 1/8 x 9 1/4. 0-486-45350-2

GEOMETRY FROM EUCLID TO KNOTS, Saul Stahl. This text provides a historical perspective on plane geometry and covers non-neutral Euclidean geometry, circles and regular polygons, projective geometry, symmetries, inversions, informal topology, and more. Includes 1,000 practice problems. Solutions available. 2003 edition. 480pp. 6 1/8 x 9 1/4. 0-486-47459-3

TOPOLOGICAL VECTOR SPACES, DISTRIBUTIONS AND KERNELS, François Trèves. Extending beyond the boundaries of Hilbert and Banach space theory, this text focuses on key aspects of functional analysis, particularly in regard to solving partial differential equations. 1967 edition. 592pp. 5 3/8 x 8 1/2. 0-486-45352-9

INTRODUCTION TO PROJECTIVE GEOMETRY, C. R. Wylie, Jr. This introductory volume offers strong reinforcement for its teachings, with detailed examples and numerous theorems, proofs, and exercises, plus complete answers to all odd-numbered end-of-chapter problems. 1970 edition. 576pp. 6 1/8 x 9 1/4. 0-486-46895-X

FOUNDATIONS OF GEOMETRY, C. R. Wylie, Jr. Geared toward students preparing to teach high school mathematics, this text explores the principles of Euclidean and non-Euclidean geometry and covers both generalities and specifics of the axiomatic method. 1964 edition. 352pp. 6 x 9. 0-486-47214-0

## Mathematics-History

THE WORKS OF ARCHIMEDES, Archimedes. Translated by Sir Thomas Heath. Complete works of ancient geometer feature such topics as the famous problems of the ratio of the areas of a cylinder and an inscribed sphere; the properties of conoids, spheroids, and spirals; more. 326pp. 5 3/8 x 8 1/2. 0-486-42084-1

THE HISTORICAL ROOTS OF ELEMENTARY MATHEMATICS, Lucas N. H. Bunt, Phillip S. Jones, and Jack D. Bedient. Exciting, hands-on approach to understanding fundamental underpinnings of modern arithmetic, algebra, geometry and number systems examines their origins in early Egyptian, Babylonian, and Greek sources. 336pp. 5 3/8 x 8 1/2. 0-486-25563-8

THE THIRTEEN BOOKS OF EUCLID'S ELEMENTS, Euclid. Contains complete English text of all 13 books of the Elements plus critical apparatus analyzing each definition, postulate, and proposition in great detail. Covers textual and linguistic matters; mathematical analyses of Euclid's ideas; classical, medieval, Renaissance and modern commentators; refutations, supports, extrapolations, reinterpretations and historical notes. 995 figures. Total of 1,425pp. All books 5 3/8 x 8 1/2.
Vol. I: 443pp. 0-486-60088-2
Vol. II: 464pp. 0-486-60089-0
Vol. III: 546pp. 0-486-60090-4

A HISTORY OF GREEK MATHEMATICS, Sir Thomas Heath. This authoritative two-volume set that covers the essentials of mathematics and features every landmark innovation and every important figure, including Euclid, Apollonius, and others. 5 3/8 x 8 1/2.
Vol. I: 461pp. 0-486-24073-8
Vol. II: 597pp. 0-486-24074-6

A MANUAL OF GREEK MATHEMATICS, Sir Thomas L. Heath. This concise but thorough history encompasses the enduring contributions of the ancient Greek mathematicians whose works form the basis of most modern mathematics. Discusses Pythagorean arithmetic, Plato, Euclid, more. 1931 edition. 576pp. 5 3/8 x 8 1/2. 0-486-43231-9

CHINESE MATHEMATICS IN THE THIRTEENTH CENTURY, Ulrich Libbrecht. An exploration of the 13th-century mathematician Ch'in, this fascinating book combines what is known of the mathematician's life with a history of his only extant work, the Shu-shu chiu-chang. 1973 edition. 592pp. 5 3/8 x 8 1/2. 0-486-44619-0

PHILOSOPHY OF MATHEMATICS AND DEDUCTIVE STRUCTURE IN EUCLID'S ELEMENTS, Ian Mueller. This text provides an understanding of the classical Greek conception of mathematics as expressed in Euclid's Elements. It focuses on philosophical, foundational, and logical questions and features helpful appendixes. 400pp. 6 1/2 x 9 1/4. 0-486-45300-6

BEYOND GEOMETRY: Classic Papers from Riemann to Einstein, Edited with an Introduction and Notes by Peter Pesic. This is the only English-language collection of these 8 accessible essays. They trace seminal ideas about the foundations of geometry that led to Einstein's general theory of relativity. 224pp. 6 1/8 x 9 1/4. 0-486-45350-2

HISTORY OF MATHEMATICS, David E. Smith. Two-volume history – from Egyptian papyri and medieval maps to modern graphs and diagrams. Non-technical chronological survey with thousands of biographical notes, critical evaluations, and contemporary opinions on over 1,100 mathematicians. 5 3/8 x 8 1/2.
Vol. I: 618pp. 0-486-20429-4
Vol. II: 736pp. 0-486-20430-8

# Physics

THEORETICAL NUCLEAR PHYSICS, John M. Blatt and Victor F. Weisskopf. An uncommonly clear and cogent investigation and correlation of key aspects of theoretical nuclear physics by leading experts: the nucleus, nuclear forces, nuclear spectroscopy, two-, three- and four-body problems, nuclear reactions, beta-decay and nuclear shell structure. 896pp. 5 3/8 x 8 1/2. 0-486-66827-4

QUANTUM THEORY, David Bohm. This advanced undergraduate-level text presents the quantum theory in terms of qualitative and imaginative concepts, followed by specific applications worked out in mathematical detail. 655pp. 5 3/8 x 8 1/2. 0-486-65969-0

ATOMIC PHYSICS AND HUMAN KNOWLEDGE, Niels Bohr. Articles and speeches by the Nobel Prize–winning physicist, dating from 1934 to 1958, offer philosophical explorations of the relevance of atomic physics to many areas of human endeavor. 1961 edition. 112pp. 5 3/8 x 8 1/2. 0-486-47928-5

COSMOLOGY, Hermann Bondi. A co-developer of the steady-state theory explores his conception of the expanding universe. This historic book was among the first to present cosmology as a separate branch of physics. 1961 edition. 192pp. 5 3/8 x 8 1/2. 0-486-47483-6

LECTURES ON QUANTUM MECHANICS, Paul A. M. Dirac. Four concise, brilliant lectures on mathematical methods in quantum mechanics from Nobel Prize-winning quantum pioneer build on idea of visualizing quantum theory through the use of classical mechanics. 96pp. 5 3/8 x 8 1/2. 0-486-41713-1

THE PRINCIPLE OF RELATIVITY, Albert Einstein and Frances A. Davis. Eleven papers that forged the general and special theories of relativity include seven papers by Einstein, two by Lorentz, and one each by Minkowski and Weyl. 1923 edition. 240pp. 5 3/8 x 8 1/2. 0-486-60081-5

PHYSICS OF WAVES, William C. Elmore and Mark A. Heald. Ideal as a classroom text or for individual study, this unique one-volume overview of classical wave theory covers wave phenomena of acoustics, optics, electromagnetic radiations, and more. 477pp. 5 3/8 x 8 1/2. 0-486-64926-1

THERMODYNAMICS, Enrico Fermi. In this classic of modern science, the Nobel Laureate presents a clear treatment of systems, the First and Second Laws of Thermodynamics, entropy, thermodynamic potentials, and much more. Calculus required. 160pp. 5 3/8 x 8 1/2. 0-486-60361-X

QUANTUM THEORY OF MANY-PARTICLE SYSTEMS, Alexander L. Fetter and John Dirk Walecka. Self-contained treatment of nonrelativistic many-particle systems discusses both formalism and applications in terms of ground-state (zero-temperature) formalism, finite-temperature formalism, canonical transformations, and applications to physical systems. 1971 edition. 640pp. 5 3/8 x 8 1/2. 0-486-42827-3

QUANTUM MECHANICS AND PATH INTEGRALS: Emended Edition, Richard P. Feynman and Albert R. Hibbs. Emended by Daniel F. Styer. The Nobel Prize–winning physicist presents unique insights into his theory and its applications. Feynman starts with fundamentals and advances to the perturbation method, quantum electrodynamics, and statistical mechanics. 1965 edition, emended in 2005. 384pp. 6 1/8 x 9 1/4. 0-486-47722-3

# Physics

INTRODUCTION TO MODERN OPTICS, Grant R. Fowles. A complete basic undergraduate course in modern optics for students in physics, technology, and engineering. The first half deals with classical physical optics; the second, quantum nature of light. Solutions. 336pp. 5 3/8 x 8 1/2. 0-486-65957-7

THE QUANTUM THEORY OF RADIATION: Third Edition, W. Heitler. The first comprehensive treatment of quantum physics in any language, this classic introduction to basic theory remains highly recommended and widely used, both as a text and as a reference. 1954 edition. 464pp. 5 3/8 x 8 1/2. 0-486-64558-4

QUANTUM FIELD THEORY, Claude Itzykson and Jean-Bernard Zuber. This comprehensive text begins with the standard quantization of electrodynamics and perturbative renormalization, advancing to functional methods, relativistic bound states, broken symmetries, nonabelian gauge fields, and asymptotic behavior. 1980 edition. 752pp. 6 1/2 x 9 1/4. 0-486-44568-2

FOUNDATIONS OF POTENTIAL THERY, Oliver D. Kellogg. Introduction to fundamentals of potential functions covers the force of gravity, fields of force, potentials, harmonic functions, electric images and Green's function, sequences of harmonic functions, fundamental existence theorems, and much more. 400pp. 5 3/8 x 8 1/2. 0-486-60144-7

FUNDAMENTALS OF MATHEMATICAL PHYSICS, Edgar A. Kraut. Indispensable for students of modern physics, this text provides the necessary background in mathematics to study the concepts of electromagnetic theory and quantum mechanics. 1967 edition. 480pp. 6 1/2 x 9 1/4. 0-486-45809-1

GEOMETRY AND LIGHT: The Science of Invisibility, Ulf Leonhardt and Thomas Philbin. Suitable for advanced undergraduate and graduate students of engineering, physics, and mathematics and scientific researchers of all types, this is the first authoritative text on invisibility and the science behind it. More than 100 full-color illustrations, plus exercises with solutions. 2010 edition. 288pp. 7 x 9 1/4. 0-486-47693-6

QUANTUM MECHANICS: New Approaches to Selected Topics, Harry J. Lipkin. Acclaimed as "excellent" (*Nature*) and "very original and refreshing" (*Physics Today*), these studies examine the Mössbauer effect, many-body quantum mechanics, scattering theory, Feynman diagrams, and relativistic quantum mechanics. 1973 edition. 480pp. 5 3/8 x 8 1/2. 0-486-45893-8

THEORY OF HEAT, James Clerk Maxwell. This classic sets forth the fundamentals of thermodynamics and kinetic theory simply enough to be understood by beginners, yet with enough subtlety to appeal to more advanced readers, too. 352pp. 5 3/8 x 8 1/2. 0-486-41735-2

QUANTUM MECHANICS, Albert Messiah. Subjects include formalism and its interpretation, analysis of simple systems, symmetries and invariance, methods of approximation, elements of relativistic quantum mechanics, much more. "Strongly recommended." – *American Journal of Physics.* 1152pp. 5 3/8 x 8 1/2. 0-486-40924-4

RELATIVISTIC QUANTUM FIELDS, Charles Nash. This graduate-level text contains techniques for performing calculations in quantum field theory. It focuses chiefly on the dimensional method and the renormalization group methods. Additional topics include functional integration and differentiation. 1978 edition. 240pp. 5 3/8 x 8 1/2. 0-486-47752-5

# Physics

MATHEMATICAL TOOLS FOR PHYSICS, James Nearing. Encouraging students' development of intuition, this original work begins with a review of basic mathematics and advances to infinite series, complex algebra, differential equations, Fourier series, and more. 2010 edition. 496pp. 6 1/8 x 9 1/4. 0-486-48212-X

TREATISE ON THERMODYNAMICS, Max Planck. Great classic, still one of the best introductions to thermodynamics. Fundamentals, first and second principles of thermodynamics, applications to special states of equilibrium, more. Numerous worked examples. 1917 edition. 297pp. 5 3/8 x 8. 0-486-66371-X

AN INTRODUCTION TO RELATIVISTIC QUANTUM FIELD THEORY, Silvan S. Schweber. Complete, systematic, and self-contained, this text introduces modern quantum field theory. "Combines thorough knowledge with a high degree of didactic ability and a delightful style." – *Mathematical Reviews.* 1961 edition. 928pp. 5 3/8 x 8 1/2. 0-486-44228-4

THE ELECTROMAGNETIC FIELD, Albert Shadowitz. Comprehensive undergraduate text covers basics of electric and magnetic fields, building up to electromagnetic theory. Related topics include relativity theory. Over 900 problems, some with solutions. 1975 edition. 768pp. 5 5/8 x 8 1/4. 0-486-65660-8

THE PRINCIPLES OF STATISTICAL MECHANICS, Richard C. Tolman. Definitive treatise offers a concise exposition of classical statistical mechanics and a thorough elucidation of quantum statistical mechanics, plus applications of statistical mechanics to thermodynamic behavior. 1930 edition. 704pp. 5 5/8 x 8 1/4. 0-486-63896-0

INTRODUCTION TO THE PHYSICS OF FLUIDS AND SOLIDS, James S. Trefil. This interesting, informative survey by a well-known science author ranges from classical physics and geophysical topics, from the rings of Saturn and the rotation of the galaxy to underground nuclear tests. 1975 edition. 320pp. 5 3/8 x 8 1/2. 0-486-47437-2

STATISTICAL PHYSICS, Gregory H. Wannier. Classic text combines thermodynamics, statistical mechanics, and kinetic theory in one unified presentation. Topics include equilibrium statistics of special systems, kinetic theory, transport coefficients, and fluctuations. Problems with solutions. 1966 edition. 532pp. 5 3/8 x 8 1/2. 0-486-65401-X

SPACE, TIME, MATTER, Hermann Weyl. Excellent introduction probes deeply into Euclidean space, Riemann's space, Einstein's general relativity, gravitational waves and energy, and laws of conservation. "A classic of physics." – *British Journal for Philosophy and Science.* 330pp. 5 3/8 x 8 1/2. 0-486-60267-2

RANDOM VIBRATIONS: Theory and Practice, Paul H. Wirsching, Thomas L. Paez and Keith Ortiz. Comprehensive text and reference covers topics in probability, statistics, and random processes, plus methods for analyzing and controlling random vibrations. Suitable for graduate students and mechanical, structural, and aerospace engineers. 1995 edition. 464pp. 5 3/8 x 8 1/2. 0-486-45015-5

PHYSICS OF SHOCK WAVES AND HIGH-TEMPERATURE HYDRO DYNAMIC PHENOMENA, Ya B. Zel'dovich and Yu P. Raizer. Physical, chemical processes in gases at high temperatures are focus of outstanding text, which combines material from gas dynamics, shock-wave theory, thermodynamics and statistical physics, other fields. 284 illustrations. 1966–1967 edition. 944pp. 6 1/8 x 9 1/4. 0-486-42002-7